D.-E. Liebscher

The Geometry of Time

Dierck-Ekkehard Liebscher

The Geometry of Time

WILEY-VCH Verlag GmbH & Co. KGaA

Author

Prof. Dr. Dierck-Ekkehard Liebscher
Sternwarte Babelsberg
Astrophysikalisches Institut Potsdam
deliebscher@aip.de

Library of Congress Card No.: Applied for
British Library Cataloging-in-Publication Data:
A catalogue record for this book is available from the British Library

Bibliographic information published by Die Deutsche Bibliothek
Die Deutsche Bibliothek lists this publication in the Deutsche Nationalbibliografie; detailed bibliographic data is available in the Internet at <http://dnb.ddb.de>.

Printed in the Federal Republic of Germany

Printed on acid-free paper

Composition Uwe Krieg, Berlin
Printing Strauss GmbH, Mörlenbach
Bookbinding Litges & Dopf Buchbinderei GmbH, Heppenheim

ISBN-13: 978- 3-527-40567-1
ISBN-10: 3-527-40567-4

Contents

Foreword

As a boy of 12, Einstein encountered the wonder of Euclidean plane geometry in a little book that he called *das heilige Geometrie-Büchlein* ("the holy geometry booklet"). Something similar happened to Dierck Liebscher—though admittedly not quite at the tender age of 12. As a physics student in Dresden, he heard lectures on projective geometry. The delight he got from these lectures has remained with him through his working life, and he has now passed on some of it in the present book.

It is a rather unusual book and all the better for it. One of the sad things about the hectic pace and competitiveness of modern scientific research is that truly beautiful discoveries and insights of earlier ages get completely forgotten. This is very true of projective geometry and the great synthesis achieved in the 19th century by Cayley and Klein, who showed that the nine consistent geometries of the plane can all be derived from a common basis by projection. When Minkowski discovered that the most basic facts of Einstein's relativity can be expressed as the pseudo-Euclidean geometry of *space and time*, Klein hailed it as a triumph of his Erlangen program for it showed that the *trigonometry* of pseudo-Euclidean space is the *kinematics* of relativity.

There is a very good reason why projective geometry is nevertheless not part of current physics courses. It can only be applied to spaces (or space–times) of constant curvature, and therefore fails in general relativity, in which the curvature in general varies from point to point. In such circumstances, one is forced (as in quantum mechanics) to use the analytical methods first introduced by Descartes. The beautiful synthetic methods of the ancient Greeks are not adequate. However, several of the most famous and important space–times that are solutions of Einstein's general relativity, notably Minkowski space and de Sitter (and anti-de Sitter) space, do have constant curvature. One of the high points of Liebscher's book is the survey of all such spaces from the unified point of view of projective geometry. It yields insights lost to the analytic approach.

Perhaps the single most important justification for this book is the advent of computer graphics and the possibility of depicting the page views of three-dimensional objects seen in perspective. Drawings and constructions may be distrusted as means to proofs, but they do give true insight that can be gained in no other way. The diagrams of this book constitute its real substance and yield totally new ways of approaching a great variety of topics in relativity and geometry. Especially interesting is the treatment of aberration, which is a vital part of relativity that gets far too little discussion in most textbooks.

This is not a textbook in any sense of the word. It is, however, a book that will instruct, deepen understanding, and open up new vistas. It will give delight to all readers prepared to make a modicum of effort. What more can one ask of a book?

Julian Barbour

South Newington, January 2004

The Geometry of Time. Dierck-E. Liebscher

ISBN: 3-527-40567-4

Preface

This is a book about geometry and physics. It tries a new approach to the interplay of the foundations of both through plane and perspective figures that are correctly constructed.

For the physicist, projective geometry is a wonderland. I entered it once through the lectures of Rudolf Bereis in Dresden, and I was captured once and for all. When I found out that projective geometry provides a really exceptional path to the geometry of relativity, to all the curious behavior of clocks and rods that takes most of the time in any popularizing attempt, the excitement grew irrevocable. Projective geometry is the unifying point of view that renders many facts in relativity because they are already familiar from the Euclidean geometry. In my book *Relativitätstheorie mit Zirkel und Lineal* this has been shown extensively. Today, figures can be drawn and varied with ease by means of computers, and it is time to present comprehensively the very wide possibilities of depicting the geometry of curved space, to include some relativistic cosmology, and to display something more of the connection between physics and geometry in general.

Famous philosophers, physicists, and mathematicians wrote about the connection between physics and geometry; so did Kant, Helmholtz, Poincaré, Einstein, and Hilbert. However, elementary illustrations of this fundamental question are rare. Here our book will enter. It considers the geometrical properties of space and time from the viewpoint of mechanics and cosmology. Concentrating on just the boundary between geometry and physics, it will not aim at a fully detailed presentation of either discipline. It will instead focus on the border region that is usually neglected in discourses on either fields. It is assumed that the reader not only has some simply college acquaintance with geometry and mechanics, but also a mind eager to be led further into the world of both topics. By looking from either side, the reader will recognize with surprise how much she or he can understand about the other side and how much each one depends on the other. Wherever possible, the text is held free of formulas. We believe the figures allow the "vide!" of Euclid. We believe that the reader will not be insensitive to the aesthetic side too. The formal aspects are offered in the appendices to the readers who wish to get a deeper understanding.

The book is not meant to give an axiomatic introduction to either mechanics or geometry. Instead, we shall try to mimic the path from the elementary experiences to the deeper ones, and not only provide the current understanding but also some of the intermediate steps. To speak with Einstein, we will first sniff with our nose on the ground before climbing the horse of generalization.

For the delight I found in writing this book, my gratitude shall cover a very wide span, beginning with the lessons in geometry I had the opportunity to take and ending with the equipment in my institute, including in between the innumerable occasions in which I enjoyed

The Geometry of Time. Dierck-E. Liebscher

ISBN: 3-527-40567-4

encouragement, discussion, and immediate help. In particular, I want to thank E. Quaisser for important advice, H.-J. Treder for many intense discussions of the fundamentals, S. Liebscher for his skills in helping with all computer work, and K. Liebscher for her support and patience. R. Schmidt studied the book as a representative of the reader. S. Antoci added some Italian spirit, which the reader will meet at many places in the volume. J.B. Barbour gave me truly necessary advice to formulate in a language that is not my own, and to clarify arguments.

Dierck-E. Liebscher

Potsdam, February 2004

The structure of the book

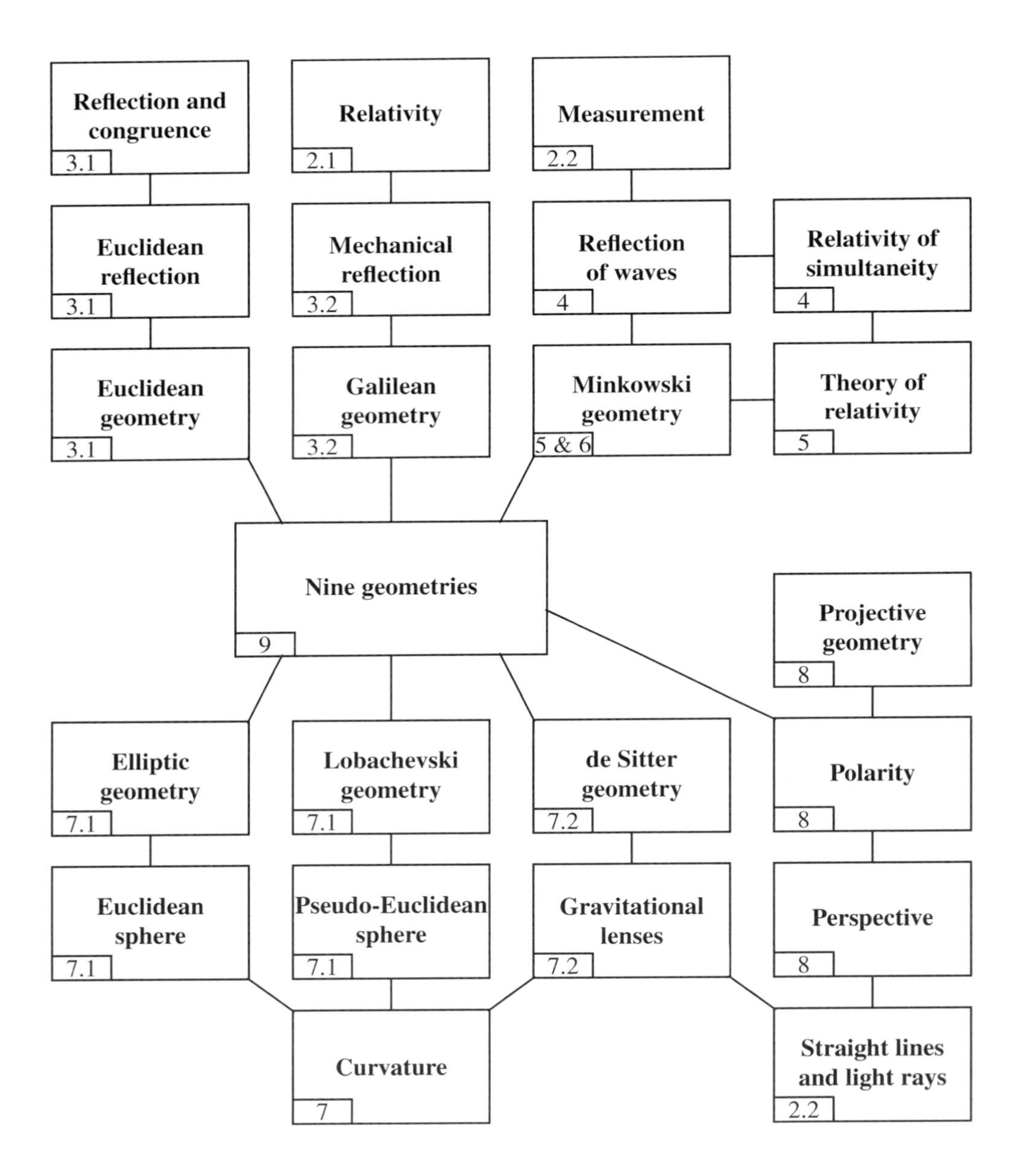

The Geometry of Time. Dierck-E. Liebscher

ISBN: 3-527-40567-4

Notation

$[\cdots]$	used for lists of coordinates, in particular the list of variables of a function, as well as for the triple product, Eq. (C.7)
$\langle\cdots\rangle$	used for the scalar product, Eq. (C.4)
$\times$	used for the cross product, Eq. (C.5)
$A \times B$, AB	used for the line connecting A and B
$g \times h$, gh	used for the point of intersection of g and h
$A \circ B$	used for the direct product, Eq. (C.10)
$A, B, \ldots$	points
$a, b, \ldots$	straight lines
$\alpha, \beta, \ldots$	planes, or angles
$\mathcal{A}, \mathcal{B}$	coefficient matrices of the absolute conic section
$\mathcal{D}$	rotation, or cross-ratio
$\mathcal{D}[A, B; E, F]$	cross-ratio
Δ	triangle, difference, increment
δ^i_k	unit matrix, zero for different, $+1$ for equal indices
$d[A, B]$	distance between the points A and B
d	infinitesimal increment
$\mathcal{E}$	unit matrix
ε^{ikl} , ε_{ikl}	permutation symbol, zero for two equal indices, -1 for odd, $+1$ for even permutations
E, F, F_1, F_2	fixed points on a straight line
$F[h]$	foot point of the line h
$\mathcal{G}$	group
$\mathcal{G}[\mathcal{A}, \boldsymbol{v}]$	element of the Galilean group, Appendix B.3
g_{ik}	metric tensor, Appendix B.3
$\mathcal{I}$	involution
$\mathcal{K}$	conic section
$k[A]$	tangent from the point A to the conic section $\mathcal{K}$
$K[g]$	point of intersection of the line g with the conic section $\mathcal{K}$
$\mathcal{L}[\mathcal{A}, \boldsymbol{v}]$	element of the Lorentz group
$\boldsymbol{n}$	vector of direction

The Geometry of Time. Dierck-E. Liebscher

ISBN: 3-527-40567-4

$\mathcal{P}$	polarity
$p[A]$	polar of the point A
p	absolute polar for all points
$\pi[A]$	polar plane to the point A in projective space
$P[g]$	pole of the straight line g
P	absolute pole for all straight lines
$\boldsymbol{p}$	momentum vector
p_k	four-momentum, Appendix B
$\Pi[a]$	circumference of a circle with a radius given by a, Appendix E
$\Sigma[\alpha]$	generalized sine, equal to the ratio of the projecting line to the projected side of an angle, Appendix E
$\mathcal{S}$	reflection, or generating system of the group of motions
s	reflecting straight line
$S[A]$	reflected point
$S_g[A]$	reflection of the point A at the straight line g
$\mathcal{T}$	transformation
u^i	four-velocity, Appendix B
$\boldsymbol{v}$	(three-dimensional) velocity vector

1 Introduction

Drawing is the first method to shape our understanding of the world, for a child, for an artist, for an engineer, and for a mathematician. At school we learn how geometry can be abstracted from the images that are meant to describe some real object, and which are studied then without respect to their content. Things in space are projected onto a plane and we learn to figure out what happens to their form. We remember the curious properties of a triangle, for instance, that we can drop perpendiculars from the vertices, and that they meet at one point, that the hypotenuse of a right-angled triangle is the diameter of a circle around the triangle, and that the square on the hypotenuse of a right-angled triangle equals the sum of the squares on the other two sides. Some of us remember the logical compactness found in the axiomatic approach. Thales, Pythagoras, and Euclid are watching us.

Time seems to be different from space. Usually, it is not mentioned in geometry, and physics produces the impression that without Leibniz's and Newton's calculus one cannot say much about it. Forms in space have an aspect of stability, time is change instead. It was Einstein's theory of relativity that demonstrated the deep connection between space and time, and between geometry and physics. It became evident that elementary geometry is to be applied to the union of space *and* time. It became equally evident that physical observation decides which geometry of space and time is to be applied to real-world phenomena, and that a careful and elementary analysis of measurements is necessary to avoid misconceptions.

Usually, one does not imagine the motions of objects as geometrical figures in the union of space and time. For the insider, it is much faster to calculate analytically. Newton already solved the geometrical problems of the *Académie Française* analytically before embedding the result in a geometrical proof. Figures are drawn as auxiliary sketches at most. The outsider understands the theory of relativity as a system of more or less complicated formulas that avoid intuition. The following will show that the foundations of the relativity theory are fully subject to geometric intuition, and that relativistic kinematics is nothing else than the elementary geometry of the union of space and time. We shall learn how to use the drawing plane and space as space–time diagrams with one or two spatial dimensions and one dimension time.

A theoretical construction represented by elementary geometry and understood as an object of immediate geometrical experience leads to a strong expectation of internal consistency, more than an analytical derivation does for the outsider. For this reason, we wish to show in this book how elementary geometry, mechanics, and fundamental properties of the universe are interconnected. We intend to do this without the rigor that may be found quite readily in the literature. Instead, we wish to expose the real constructions and the relationships that produce the often aesthetically striking character of geometry. That is, we intend to fall in

The Geometry of Time. Dierck-E. Liebscher

ISBN: 3-527-40567-4

between all the stools available. However, we will discover many unexpected and astonishing relationships and associations. We shall consider the geometry of space and time and demonstrate by elementary means

- how physically elementary experiments receive a geometrical interpretation,
- how physical experiments restrict the properties of applicable geometries, and
- how geometrical properties determine correct physical formulations.

The figures of this book are produced with IDL. The programs can be requested by e-mail *deliebscher@aip.de*.

In Chapter 2 we introduce the notion of timetables as elementary representations of space–times. We shall learn the first means to draw in a space–time plane. The question of the definition of distances in timetables is left open here. Chapter 3 introduces the fundamental role of reflections. This role is a bit surprising because *real* motions are split into two reflections that produce only *virtual* images. However, in our timetables reflections are real and much simpler than other motions. We use this to get a first notion of the strangeness of the geometry in a timetable. Chapter 4 presents the central problem of Einstein's (special) theory of relativity. This was the first occasion to consider geometries different from the Euclidean geometry of space in the framework of physics. We correct the reflection procedure of Chapter 3 to solve the central problem and obtain the geometry of the space–time called the Minkowski geometry. The relativity theory and its paradoxes are considered in Chapter 5 with the help of this geometry. The elementary metric properties of the Minkowski geometry are compared with their Euclidean analogs in Chapter 6. Chapter 7 extends the relation between the Euclidean and Minkowski geometries of the plane to homogeneously curved surfaces, always trying to keep contact with physical examples. We obtain new, but characteristically similar, geometries. Chapter 8 presents the initial notions of projective geometry, which in Chapter 9 unites the geometries in one family, i.e., the Cayley–Klein geometries. This family can be characterized axiomatically as one expects for geometry. Chapter 10 deals with some general questions connected with the physical interpretation of these geometries.

All the notions explained in this volume are the subject of well-founded and strictly defined and formalized theories. It is not our aim to repeat these here, because we are interested in the interface, where these notions sometimes have to be unsharp enough to see that they fit. The necessary formal background for geometry is given in the appendices. Appendix A explains groups of motions and their generation by sets of generating elements interpreted as reflections. Appendix B considers questions connected with the physical introduction of coordinate systems, which, since the time of Descartes, have permitted the application of arithmetic methods to calculate and prove geometrical results. It explains in detail the transformations connected with changes in reference and introduced in the Riemannian geometry as far as these notions are concerned. Appendices C and D formalize the notions of projective and projective–metric geometry used in Chapters 8 and 9. Appendix E formalizes the classification of the Cayley–Klein geometries and, finally, gives the formal representation of the metric in projective metric spaces. In order to provide for a rapid access to definitions of the various notions used or touched in the book, a glossary is given instead of an index.

You will find many books about geometry or theory of relativity. Here only that part is cited that has some connection with our topic. Geometric and graphic presentation of the theory of relativity can be found in [1–7]. There are elementary [8–11] and less elementary [12, 13] introductions to the theory of relativity, in the general theory [14] and cosmology [15–18]. The descriptive and projective geometry can be learned in older [19–24] and more recent books [25]. Detailed information about the non-Euclidean geometry can be found in [26, 27]. General introduction to geometry is provided in [28–31]. The spatial imagination is trained in [32, 33]. And [34] is dedicated to computer graphics in our context.

2 The World of Space and Time

2.1 Timetables

The investigation of any motion turns out to be the examination of a line (in general, a figure) in *space–time*, the product of space and time. We will also refer to space–time as the world. Any point of space–time is characterized by its position in space and its moment of time. We call such a point an event. The history of the motion of a material point is called a world-line.

The world-lines that represent the history of material points represent a kind of timetable of motion. Such a timetable is the geometrical representation of motion. The simplest motion is that which is in only one spatial direction. We can understand the geometrical representation as registration on a strip like the strip registration of a seismogram, an electrocardiogram, or an electroencephalogram. In principle, we control the rate of unrolling the strip by Newton's first law. Free motion shall draw straight lines in the familiar sense (Figure 2.1). Accelerated motion will draw lines that deviate from straight ones (Figure 2.2).

It is necessary here to have a look at Newton's first law. We take a modern form that avoids any trouble with the definition and realisation of systems of reference:

> **The set of world-lines of bodies that are not influenced by other objects is a set of straight lines.**

Here, a set of lines is a set of *straight* lines if two of them have at most one intersection and if two points (events) have at most one connecting line. Straight lines are *not* defined by linear relations of coordinates that would need prescriptions of constructions beforehand. A lone line can always be straight, i.e., embedded in a set of lines that fulfills the conditions of a set of straight lines. It is now mathematics to show that a set of straight lines allows us to construct coordinates in which the members can be characterized through linear relations. These coordinates constitute the linear reference systems that we are using. Newton's first law implies that we can define coordinates in such a way that a force-free motion can be described through linear relations in space and time, i.e., through familiar straight lines.

The motion in two spatial directions will be shown in projections of three-dimensional space–times. As an illustration, we show the timetable of a train (Figure 2.3), the timetable of the earth–moon system (Figure 2.4), the timetable of a three-body system (Figure 2.5), and the timetable of a collision visualized best on a billiard table (Figure 2.6). Collisions are evaluated in the study of elementary particles (Figure 2.7). Space–time pictures of moving surfaces are indicated in Figures 2.8 and 2.9.

The Geometry of Time. Dierck-E. Liebscher

ISBN: 3-527-40567-4

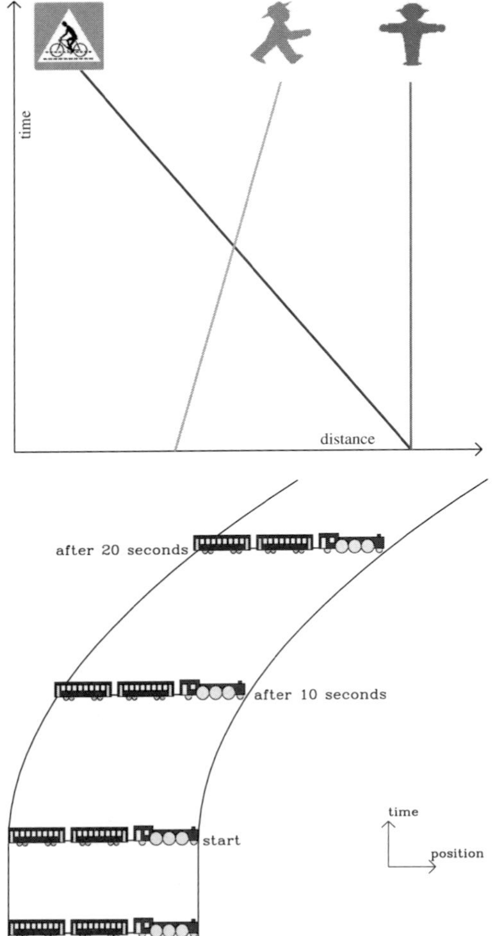

Figure 2.1: *The registration strip.*

The simplest example for a geometrical timetable is the registration strip. The motion proceeds on the upper horizontal line, and the strip is rolled off downward. At any given instant, we obtain registration lines that hang down from the present position of the registered objects.

Figure 2.2: *Timetable of an accelerating train.*

We draw world-lines on a two-dimensional space–time plane. The time axis is taken vertical. The slope of the world-line with respect to the vertical is the velocity of the object. Motion with a constant acceleration produces a parabola for the world-line. This parabola is analogous to the parabola of a thrown object for which the horizontal coordinate plays the part of time, because the horizontal component of the velocity remains constant.

When we intend to pass from the geometry of space to the geometry of space–time, we would first expect the reorientations and translations in space–time to be constructed byreflections just as their counterparts in space. Reflections can be efficiently realized by mechanical means and can be determined by observation. After doing this we can develop the geometry. Reflection at a straight line in space–time means reflection at some unaccelerated motion. If we descend for the moment from the Pegasus of phantasy,[1] and do not ask for all possible geometries of the world, but only for the physically observable geometry, we have to investigate the reflection of mechanical motion. With that it will be seen in Chapter 5 that wave phenomena when explained by mechanics seem to provide a curious absolute ori-

[1] *Wenn wir an etwas arbeiten, dann steigen wir vom hohen logischen Ross herunter und schnüffeln am Boden mit der Nase herum. Danach verwischen wir unsere Spuren wieder, um die Gottähnlichkeit zu erhöhen.* (In working on something, we descend from the high horse of logic and sniff at the ground with our nose. Afterward, we hide our tracks in order to increase our similarity to God: A. Einstein, cited in [35].) G.K. Chesterton puts it still stronger: *You can only find truth with logic if you have already found truth without it.*

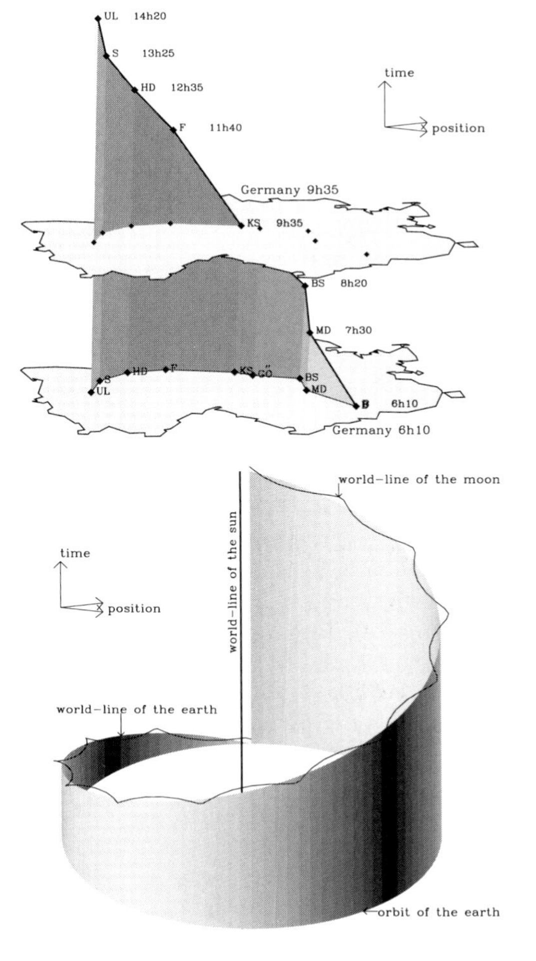

Figure 2.3: *World, world-line, timetable.*

The world is a product of space and time. We represent space by a plane in which the border of Germany and the railway from Berlin to Ulm are indicated. The history of a moving point, i.e., a world-line, is exemplified here by the timetable of a train from Berlin to Ulm. The elapsed time is represented by the height over the fundamental plane. The slower the train, the steeper the world-line. The world-line will never be horizontal, because that would have to be interpreted as the train being at different places at the same time. The world-line can be vertical: In this case, the train has stopped.

Figure 2.4: *The motion of the earth and the moon as a timetable.*

The world-line of the earth is a spiral. In the figure it is represented by the upper edge of a palisade showing the projection onto the space represented by the fundamental plane. This projection is a Kepler ellipse, as we know. After 1 year, the earth is at the same place in the solar system. The curve is drawn up to this moment. In contrast to its projection, the world-line is not closed. The world-line of the moon (indicated as a dark line) winds around the earth's world-line.

entation in space–time. The questions connected with this discovery (whose basic features will be explained later) finally led Einstein to the theory of relativity in 1905. H. Minkowski established the geometry of space–time on this foundation. F. Klein identified this geometry as the geometry of a whole family that we will illustrate in the space–time plane: It is the family of the nine aforementioned geometries of plane (Chapter 9, [36]). Before explaining the characteristic features of this family, we must add some more of the physical background.

2.2 Surveying Space–Time

How should we characterize the fundamental methods for surveying space–time? Roughly speaking, there are three different types of procedures that, when combined, allow one to fix the position of an event in space and time. These are *sighting*, *application of rulers* with meter sticks, and *measurement of duration*.

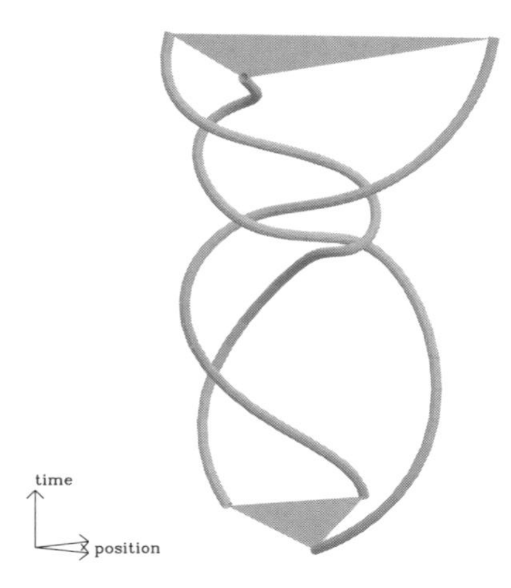

Figure 2.5: *The timetable of three bodies in plane motion.*

Three bodies move in a plane. The world-lines are given together with the triangles of the initial and the final configuration [48].

In *sighting* we try to position a cross hair in such a way that the light ray from the object to the eye passes through the cross hair. Two points (the eye and the object) define a straight line. The possible statement is about the lining up (the incidence) of a third point (the cross hair) with the previous two. Of course, the identity of the light ray with a straight line has to be supposed. The relation between straight lines and light rays was an important subject of Greek philosophy. In his celebrated parabola about prisoners sitting in a cave, Platon assumes such a relation. So does Parmenides in deriving the spherical shape of the earth, and also Anaxagoras in determining the distance to the sun. Exaggerating a bit, we can say that the notion of a straight line is more important than that of a point: For Platon the latter is merely the intersection of two rays. In Chapter 3 and Appendix A we discover this in a new guise when we discuss abstract groups generated by reflections.

Euclid stated explicitly that it is light propagation that defines straight lines physically. One can formulate this as an integral principle by requiring that light rays be the shortest lines (in space measured by the comparison with meter sticks). However, the effective length of a line segment may differ from that expected geometrically if we must take into account a refractive index. In this case, the apparent geometric length has to be multiplied with the refractive index or divided by the local phase velocity of light. The effective length of a path is here the time that the phase of the light needs to pass through. We then obtain Fermat's principle. The stretched rope used by a gardener to find a straight line implements such an integral principle, too: The straight line turns out to be consist of a minimal number of atomic distances.[2] Here, we use the light ray for constructing straight lines. Figure 2.10 shows a device that reduces the apparent position of a star to materially measurable angles.

[2]The fact that light rays and ropes can sometimes be used to define straight lines with a certain degree of mutual consistency looks to the devil's advocate like a very curious accident if not a miracle [37]. It is the miracle of the possibility of defining a geometry. Of course, the rope should not be stretched too much; otherwise we would find also the "interatomic distances" to be stretched, so, by pulling more and more, the distance measured by counting the atoms gets smaller and smaller, without reaching a minimum, until the rope breaks [37]. The maximum distance found by a stretched rope is the idealization of the independent measure.

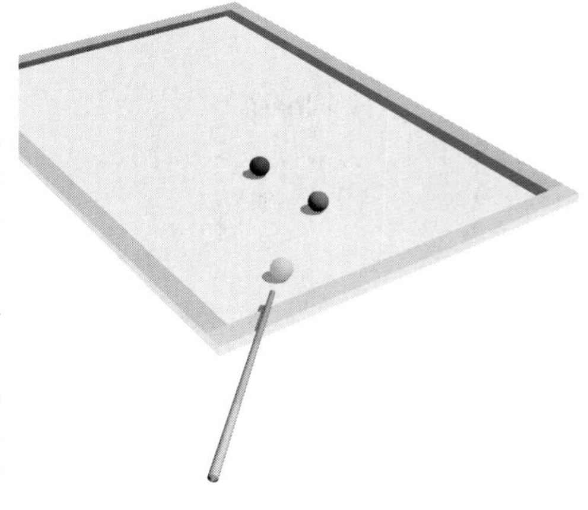

Figure 2.6: *The billiard table.*

The billiard table is our model of the two-dimensional space in which point-like masses interact only by collision. All the nice tricks of the true billiard game that use the finite size of the balls are neglected. The ball in front is shot against the other ones that are at rest. The more centrally they are hit, the more motion is transferred. In the approximation that neglects the rotational motion of the balls, the two balls advance after the hit in right angles. In the lower-left figure, these positions are shown. On the right, the four instants are stacked, and we obtain the timetable of the collision with the world-lines of the three bodies.

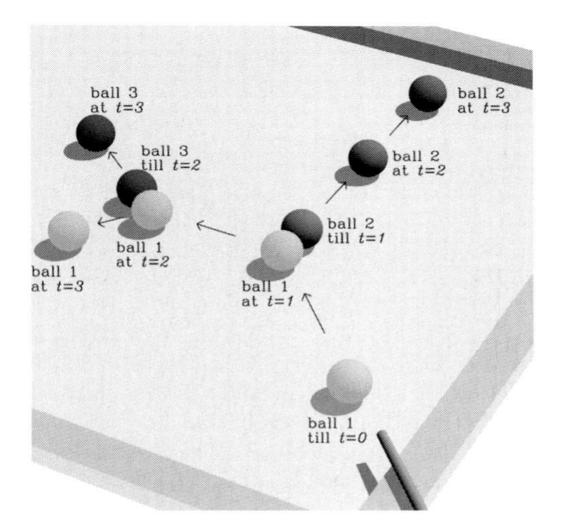

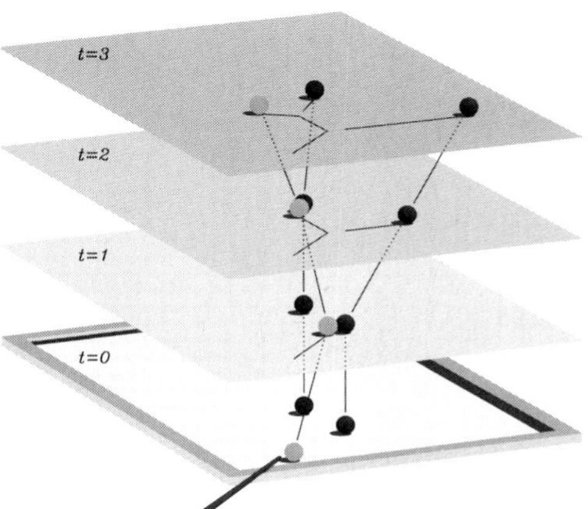

No other principles have been as successful in physics as the extremum principles. In general, they deal with extrema of values attributed to paths through abstract configuration spaces (in which each degree of freedom adds one dimension and the equations of motion are of second order) or phase spaces (in which each degree of freedom adds two dimensions and the equations of motion are of first order). The value attributed to a path is given, in general, by an integral called the action integral, because it sums up products with the physical dimension $energy \times time$. The physical processes describe curves in configuration or phase space. The equations of motion are derived as a necessary condition for an extremum of the action integral.

In *applying rulers* we suppose that they can be moved without being changed (Figure 2.11). In moving the meter stick after its calibration no change of the stick should be allowed. In any case, such a change would be detected only in comparing different sticks.

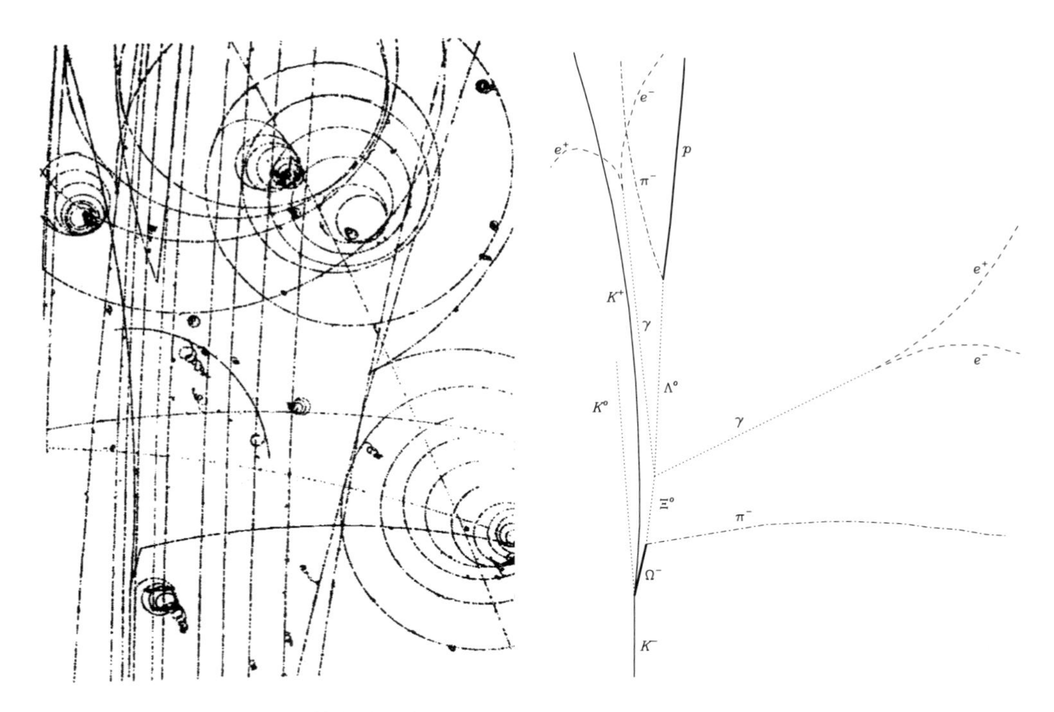

Figure 2.7: *Tracks in the bubble chamber.*

In the bubble chamber, we observe the billiard of elementary particles that leave tracks if electrically charged. The curvature of the tracks is produced by an external magnetic field and allows us to determine the momentum of the particles. The thickness of the tracks allows us to calculate the values of mass and energy. Using the corresponding conservation laws, we can characterize also the neutral particles that do not leave visible traces. Here, we show the photo of the famous event in which the Ω^- hyperon could be measured [38] together with a sketch of the process chain. A K^- meson from the accelerator (coming from below) collides with a proton of the chamber gas and produces an Ω^- hyperon together with a neutral K^o and a K^+ meson with a track leaving the picture at the upper left. The Ω^- leaves a short track showing that its lifetime is much longer (about 10^{-11} s) than that of ordinary intermediate resonances. It decays into a neutral Ξ^o hyperon and a π^- meson that leaves a track to the right border. The trackless Ξ^o hyperon decays into three other neutral particles, a Λ^o hyperon and two photons (γ). These three particles leave no track themselves but their decay products do. The Λ^o decays into a characteristic proton–π^- pair. In colliding with other protons of the chamber gas, the photons produce electron–positron pairs that draw characteristic pairs of spirals. By plain luck even these particles of the fourth generation are seen, and the process can be reconstructed in total

We should expect such changes only if the forces necessary to move the stick are comparable to the internal forces necessary to change the structure and the distance between the marks permanently. This requirement is motivated from the physical point of view. After all, we know of a theoretical construction due to H. Weyl in which a hypothetical dependence of some scale on history is exploited to represent the electromagnetic field, and we know of the objection, raised by Einstein, that such a dependence on history is excluded by the observed

Figure 2.8: *Timetable of an explosion.*

At a $t = 0$, a charge explodes whose fragments move in all directions with equal speed. Their distance from the center is proportional to the time elapsed. Therefore, the world-lines form the mantle of a circular cone. If we instead consider a flash of light occurring at some point in space, it sends a light signal propagating as a wavefront. At any given time the wavefront forms a sphere (in our plane of positions a circle) whose radius is proportional to the elapsed time. The wavefronts at different times form a circular cone apparently identical with the explosion cone. The mantle surface is formed by the world-lines of the light signals.

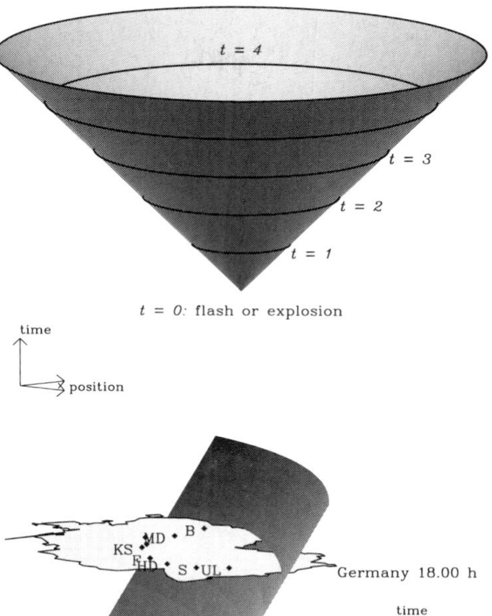

Figure 2.9: *Timetable of a weather front.*

A weather front passes Germany from west to east. Its timetable is a surface. If the form of the front does not change and the motion is uniform, the surface is a cylinder. If the front is straight, the surface is a plane. Weather fronts and wavefronts differ mainly in velocity, i.e., in the inclination of the surface against the vertical time axis.

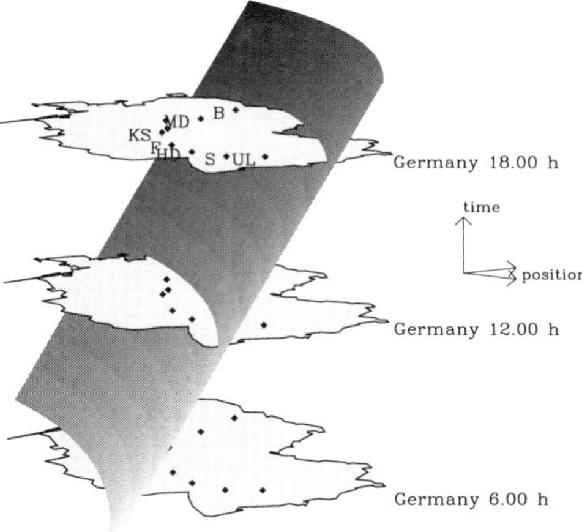

narrowness of certain spectral lines of cosmic objects. Another typical necessity consists in knowing what is simultaneous at different points of space. Simultaneity is decisive for any length measurement of moving objects (Figure 2.12). After all, we must read off on both ends of an interval simultaneously to get a sensible result. This requirement is the space–time analog of the equally obvious requirement that the object in question and the measuring rod must be parallel if the measurement is performed at a distance, for instance, between the parallel jaws of a sliding rule or the parallel light rays in long-jump measurements.

To illustrate a *measurement of duration* let us imagine an observer inside a ballistic rocket with closed windows. If the motion is inertial, the only physical occurrence that he or she can observe is the very flow of time, and he or she can obtain a measure of this flow by counting the ticks of his or her wristwatch. As previously assumed for the length-measuring devices, the timepiece should not change its rate while moving with the rocket. Again, deviations from this property show up only by comparing different timepieces, and deviations can be excluded

Figure 2.10: *Sighting.*

The picture taken from the first volume of the *Machina coelestis* of Hevelius ([39], Bibliothek der kgl. preuss. Sternwarte Berlin) shows the use of a quadrant. After sighting, height and azimuth of the star can be read from the graduation

only if the external forces accelerating the watch are very small compared to the internal forces governing the periodic process that constitutes the timepiece. Anyone who tries to displace a pendulum clock without stopping its motion can grasp the importance of this condition. For an object moving through our laboratory, we either need more than one clock to follow its motion or measure time from a distance. In both cases, projection effects similar to these in length measurements have to be taken into account. We shall meet such effects in discussing the paradoxes of Einstein's theory of relativity (Chapter 6).

In *sounding* all three methods are combined. If a sound signal propagating through the surrounding medium can be echoed back by an object, the position of the latter at the moment

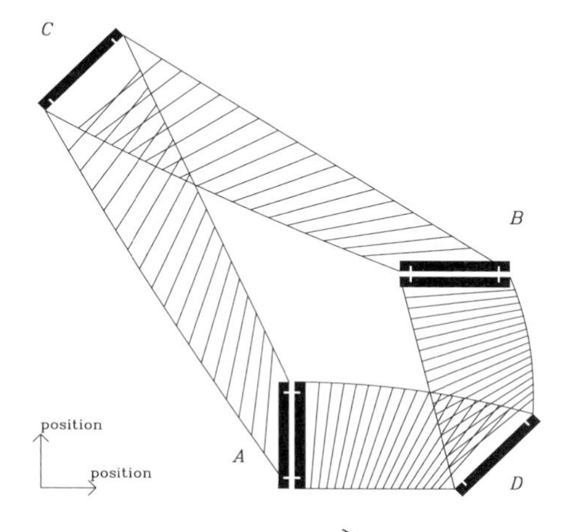

Figure 2.11: *The application of rulers.*

The distance of two points on a rigid body can be determined by applying a ruler, because meter sticks of different construction do not show variations if moved through space carefully enough. Two meter sticks are compared at A and moved on different paths (one passing through C, the other through D to B) and compared again. What happens in between? The more rigid the sticks are, i.e., the larger the internal forces are compared to the external inertial forces, the less can happen and the more accurately the new comparison will produce the same result as the former one.

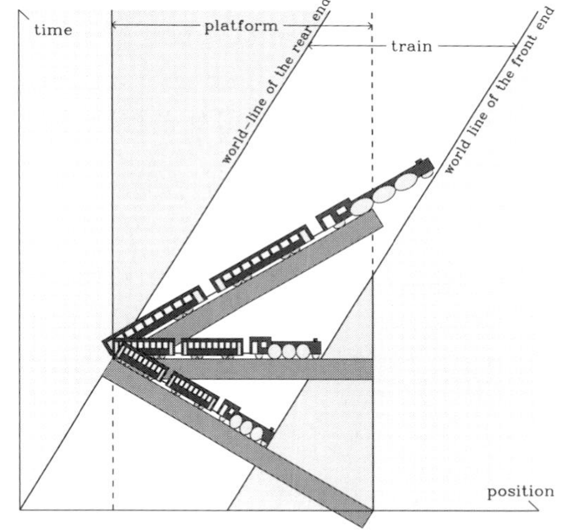

Figure 2.12: *Simultaneity and length measurement.*

If we try the fit of a measuring rod we must read off at both ends simultaneously, if the object of the measurement is moving. If we read off at the front end of a train too early the result is too small, and if too late, the result is too large.

when it is reached by the signal can be calculated (Figure 2.13). However, we need to know the velocity of sound relative to the measuring device when it propagates in both directions. As we will see later (Chapter 5, Figure 4.9), the use of electromagnetic waves relieves us from this task since the speed of light is *not* changed by composition with other velocities (except for aberration). If we can detect an object using radar, we obtain four data about the event: the time when the signal was emitted, the duration of its back and forth trip, and the direction of the returning signal. Under the ideal conditions of light propagation, the measurement of the propagation time of a signal allows not only the determination of distance but also, together with the measurement of direction and time, a reconstruction of the complete space–time coordinates of the event "reflection of the signal" [40, 41].

The echo sounder can be used to infer velocities. Let us assume that the sounder emits signals with a certain period. The signals in turn are reflected by the mirror. If the mirror has a constant distance from the sounder, the period of the reflected signals is equal to that of the

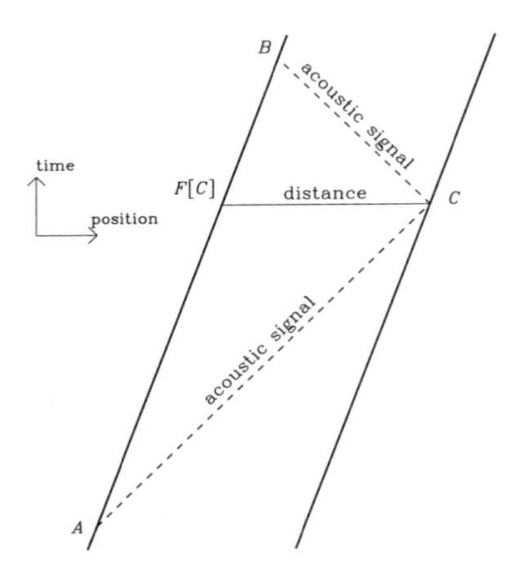

Figure 2.13: *Echo sounding as determination of position.*

The sketch shows the world-lines of an observer and an object at a fixed distance, both in motion through the medium that is here at rest. We can then draw the world-lines of the acoustic signals in both directions with the same slope. The calculation of the distance $d[C,F]$ by the time difference t_{AB} requires knowledge of the magnitudes v_+ and v_- of the relative velocities of the signals: $t_{AB} = d[C,F]\,(\frac{1}{v_+} + \frac{1}{v_-})$.

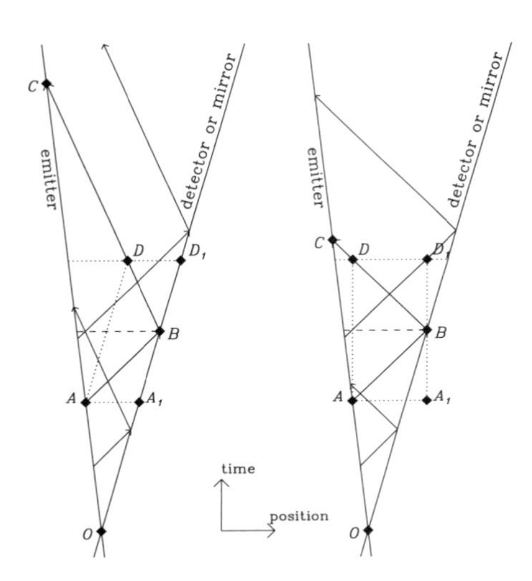

Figure 2.14: *The Doppler effect.*

We see the world-lines of a sender and a reflector in relative motion. On the left-hand side, the signals consist of particles. At reflection, only the sign of their velocity *relative to the mirror* is changed: The parallelogram AA_1D_1D is adjusted through the world-line of the mirror. On the right-hand side, the signal consists of wavefronts or groups. In this case the sign of the velocity *with respect to the carrier* is changed, and the motion of the mirror does not enter. The parallelogram AA_1D_1D is adjusted through the velocity of the carrier (assumed to be at rest in our frame). The change in period is given by $(t_C - t_O) : (t_A - t_O)$. If we now assume the existence of a universal time, we can compare the period on the mirror with that at the sender too. The change in period, $(t_B - t_O) : (t_A - t_O)$, is called the *Doppler effect*.

sounder. If the mirror moves, we observe a change in period (Figure 2.14). In the evaluation of the figure, we count velocities positive if the object moves in the direction of the emitted signal. We obtain

$$\frac{t_C - t_O}{t_A - t_O} = \frac{v_{\text{signal}} - v_{\text{emitter}}}{v_{\text{signal}} - v_{\text{reflector}}} \, \frac{v_{\text{reflector}} - v_{\text{refl. signal}}}{v_{\text{reflector}} - v_{\text{emitter}}} .$$

We must distinguish between two cases, i.e., that of signals in the form of emanated particles and that of signals in the form of wave pulses. In the case of signals consisting of particles moving with a given velocity with respect to the emitter, we must calculate with a reflection subject to Huygens' law ($v_{\text{refl. signal}} = 2v_{\text{reflector}} - v_{\text{signal}}$, Figure 3.15). Then the change in period depends only on the radial component of the relative velocity between emitter and reflector. For technical application, acoustical or electromagnetic waves are used. The change in period is measured as the change in frequency. Now the velocity of the signal is given with respect to the carrier medium ($v_{\text{refl. signal}} = 2v_{\text{medium}} - v_{\text{signal}}$), and it must not be related to the emitter or the reflector. Hence, the effect depends on the velocities of the emitter and reflector with respect to the medium separately.

Up to this point, only one clock at the position of the sender is needed to establish the effect. If we now assume the existence of a universal time, we can compare the period on the mirror with that at the sender too. We obtain

$$\frac{t_B - t_O}{t_A - t_O} = \frac{v_{\text{signal}} - v_{\text{emitter}}}{v_{\text{signal}} - v_{\text{reflector}}}.$$

This is called the *Doppler effect*. In the case of particle ejection, $|v_{\text{signal}} - v_{\text{emitter}}|$ should be considered a given constant, in the case of waves $|v_{\text{signal}}|$ itself. The Doppler effect shows in both cases that the propagation velocity of the signal is finite, i.e., the propagation is *not* instantaneous. The acoustic Doppler effect depends on the velocities relative to the medium. The optical Doppler effect obtains its final form in the relativity theory (Figure 5.11).

The classical standard for length is a rigid body with marks on its quasicrystalline microscopic structure. The characteristic distances in this structure are determined by quantum mechanics, whose natural unit is the Bohr radius of the hydrogen atom. Comparison with a rigid body implies comparison with this radius. We expect it to remain the same if the factors that determine it do not change with time and position. Thus, the same laws that identify the Bohr radius as the atomic unit of length determine the structure of a rigid body. If one could vary the factors determining the Bohr radius, the size of any rigid body would change accordingly.

The classical standard for time measurement is the course of the planets, i.e., the ephemeris time. It is determined by Kepler's third law, and has to account for all perturbations and uncertainties of the solar system. The establishment of atomic time provided a microscopic unit. The transitions between bound states of the atom that produce the spectral lines have a common measure, the Rydberg constant, which is a frequency and a time normalization. The stability of the atoms provides the stability of this time unit.

The microscopic lengths and time units are determined by the forces involved and by the inertial masses defined by Newton's laws. They are so readily available because stationary states cannot continuously vary and are therefore stable to a certain extent. This is due to the laws of quantum mechanics. We consider space as isotropic if the virtual orientation dependence of inertial mass is compensated by that of the forces. A sphere is defined on one hand by the final positions of the partners after a symmetric collision, and, on the other hand, by the equipotential surface of the gravitational or the electrostatic field,

$$\text{distance} \propto \frac{1}{\sqrt{\text{field strength}}},$$

for which only the structure of the source can lead to systematic errors, or by the surface of

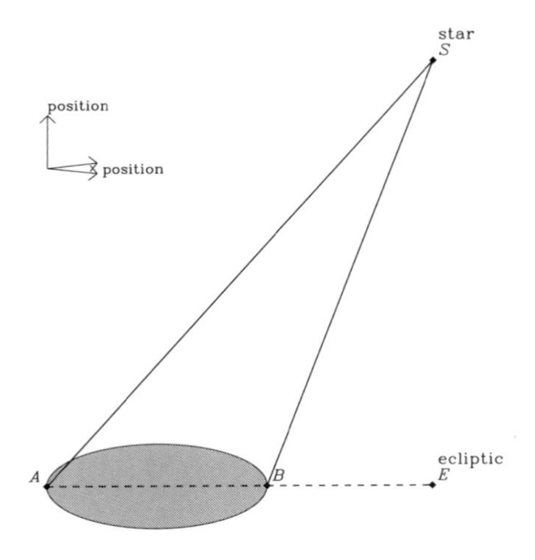

Figure 2.15: *Trigonometric parallax.*

The figure shows the orbit of the earth in the ecliptic plane and a star S above this plane. The determination of the trigonometric parallax presupposes that the sum of the angles of a triangle yields a flat angle. The two angles $\angle EAS$ and $\angle EBS$ are observed as apparent heights above the ecliptic; hence all the elements of the triangle are determined, since the size AB of the earth's orbit is known. Bessel developed this method, and the first star to have its distance measured was 61 Cygni. Its angle $\angle ASB$ is about 0.3 arcsec, and its distance 3.4 pc $\approx 10^{14}$ km.

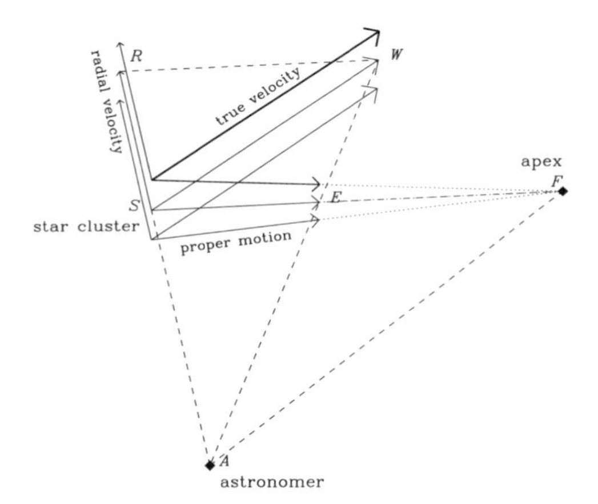

Figure 2.16: *Moving-cluster parallax.*

The observer sees the proper motion $\vec{SE}$ as the projection of the true motion $\vec{SW}$ of a stellar cluster onto the sky (perpendicular to the line of sight) and determines an apex (*vanishing point*) F. The angle $\angle SAF$ is equal to the angle $\angle RSW$ of the radial component (radial velocity) with the true motion. The angle $\angle RSE$ is right. Hence it yields $SE = SR \tan(\angle RSW)$. We now measure the proper motion as an angle, and the radial velocity as a true length per unit time. The ratio of these two provides the distance.

constant intensity of a symmetric source,

$$\text{distance} \propto \frac{1}{\sqrt{\text{intensity}}}.$$

On the earth we can test at least, in principle, the appropriateness of geometrical theorems. Measuring in the universe, we must presuppose applicability at least to a large extent. For instance, in observing parallaxes we must presuppose the validity of our geometrical conceptions, and they identify the quantities to be calculated from the measurement.[3] That is, the geometrical relations must be known a priori in order to interpret the observations. Determining the trigonometric parallax, we use the diameter of the orbit of the earth around the sun as a base. Then we measure two angles: the maximal and the minimal height of the star

[3]Speaking with Einstein, the theory *always* determines what is measured, but our analysis will not dig that deep.

Figure 2.17: *Apparent size and distance.*

The observation of the reflection of the burst of light of a supernova allows us to conclude the physical size. When we compare this size with the apparent size (the angle), we obtain the distance as in the case of the moving-cluster parallax. The distance to the supernova 1987A in the Large Magellanic Cloud can be found in this way because the size R of the observed ring is proportional to the time t elapsed since the explosion of the supernova S.

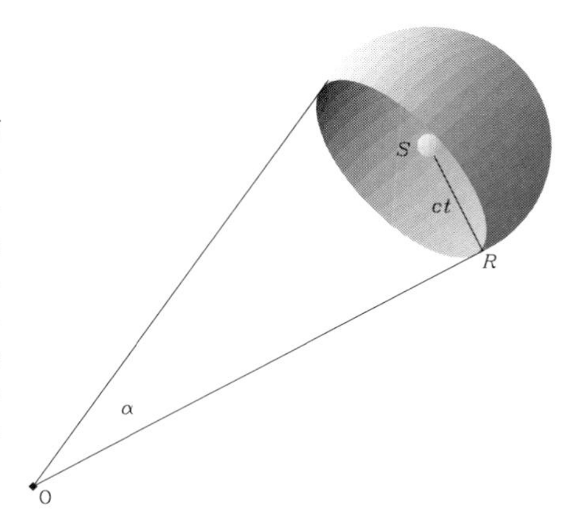

Figure 2.18: *Intensity and distance.*

We show three concentric spheres around a source with cuts bounded by rays from the center, in which we imagine a source. The cuts grow with the size of the spheres. When the distance from the source is increased, a detector of fixed size (indicated by the small black squares) catches a decreasing part of the emitted power and we can thereby determine the change in its distance from the source. The detector receives the fraction of the power corresponding to the ratio of the detector area to the total surface area. Consequently, this ratio is inversely proportional to the measured intensity.

in question above the ecliptic (Figure 2.15). From the base AB and the two angles $\angle SAE$ and $\angle SBE$, the triangle ΔABS can be constructed and evaluated. In determining a moving-cluster parallax (Figure 2.16), the radial velocity of a stellar cluster is the basic quantity that can be translated into a true proper motion by an angle that can also be observed. The apparent proper motion being measured; its relation to the calculated true proper motion yields the distance. Another opportunity to observe the apparent size of objects with given physical size is explained in Figure 2.17.

In addition to the apparent size, another important method for characterizing distances is determination of the apparent magnitude for sources of known luminosity. Here, the base of the triangulation is the size of the observer's detector. The fraction of the energy flowing through its area is inversely proportional to the surface of the sphere that can be imagined around the source and passing through the detector (Figure 2.18). The surface of the sphere collects all the power emitted by the source independently of its size. Distance measurements on the basis of apparent magnitudes or intensities may be interpreted as determination of surfaces of virtual spheres imagined around the source and passing through the observer (or vice versa). Therefore, they are important for cosmology because, due to the possible curvature of space, the square of distance should not be expected to be simply proportional to the spherical surface area. In cosmology, the other basic measurement is the volume that is determined by counting objects of a given class assumed to be homogeneously distributed. In the calculation, the Einstein equations for the universe (Friedmann equations) and the cosmological red-shift (the expression of the overall expansion of the universe) must be taken into account, of course.

2.3 Physical Prerequisites of Geometry

In this section, we shall investigate some deeper problems of the connection between physics and elementary geometry, and the impatient reader may jump to the next chapter.

The investigation of the laws of forms requires stability of these forms. Therefore, we draw geometrical figures on rigid bodies in which the coordination of atoms and molecules is permanent and does not change by manipulations such as rotation and displacement.[4] We see immediately many properties that can be compared independently of position, orientation, or history: first of all the shape of a rigid body. Furthermore, we find out that one has to measure with high precision in order to detect any dependence of the properties of a body or of a process on the place and the time of its preparation. That is, the foundation of the point of view that observed relationships is not due to the absolute position and orientation of objects in space–time, but to as yet unspecified interactions with other objects. In this way, we arrive at the first relativity principle.

> **Position and orientation of an object can only be determined in relation to other objects. Two objects differing only in position and orientation are identical. If these objects are geometrical figures, we call it congruence.**

It is difficult to imagine how a geometrical system could be found without this physical phenomenon. Nevertheless, after learning the constructional features of geometrical relations, we can imagine a universe in which the principle of relativity formulated above did not hold. Comparing with our experience, one would call such a space inhomogeneous. Physics could be like that. Discovering that the first principle of relativity is only an approximation would force us to look for physical reasons for such an inhomogeneity just as the structure and the internal motion of a nearly rigid body is subject to explanation.[5]

A rigid body can be used as a measuring rod as far as its structure and size are guaranteed by the stability of both the structure and interaction of its constituent atoms. Thus, the standard

---

[4]One could object that the argument is circuitous because the stability of the microscopic entities is found and formulated by macroscopic observations that need stable macroscopic bodies to be set up [37]. However, this situation is not unusual in physics and shows that one can only find either consistency or contradiction. It might be that there exists more than one consistent and applicable description, but so far we are happy to have at least one.

[5]The simplest example of a position measurement without obvious relation to a distant object is the measurement of height by a barometer. Apparently, we find one component of the position without seeing the ground. In fact, it is not seen directly, but its presence at a certain distance is inferred from the state of the atmosphere. This atmosphere assumes the role of the external object. We can see that the height measurement by a barometer is also a measurement in relation to an external object, not an absolute determination.

meter of Paris fixes the length unit through the characteristic distance of the atoms in the metallic structure. Macroscopically speaking, this characteristic distance is determined in turn by the equilibrium of different forces.[6] We are ready to accept that when their source can be approximated by a geometrical point the most important forces (gravitation, Coulomb force) depend only on distance, and that the equipotential surfaces are spheres. However, one should see that the sphere is defined physically only by the equilibrium of just these forces, which renders the structure of an isolated body indifferent to reorientation.[7] It is a reassuring observation that at least these forces agree to conspire in favor of geometry, and that spheres can be defined. We should remember that part of every notion is always a mere convention, while the remainder depends on its consistent applicability [43].

Let us imagine that there are forces that do not conspire to create geometry. Then objects of different compositions can be conceived that change their form relative to each other because the mean distances of atoms (of the one body relative to those of the other) change on reorientation or repositioning. Then any immediate development of a notion of congruence would be excluded. If two rigid bodies may be compared independently of their orientation and position in space, geometry and length in particular are defined. With respect to this length, equipotential surfaces are necessarily spheres, and the space appears necessarily homogeneous and isotropic, i.e., without privileged direction. Let us for the moment imagine an absolute space without relation to immersed objects in which physical bodies would expand if turned into a certain direction.[8] If such a dilation were to affect the individual bodies by different amounts, absolute directions would become observable, but the notion of congruency would be found inappropriate.

We must acknowledge that microscopic precision of measurement does not necessarily improve the visibility of geometrical properties. The inhomogeneity of the matter distribution on microscopic scales can make it more difficult to see the global relations. Roughness of measurement averages microscopic peculiarities and is thus required for many intuitive considerations to make sense.[9] For example, Galileo's statement that all objects fall with the same acceleration can only be demonstrated to rough precision in a casually designed experiment. Considerable effort in preparation of the experiment is necessary to verify that the accelerations are identical to high accuracy [44, 45]. Observation tests applicability, not the law itself. Therefore, we can use inaccurate preparations of our experiments if we do not exaggerate the individual result. Euclid, as it is told, drew his figures in the sand.[10]

---

[6]Newton's second law requires the explanation of any acceleration as due to interaction with other physical objects, which is interpreted as a force at the position of the accelerated object. Newton identified the *gravitational force* with its famous dependence on the inverse square of the distance as the cause of the orbits of planets and moons around massive celestial bodies. Later it was found that the electrostatic force also obeys the same law of dependence on distance, and that one may represent these forces as gradients of appropriate potentials. Euclidean geometry is connected with the empirical fact that these potentials depend only on the distance to the central source of force. Anything else would not only complicate physics but also empirical geometry.

[7]If only one force existed, that would be trivial. The point is again that it is the conspiracy of the different forces and the consistency that matter.

[8]In Aristotelian physics, the vertical to the surface of the earth could produce such an effect.

[9]Another side of this fact is that averaging is deliberately used by experimentalists to improve the accuracy of a macroscopic measurement by smoothing out short-range noise.

[10]This emphasizes the necessity of independent proofs and not a fundamental lack of precision of the method. In contrast, the construction with ruler and compass was the most precise method of calculation till the advent of the logarithm tables [81].

Geometry is abstracted from observations despite the fact that congruence of rigid bodies is only approximately valid, mainly because ultimately rigid bodies cannot exist: First, all bodies allow internal motions (acoustic waves) and even plastic deformations. As one learns in thermodynamics, the amount of internal motion increases with temperature. Since the advent of quantum mechanics, we have also learned that the atoms will not be at rest with respect to each other even at zero absolute temperature. Moreover, one learns from general relativity that the gravitational field must be interpreted as curvature of space–time varying from point to point in such a way that motion of a completely rigid body is impossible in principle as well as in practice. However, the only approximate congruence of solid bodies is already sufficient for the existence of a geometry.

There exists a similarly important restriction when we compare light rays and straight lines. A light ray always has a nonzero divergence (due to the second law of thermodynamics) and an uncertainty of position and direction (produced by diffraction on measured and measuring objects). Together with Platon's absolute identification of the light ray with a straight line, one has to keep in mind Aristotle's objection that the straight line of geometry can never exactly coincide with anything in reality.[11] Nevertheless light has a very exceptional role in both the special and general theory of relativity: one can build their theories of measurements merely by using light and a standard length or standard (atomic) clock at one event only [41, 46, 47]. Of course, the atomic clock itself is a complicated object. In addition, it is much more recent than the concepts of mechanics that we shall use here. We have the impression of a deeply rooted conspiracy among the observed motions that allow a simple geometric understanding of time [48]. Newton's first law is a statement about such a conspiracy of free motions (Section 2.1).

Just as changes in position and form of different bodies are measurable only in relation to each other and therefore allow a geometry of space, we measure the course of different *motions* relative to each other and thus perceive time. Again, in order to prepare the physical notion of time we must stipulate by convention that a periodic system defining the unit produces equal units independent of where and when it is started. Without such a general independence of time itself (at least in a first approximation) it would be difficult to define a measurable concept of time at all. For the moment, the notion of time seems to be independent of the experience of space. Ideal clocks are not changed by reorientation and repositioning; even motion does not seem to change them as long as the accelerations do not produce perturbing inertial forces that are too strong. Apparently, one can transport a normal clock in order to synchronize all other clocks by comparison and to get an absolute time by this procedure. Absolute time includes the following: Whether or not two events are simultaneous seems to be a question that can be definitively decided by only one measurement.

All this discussion shows how necessary it is to get a point to start from that does not refer to all these complicated notions of real bodies, sticks and clocks. It is just dangerous to be satisfied with objects that do not explain but are to be explained instead. This is the reason why we start from axioms such as Newton's first law (Section 2.1).

[11] "It is not even true that geodesy considers only seizable and transitory quantities: It would perish together with them. Astronomy as well does not only deal with sensual quantities and the given sky. The sensual lines are not the lines considered by the geometer." (Aristotle, *Metaphysics*, volume B.2).

3 Reflection and Collision

3.1 Geometry and Reflection

In this section, we leave physics for the moment and introduce some elementary notions of geometry. We shall find out which construction we need in a space–time in order to generate a geometry.

Habitually, we understand a *geometry* as the complex of relations that result from a convention, namely, which figures we should accept as having the same shape, i.e., being congruent. We have already seen that having the same shape is in geometry a property more general than is suggested by everyday use and refers directly to the transformations that we have agreed upon to accept as allowed. For instance, by enlarging the Euclidean convention we can regard as congruent figures that are only similar provided that the dilations are included in the allowed transformations (which is not done in the Euclidean geometry, of course).

In the Euclidean geometry, the notion of congruence can be reexamined in mechanics when we try to move congruent figures into an identical position by a combination of consecutive translations and rotations. Above all, we intend to translate and rotate material bodies, rigid bodies. The figures in question are drawn on their surfaces. Consequently, what is congruent in practice depends on the laws of physics that determine the real motion and formation of a rigid body. In mathematics, we abstract from such models and call any change that puts congruent figures into an identical position a motion. Congruence means equivalence. Hence, the motions must form a group:[1] The composition of two motions is again a motion. If not, we could not speak of equivalence. The trivial motion changes nothing at all. We take it as an identity (neutral element) of the group. Successive motions can be combined at will if their order is not changed. The motion back is included too; it serves as the inverse motion that after composition with the original one always yields the original state. In this conception, geometry represents the possibility of separating external properties of a figure (position and orientation) from its internal ones. To compare with physical terms again, we must identify operations that leave invariant some set of properties of our bodies. This set constitutes the internal properties (in the simplest version, the shape of a body). Equality of internal properties corresponds to congruence; the operations form the group of motions. In the following, we denote as motions the translations, rotations, and their combinations (screwing motions) in space. Physical experience should reveal how rotations in a world of space and time look like. We must find a method for constructing in physics rotations and translations in order to

[1]The formal aspect can be found in Appendix A.

The Geometry of Time. Dierck-E. Liebscher

ISBN: 3-527-40567-4

obtain an impression how to represent motions. After this is done, we can look for abstract definitions.

We learn at school already that (Euclidean) translations and rotations can be composed of two reflections.[2] This is important for our purposes, because we can easily construct mechanical realizations of reflections and because the totality of reflections is simpler than that of rotations in this respect. Nevertheless, the reflection on a line (in space on a plane) is not a motion composed of translations and rotations. In the space of everyday experience, a reflection always produces virtual, intangible images. The reduction of real motions to virtual reflections seems to be only an abstraction. However, we obtain through reflections the form of motions (rotations in particular) in a world of space *and* time.

The reflections are particularly simple because they are their own inverse. The same reflection that produces the virtual image reflects this image into the position identical to the imaged object. In practice, stating the congruence of objects depends on the correct comprehension and combination of the properties of reflections. Everybody can see in a double mirror that an image reflected a second time has no longer permuted sides but is only rotated (Figures 3.1 and 3.2). When we stand between two parallel reflecting walls we see ourselves replicated in a long row of shifted images alternately displaying permuted sides and regular ones. Double reflection on parallel mirrors yields translation. This completes the property of reflections of generating rotations and translations. With these rotations and translations, we can now check our conception of congruence: Two figures are congruent if they coincide when they are brought to the identical position by a combination of rotations and translations. The reflections generate all motions [49, 50]. Consequently, reflections define the comparison of lengths and angles by congruence of finite lines and angles (Figure 3.3).

The most important angle is the right angle. A line is *orthogonal* to a mirror $\mathcal{S}$ if it coincides with its reflected image. A right angle reflected on one of its legs is complemented by its image to a flat angle. The line joining a point A to its reflected image $S[A]$ is the perpendicular from A to the reflecting line. The reflecting line is the locus of all points equally distant from A and its image $S[A]$. In addition, the reflecting line divides the angles $\angle AQS[A]$ into two equal parts. We will illustrate these definitions many times.[3] This is the factual definition of the comparison of lengths and angles. The main point is that it is impossible to describe a right angle without defining a reflection or to describe a reflection without defining perpendicularity before. In the particular cases, one of the two must be defined explicitly. We shall start with the definition of reflections.

We now note that the usual construction of a perpendicular takes just the opposite way: One starts with the metric property, takes the compass, and determines the intersections of circles (Figure 3.4). Here, we do not proceed this way. Just because we intend to derive the motions from the reflections (no other means being in sight) length and angle are derived concepts. The circle will be such a *derived* concept, too. Remembering the fact that the

[2] In the plane, the mirrors are meant to be straight lines. If we emphasize the *map* aspect (in which a reflection is simply a nontrivial map that is its own inverse), there also exist other constructions (for example the inversions on the circle). Here, reflections on points become important. In the plane, the reflection on a point can be regarded as a rotation too, the angle of rotation being the flat angle. Consequently, point reflections are the product of two reflections about straight lines that both pass through the point in question and are orthogonal to each other (Appendix A).

[3] In spite of the fact that some readers may find this axiomatic language difficult, they are asked to have patience. The necessity of the given abstraction will become clear by practice.

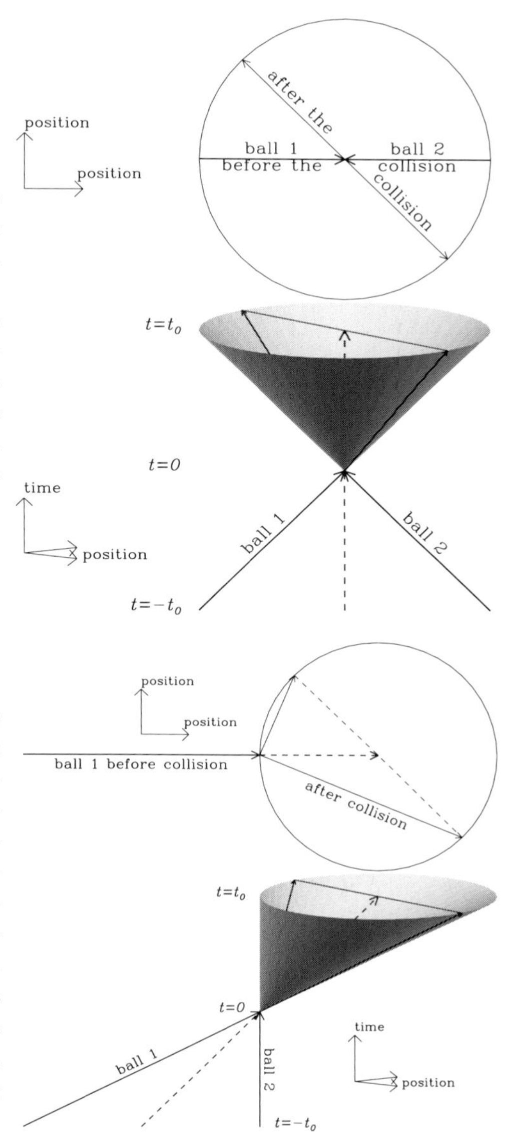

Figure 3.7: *Timetable of a symmetric collision.*

Two billiard balls collide with equal but opposite velocity. After the collision (at $t = 0$) they leave again with equal and opposite velocity. The locus of positions reached after some given time $t = t_0$ forms a circle. If no kinetic energy is lost, the diameter of this circle is equal to the distance of the colliding balls at the time $t = -t_0$ before the collision. In the upper part of the figure we see the ground view of the cone with a vector pair representing the equal and opposite velocity of the partners after the collision.

Figure 3.8: *Timetable of a billiard collision.*

We illustrate the collision of a moving ball with one ball at rest. After the collision, the balls move with equal and opposite velocity relative to the common center of mass that is in motion itself. Therefore, after a given lapse of time the possible worldlines reach a circle again (i.e., a sphere in three-dimensional space). The upper part shows the vector pair in the upper part of Figure 3.7, now forming a right-angle triangle. The third edge passes through the center of the circle. This figure may be transformed from Figure 3.7 by some shear. The sections parallel to the spatial plane are identical, only shifted to the right by an amount increasing linearly with time.

combined with the velocities measured by the observer in the one system to yield the velocities measured by the observer in the other one. Huygens supposes an additive composition of velocities, quite expected by common sense.

This is the first example of how we conclude from mechanical laws geometrical relations and of how these relations turn out to be very different from the Euclidean geometry. After all, in the expected geometry of timetables Figures 3.7 and 3.8 are congruent: Up to the orientation in space–time (i.e., up to a common velocity), they describe the same physical object. In fact,

Figure 3.9: *Relativity of velocity.*

This is Huygens' famous sketch of the comparison of two observers in relative motion: One man on the bank of the river and one in the boat passing by. After Galileo, Huygens was the first to argue for a universal subtraction of the relative velocity of the observer if a given motion is described from a point moving itself. Using his argument, we derive Figure 3.8 from Figure 3.7. If the man standing on the bank moves the two balls against each other with equal velocity in such a way that one is moving with the velocity of the boat, the man on the bank observes the situation shown in Figure 3.7, and the man in the boat observes that shown in Figure 3.8.

the cone in Figure 3.7 is realized in total if the collision experiment is repeated often enough. In a single-collision experiment, one finds a final motion of the partners that is represented by opposite mantle lines of the cones in the timetable. Equivalently, the cone may be interpreted as the totality of the world-lines of fragments of an explosion, if all these fragments leave the position of the explosion with the same speed (Figure 2.8).

We easily conceive the velocities $\boldsymbol{v}$ of objects before and after a collision.[7] The attempt to state a balance of the total velocities fails. The sum of the velocities before the collision *differs* in general from the sum after the collision. The curious and basic experience is the observation that one finds equal sums after the velocities $\boldsymbol{v}$ are multiplied with some weight factors. These factors are called *masses*, more precisely inertial masses, in order to distinguish them from the gravitational charge.

[7] Velocities are usually described by the three components along the different directions of space. Such a quantity is called a vector. As usual, a boldface letter stands for all three components of a vector such as a velocity.

If one adds the velocities that the bodies possess before the collision, each weighed with the corresponding inertial mass, one finds the same result as can be obtained by weighing the velocities of the collision products with their inertial masses after the collision. The sum of the inertial masses after the collision is equal to the sum before.

Moreover, the masses, being independent of the particular circumstances of the collision, appear related only to internal properties of the objects. We call momentum the product of velocity $\boldsymbol{v}$ and inertial mass m. Momenta can be added simultaneously with the masses, and multiplying the mass means multiplying the momentum too. Therefore, we form out of the mass and the three spatial components of the momentum a four-component quantity. It is called *four-momentum* in order to distinguish it from the momentum in space introduced above. In a general collision, the sum of the four momenta after the collision is equal to the sum before the collision. This *momentum conservation theorem* is the foundation of dynamics. Usually, it is understood as a corollary to Newton's third law, in a form given by Huygens.[8] We construct the total four-momentum (constant in the collision) by a momentum parallelogram (Figures 3.10 and 3.11). At this point, it is important to note that the diagrams in the mv–m plane (Figure 3.11) *cannot be but similar* to those in the x–t plane. This is true in the four-dimensional case too, because the masses do not depend on orientation. *Any reflection in the space–time is related to the corresponding reflection in the momentum space.* In addition, the momentum diagram (Figure 3.11) constructed from the registered velocities (Figure 3.10) shows immediately that the construction of weights in order to get conserved quantities includes the conservation of these weights themselves.

The general theorem of conservation of momentum shall be exemplified now through the collision of two bodies. First we consider the totally inelastic case. Here, the collision binds all partners together, and they form a common object. What is its velocity? The third law, in Huygens' terms, expresses the experimental finding that there exists a mixing rule. The velocity $\boldsymbol{V}$ of the object formed in the totally inelastic collision by the two colliding particles is the weighted average of their velocities $\boldsymbol{v}_1$ and $\boldsymbol{v}_2$ before the collision,

$$\boldsymbol{V} = \frac{m_1\,\boldsymbol{v}_1 \;+\; m_2\,\boldsymbol{v}_2}{m_1 + m_2}, \qquad M = m_1 + m_2. \tag{3.1}$$

Again, if one adds the velocities that two bodies possess before the collision, each weighed with the corresponding inertial mass, one finds the same result as can be obtained by weighing the velocity of the newly formed body after the collision with its mass, which is the sum of the two initial masses. In some sense, the totally inelastic collision is the opposite of the perfectly elastic collision. In perfectly elastic collisions, the partners preserve all their internal properties (in particular, their internal energies) and only change their velocity. The general energy conservation takes the form of the conservation of the sum of the kinetic energies $\frac{1}{2}m\boldsymbol{v}^2$. The velocity $\boldsymbol{V}$ that is given by Eq. (3.1) is the velocity of the center of mass. If we refer to it, we obtain for the velocities $\boldsymbol{v}_k{}'$ after a perfectly elastic collision simply

$$\boldsymbol{v}_1{}' - \boldsymbol{V} = -(\boldsymbol{v}_1 - \boldsymbol{V}), \qquad \boldsymbol{v}_2{}' - \boldsymbol{V} = -(\boldsymbol{v}_2 - \boldsymbol{V}). \tag{3.2}$$

[8] In this form it could be named Huygens' law. It is published extensively in the posthumous writings [54]. The concept is older, however, than Newton's *Principia*. Most important, it does *not* need the notion of force. Hence, its place should next to the first axiom.

It is the relative velocities that change their sign.[9] Figure 3.12 shows momentum diagrams of such elastic collisions for four different ratios of the masses, but with equal initial velocities. After the collision, the possible motions of each partner form an oblique circular cone about the world-line of the center of mass. The vertex angle is inversely proportional to the inertial mass itself. The undisturbed motion always yields a generatrix of the cone. If the mass of one partner is very large compared to the mass of the other, it acts as a perfect mirror: The change of its own velocity can be neglected, and the other partner changes the sign of its relative velocity. Figure 3.13 shows the situation in two spatial dimensions.

After this preparation, we consider a swarm of particles that simultaneously begin to move, as in an explosion, in all directions with the same speed at some event E. Their world-lines are all straight and have equal inclination to the time axis. They form a cone (Figure 2.8). If we erect a mirror in the path of these particles we obtain a preliminary physical method to construct reflections in a timetable (Figure 3.14). To construct a full geometry, we now have to ask for a reflection on arbitrary planes, i.e., on a mirror moving uniformly but with arbitrary velocity. We consider such a moving mirror now. The momentum balance yields again a cone of reflected world-lines, but this cone is now oblique (Figure 3.15). Nevertheless, its vertex is well defined and yields the reflected image $S[E]$ of the explosion E. This is the physical construction of the reflection in a space–time. The result $S[E]$ does not depend on the speed chosen for the fragments of the explosion. If we consider a timetable of two dimensions only, we use the fact that the image of an event that is reflected on a world-line does not depend on the speed of the reflected real particles that are emanating from the event (Figure 3.16). We obtain the reflection prescription by considering two particles of different velocities passing through the event in question. They are both reflected as usual. Their new world-lines are followed back to their intersection point. This is the event reflected by the world-line of the mirror. It is obvious that the reflection produces an image event *simultaneous* with the original event. The usual distances to the mirror are equal too (Figure 3.17). This is a situation that we expect as "natural." Indeed, we always see our image in a mirror just at the moment when we look at it, and it recedes into the depth of the mirror in just the degree we recede from it. Nobody observes any retardation in the motions of the image as compared with her or his own motions.

The first consequence of this absolute simultaneity is a rather strange kind of distance in space–time. *This space–time distance is the pure time lapse.* Let us suppose two events O and A, and mirrors passing through the event O with arbitrary velocity. All reflected images $S[A]$ are simultaneous with A. However, the distance has to be defined in such a way that all images $S[A]$ have the same distance to O as A. Consequently, the space–time distance depends only on the time lapse between the events O and A, or O and $S[A]$. The relative spatial position of the different images $S[A]$ does not influence it.

It follows that distinct but simultaneous events (with no time lapse in between) have space–time distance zero. Compared with the Euclidean experience in space, this is something completely new and unexpected. For *space–time*, it is typical, and the following chapters will provide more examples. What about the *ordinary* distance of simultaneous events? The spatial distance is related to an angle in this construction, i.e., it provides a measure for the difference in space–time directions. Angles are formed by two world-lines passing through the same

[9]Newton used the word *reflection* for collisions in general.

Figure 3.10: *Conservation of momentum and inertial mass. I.*

We construct the timetable for a general one-dimensional (central) collision of two billiard balls of different masses. Ball 2 is at rest before the collision with ball 1. After the collision, ball 2 will move in the direction in which it was pushed. Ball 1 will, depending on its mass, follow it, come to rest, or spring back. For the moment, the ordinate of the velocity vector only denotes the trivial value 1 for the clock rate. Now we try to find the straight line that divides the attained positions at fixed times before and after the collision in the same ratio. The line that we obtain is the world-line of the center of mass. The individual masses are inversely proportional to the distances from this line.

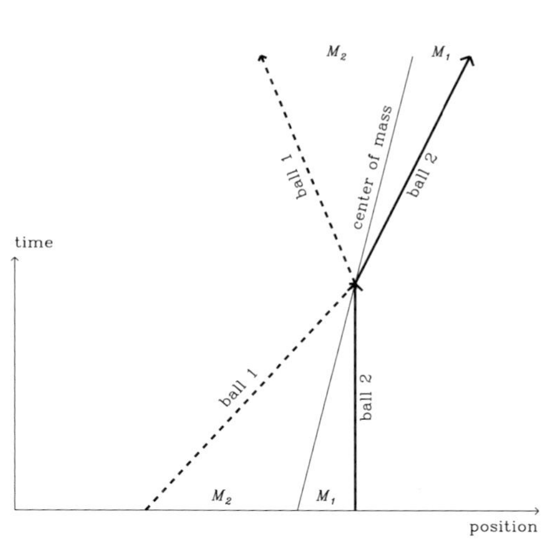

Figure 3.11: *Conservation of momentum and inertial mass. II.*

For Figure 3.10, we draw the sum of the velocities before and after the collision. The two sums differ in general, i.e., the sum of the velocities is not conserved in the collision. We get conservation if the velocities are weighted, i.e., stretched or shortened corresponding to their weight. The weighted velocities are the momenta; the weights themselves are the inertial masses. If the ordinate of the velocity is the clock rate, the ordinate (time component) of the momentum is the mass itself. The total momentum has the direction of the world-line of the center of mass. This center of mass is, at every instant of time, the average of the positions weighted with the obtained masses. We show the case of a perfectly elastic collision: The distances AS and SB are equal. $(BS)^2$ and $(AS)^2$ are proportional to the kinetic energies before and after the collision.

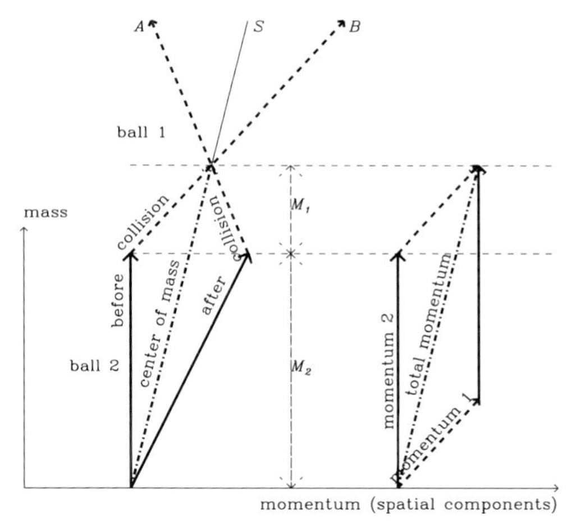

event, which is the vertex of the angle. Two angles at some event are equal if their sides cut off segments of a horizontal plane that are equal in the usual sense. If the angles are carried by different events, the lengths of a segment must be divided by the space–time length of the sides of the angle. This last is a time: In such a way, angles in space–time must be interpreted physically as relative velocities.

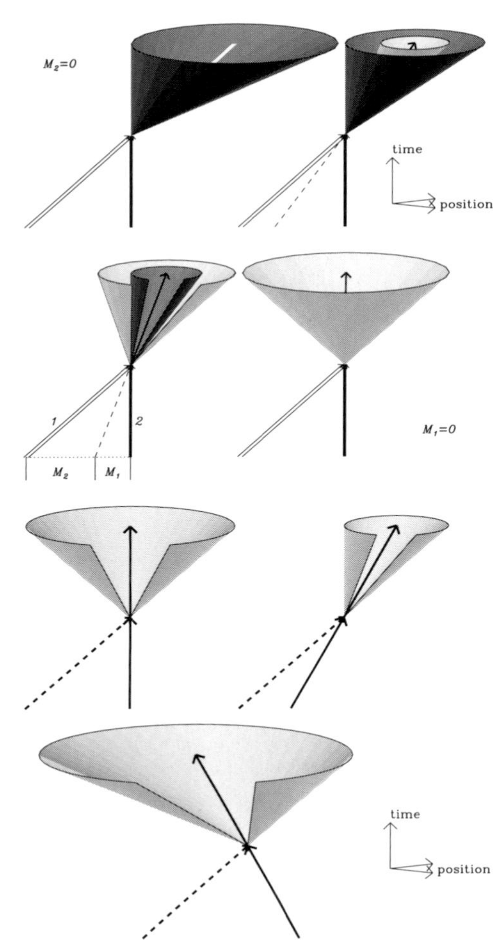

Figure 3.12: *The elastic collision with different masses.*

We first construct the total momentum and obtain the world-line of the center of mass. Now the collision is the physical reflection on this world-line. In the lower left, we recognize the situation shown in both the previous figures. The cones are cut open in order to facilitate the comparison. In the lower right, the pushing mass is much smaller than the pushed one; in the upper left we see the opposite case. In the upper part, the pushing mass is heavier than the pushed one; in the lower part, this relation is inverted.

Figure 3.13: *The elastic collision with a very heavy target.*

We consider the lower right of Figure 3.12 and generalize it for the case when the pushed ball moves itself. The physical reflection is found by marking the relative speed before the collision in all directions after the collision.

The geometry that we have constructed shows two facts. First, it is physics that can lead us to find the appropriate geometry for space–time. Second, the geometry of space *and* time is rather different from the geometry of space alone. However, the two geometries are structurally similar with respect to the existence of points, straight lines, angles, distances, and unique parallels. We developed the geometry by considering the mechanics found by Galileo and Newton. It is called the *Galilean geometry*.[10] This geometry admits for generic cases the desired comparison of lengths and angles, mediated by reflection. The horizontal plays a distinguished role that is the immediate expression of absolute simultaneity. The space–time distance between two events is given by the time interval and does not depend on the separa-

[10]Galileo would certainly have refused to accept this geometry as *Galilean*. We use the notion as an abbreviation for the geometry of space–time induced by Galilean relativity. In fact, seeing a geometry of space *and* time behind mechanics is the consequence of the Einstein's relativity [37]. More extended considerations of the Galilean geometry are found in [36].

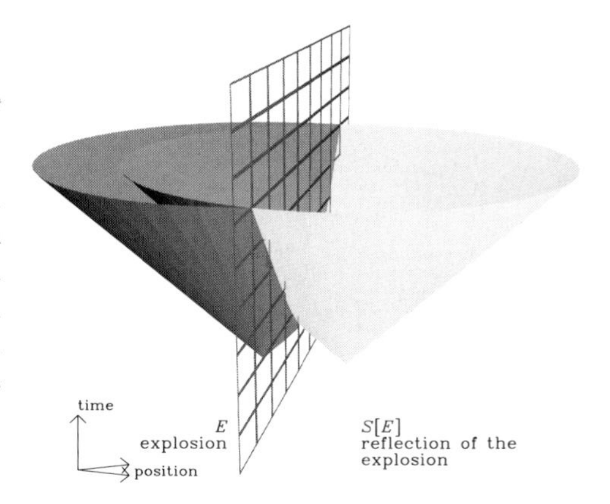

Figure 3.14: *The reflection of an explosion cone at a fixed reflector.*

The world-lines of the fragments produced in an explosion E, if they all have equal speed, form a cone. The world-lines of the fragments reflected by the mirror form part of a second cone, which can be completed in order to find its vertex $S[E]$. This vertex is the event that has to be taken as the reflected image of the explosion E.

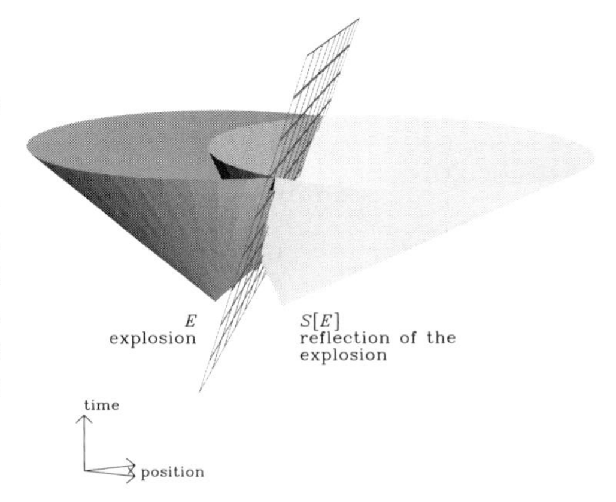

Figure 3.15: *The reflection of an explosion cone at a reflector in motion.*

If the mirror is itself moving, the reflected fragments form an oblique circular cone, which can be completed to show its vertex $S[E]$. This event is the reflection image of E by the moving cone. Comparing with Figure 3.14, we can see that the simultaneity of E and $S[E]$ does not depend on the motion of the mirror.

tion in space. An event D can be conceived as a reflection image $S[C]$ of some other event C if and only if C and D are simultaneous. All points on a horizontal have the space–time distance zero (Figure 3.17). Independently of the inclination of a line in the timetable, the perpendiculars (given by the lines connecting a point A to its image $S[A]$) are always horizontal. Conversely, the only exception is the horizontal itself. All other straight lines are perpendicular to it. This state of affairs is curious but perfectly consistent.

It is necessary to note that there remains a difference between the individual physical process of the reflection of a real object by a real mirror on the one hand and the abstract reflection by an abstract line in a timetable on the other. We intend to take the world-line of the *real* object (after its reflection by the collision with the real mirror) as *abstract* reflection of the virtually undisturbed original world-line on the world-line of the mirror. This identification allows us to use *real* reflections to extract the abstract definition for an operation that produces *virtual* images, as we know from Figure 3.1.

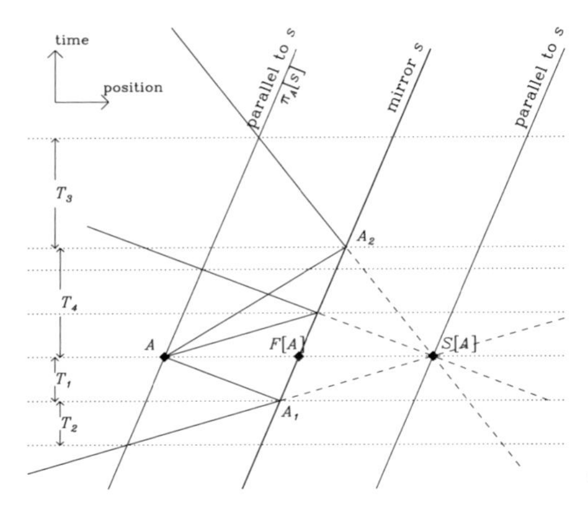

Figure 3.16: *Construction of the Galilean reflection in the space–time plane.*

This is a kind of vertical cut through Figure 3.15. The event A is seen in the moving mirror s. We draw the parallel $\pi_A[s]$ to s through A. Then a particle coming from A will be reflected at the mirror in such a way that it needs the same time to reach there and back: $T_1 = t[A, s] = t[s, \pi_A[s]] = T_2$. If we follow the reflected world-lines behind the mirror, all meet at one world-point, the event $S[A]$ that is the reflected image of A. Moreover, a particle mirrored to pass through A would pass through $S[A]$ if the mirror did not reflect it.

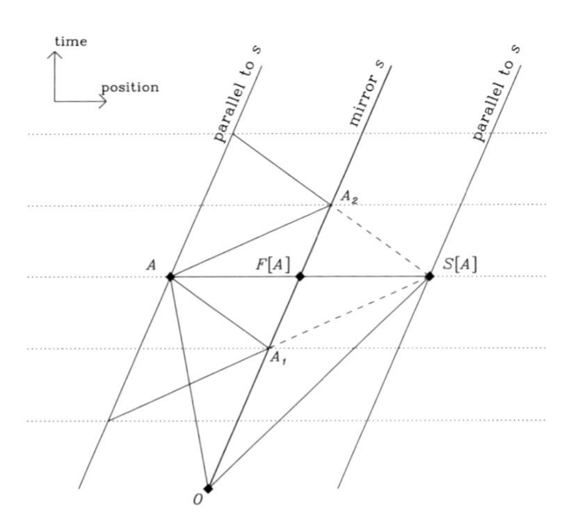

Figure 3.17: *Length and angle in the Galilean geometry.*

Reflections are to determine the comparison of lengths and angles. A reflection ought not alter a length or an angle (except for the sign). This means for the Galilean geometry that the length of chord is given by its time component $d[O, A] = d[O, S[A]]$ and that the spatial distance becomes a measure of the angle, $\angle AOF[A] = \angle S[A]OF[A]$. Equal angles signify equal ratios of spatial distance to time interval, i.e., equal relative velocities. In the end, we can interpret the construction with the same wording as in Figure 3.4, only the measure of distance has changed.

The advent of the theory of electromagnetic phenomena including light changed the picture again. It showed that the simple composition of velocities that we used invoking Huygens' cannot be applied for large velocities, that is, for velocities that are a significant fraction of the speed of light.[11] Nevertheless, we shall again be able to construct a geometry on the basis of this experimental fact. This time, the consequences are even more surprising.

[11] We remember that *being large* presupposes the existence of a comparison *standard*. The standard that can be reproduced best is nowadays the speed of light. As regards relativity, the speed of light is fundamental for its universally isotropic propagation.

4 The Relativity Principle of Mechanics and Wave Propagation

When two figures with different Galilean orientations are congruent, all velocities differ by a common *relative velocity*. This relative velocity is simply the difference of corresponding velocities here. Congruence means that internal properties of the congruent figures do not allow us to determine position and orientation (in space–time, reorientation includes composition with a common velocity). Reorientation and translation produces congruent figures, i.e., figures with equal internal properties. In the language of mechanics, the observation of the internal relations in some process of motion does not suffice to determine a common velocity without reference to external objects. In mechanics, only relative velocities matter. Galileo describes this observation in his beautiful language[1] in his Dialogs. He considers flies, but-

[1] *Salviati.* Riserratevi con qualche amico nella maggiore stanza che sia sotto coverta di alcun gran naviglio, e quivi fate d'aver mosche, farfalle e simili animaletti volanti; siavi anco un gran vaso d'acqua, e dentrovi de' pescetti; sospendasi anco in alto qualche secchiello, che a goccia a goccia vada versando dell'acqua in un altro vaso di angusta bocca che sia posto a basso; e stando ferma la nave, osservate deligentemente come quelli animaletti volanti con pari velocità vanno verso tutte le parti della stanza. I pesci si vedranno andar notando indifferentemente per tutti i versi, le stille cadenti entreranno tutte nel vaso sottoposto; e voi gettando all'amico alcuna cosa, non più gagliardamente la dovrete gettare verso quella parte che verso questa, quando le lontananze sieno eguali; e saltando voi, come si dice, a piè giunti, eguali spazii passerete verso tutti le parti. Osservate che avrete diligentemente tutte queste cose, benché niun dubbio ci sia che mentre il vascello sta fermo non debbano succeder così: fate muover la nave con quanta si voglia velocità; ché (pur che il moto sia uniforme e non fluttuante in qua e in là) voi non riconoscerete una minima mutazione in tutti li nominati effetti; e da alcuno di quelli potrete comprender se la nave cammina, o pure sta ferma. Voi saltando passerete nel tavolato i medesimi spazii che prima; né perché la nave si muova velocissimamente, farete maggior salti verso la poppa che verso la prua, benché nel tempo che voi state in aria il tavolato sottopostovi scorra verso la parte contraria al vostro salto; e gettando alcuna cosa al compagno, non con più forza bisognerà tirarla per arrivarlo, se egli sarà verso la prua e voi verso poppa che se voi foste situati per l'opposito: le gocciole cadranno come prima nel vaso inferiore senza caderne pur una verso poppa benché mentra la gocciola è per aria, la nave scorra molto palmi; i pesci nella loro acqua non con più fatica noteranno verso la precedente che verso la susseguente parte del vaso; ma con pari agevolezza verranno al cibo qualsivoglia luogo dell'orlo del vaso; e finalmente le farfalle e le mosche continueranno i lor voli indifferentemente verso tutte le parti; ne mai accaderà che si riduchino verso la parete che riguarda la poppa, quasi che fussero stracche in tener dietro al veloce corso della nave, dalla quale per lungo tempo trattenendosi per aria saranno state separate; e se, abbrucciando alcuna lagrima d'incenso, si farà un poco di fumo, vedrassi ascendere in alto, ed in guisa di nugoletta trattenervisi, e indifferentemente muoversi non più verso questa che quella parte; e di tutta questa corrispondenza d'effetti ne è cagione l'esser il moto della nave comune a tutte le cose contenute in essa, ed all'aria ancora; che perciò dissi io che si stesse sotto coverta, che quando si stesse di sopra e nell'aria aperta e non seguace del corso della nave, differenze più o meno notabili si vederebbero in alcuni degli effetti nominati; e non è dubbio che il fumo resterebbe in dietro quanto l'aria stessa, le mosche parimenti e le farfalle, impedite dall'aria, non potrebber seguire il moto della nave, quando da essa per ispazio assai notabile si separassero, ma trattenendovisi vicine perché la nave stessa, come di fabbrica anfrattuosa, porta seco parte dell'aria sua prossima, senza intoppo o fatica seguirebbon la nave; e per simil cagione veggiamo tal volta nel correr la posta le mosche importune e i tafani seguir i cavalli, volandogli ora in questa ed ora in quella parte del corpo; ma nelle gocciole cadenti pochissima sarebbe la differenza, e nei salti e nei proietti gravi del tutto impercettibile.

The Geometry of Time. Dierck-E. Liebscher

ISBN: 3-527-40567-4

terflies, fishes, men jumping and throwing balls, all in a room under deck of a sailing ship. None of these things can reveal the motion of the ship as long as its motion is uniform (and not rotating).

Not only position and orientation in space, velocity too can only be determined relative to external objects.

Galileo's conjecture (that there is no internal means to establish uniform motion of an isolated, and screened from the outside room) is universally valid. We call it the *principle of relativity of mechanics* because the description of mechanical motion must always respect this requirement. It is, however, nothing else than the application of the relativity principle of geometry (page 18) to the particular case of space–time, and it allows us to apply geometrical reasoning to mechanics.

When we interpret the figures in the Galilean geometry as sketches of mechanical motion, the complement of congruency is *relativity*. Two congruent figures, as we know, differ only in *space–time* position and orientation: This includes position in space and time and orientation in space as well as velocity. All properties that depend on position and orientation are relative unless they refer to other objects; they have no objective meaning. One such relative property is the apparent[2] shape of a distant object: Its apparent form depends on the relative orientation to the observer; its apparent size can also depend on the distance. Our point is that the notion of orientation in a world of space and time includes velocity too. The already known relativity of velocity means in geometry the relativity of orientation in space–time. Also in space–time, orientation obtains an objective meaning only if referred to other objects.

We proceed one step further. The apparent length of a stick in space depends on its orientation with respect to us. It can be oriented transversally and show us a maximum extension, and it can be oriented radially and show us only its thickness. In a space–time, time is a coordinate together with the space coordinates. We should expect us to be forced to distinguish apparent time intervals measured by a distant observer from time intervals measured in the observed object itself. The length of the apparent time intervals should be expected to depend on the space–time orientation, i.e., on the relative velocity of the object. Newtonian mechanics does not contain such an effect. Earlier physicists concluded that time is absolute and believed that time intervals do not depend on the state of the observer. After having analyzed the propagation of waves, we shall be forced to revise this geometry of space–time. One of the consequences will be that the apparent flow of time becomes relative, too. The surprising consequences of such a relativity are our subject now.

In Newtonian mechanics, this principle is realized by the unchanged form of the equations of motion when we change the reference frame. These changes are subject to absolute simultaneity and additive composition of velocities. They are called the Galilean transformations.[3]

Sagredo. Queste osservazioni, ancorché navigando non mi sia venuto in mente di farle a posta, tuttavia son più che sicuro che succederanno nella maniera raccontata; in confermazione di che mi ricordo essermi cento volte trovato, essendo nella mia camera, a domandar se la nave camminava o stava ferma; e talvolta, essendo sopra fantasia, ho creduto che ella andasse per un verso, mentre il moto era al contrario [55] (English translation in [56]).

[2]The attribute *apparent* is used exclusively as it is used in astronomy. In a space, it denotes the projection onto the field of view, in astronomy, onto the apparent sky. The projection center is the eye of the observer. In a space–time, it also denotes projection onto the local flow of time and the locally simultaneous space. The attribute *apparent* will never indicate deception or illusion.

[3]For the formal aspect, see Appendix B.

These transformations are subject to a twofold interpretation. First, they are understood as reconstruction of the same scenario for another event, with another orientation in space and with another common basic velocity. Being an active procedure, this is a difficult demand on the experimenting observer. Second, the description of the second experiment can be obtained from that of the first by the pure mathematical substitution of new coordinates for the old ones, i.e., as an operation on a sheet of paper . Relativity means that both ends in the same result. The simplest example is the translation in time. Such a relativity of location in time exists if the result of a preparation at a later time yields the same result as the substitution of the later time in the description of the first experiment. The space–time kinship of translations in space with translations in time is the same as the kinship of velocity and orientation.

The additive composition of velocities led us to see the relativity of velocity. In addition, it seems obvious and out of question that velocities must be composed additively. This picture changes when we consider the propagation of waves. In mechanics, the prototype is sound waves. In contrast to particles, the propagation speed of waves is not connected with the amount of energy or momentum that is transported. Waves seem to be determined by the properties of a carrier medium. In particular, the propagation speed is determined by and refers to the carrier medium. No transport of particles is included. It is the atomistic structure of matter that allows us to make a mechanical model for the propagation of waves. In particular, sound waves are described adequately by mechanical waves, i.e. waves in the motion of the components of a material. The atoms in the lattice of a solid body, as well as the molecules of a gas, are more or less bound to a given position. The particles may, however, push each other around a bit, and it is in this way that energy is transported, without each individual particle straying much from its own territory. The totality of the particles, which do not matter individually, constitutes the medium of sound propagation. It can be considered as the *continuum* or even *plenum* of Descartes,[4] but for us it is important to understand the completely mechanical character of its motions. Relative to the otherwise structureless medium, sound propagates with a speed that in general is independent of direction. The propagation of a sound signal is also depicted by a straight circular cone in our timetable.

Now we interrupt the undisturbed propagation by a reflecting mirror. As long as the mirror is at rest in the medium, the picture of the explosion cone (Figure 3.15) does not change. If the mirror moves, the reflected cone is now part of a cone that is straight and again circular. However, its vertex event $S[E]$ is no longer simultaneous with the explosion at E (Figure 4.1). Supported by our knowledge of the geometry of Newtonian mechanics, we declare that Figures 3.14 and 3.15 are congruent. It follows that the scene of Figure 4.1 is not congruent to them. However, this need not be a defect, because the latter scene contains a relative velocity that must be taken into account: that of the mirror with respect to the medium. After subtraction of the mirror's velocity in Figure 4.1, we obtain an oblique cone for both the direct and the reflected propagation[5] (Figure 4.2). The obliqueness of the cones expresses the fact that the sound propagation is no longer independent of its direction. The medium, at rest in

[4]Before Newton stated gravitation to be an action at a distance, the carrier of any action was believed to be a hypothetical plenum or continuum. The argument in favor of such a construction was answered and founded by particle mechanics in analyzing the atomic structure.

[5]Of course, we have to idealize our mirrors in order to keep them reflecting while the medium can pass freely. For instance, mirrors can be made out of nets. In the discussion of the Michelson experiment, the question was how to enable the assumed *aether* to flow freely through the experimental setup (Figure 4.4).

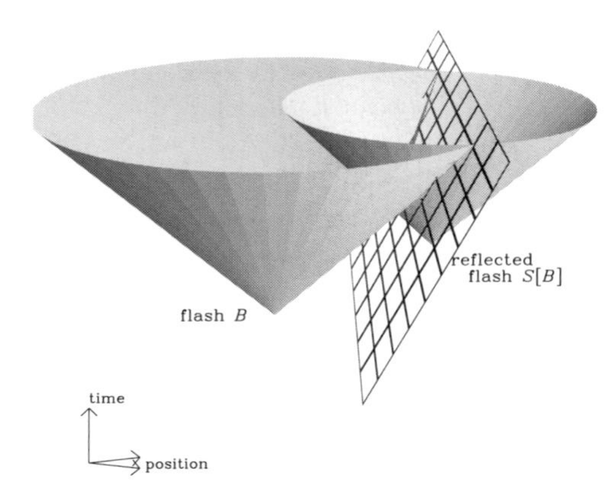

Figure 4.1: *Wave propagation and reflector in motion.*

In contrast to the explosion cone for which the speed relative to the mirror changes its sign but not its magnitude, we now assume that the propagation speed is always independent of direction. The reflection changes the sign of the velocity relative to the medium independently of the motion of any mirror. The reflected wave propagation is now part of a straight circular cone. Remarkably, its vertex is *not* simultaneous with that of the unreflected cone. B and $S[B]$ are no longer simultaneous. In addition, the time lapse between B and $S[B]$ depends on the motion of the mirror. If we ever define space–time reflections with the demand that our figure is to be congruent to Figure 3.14, B and $S[B]$ must be simultaneous. In addition, simultaneity will become dependent on the motion of the observer (plane of drawing or mirror, resp.).

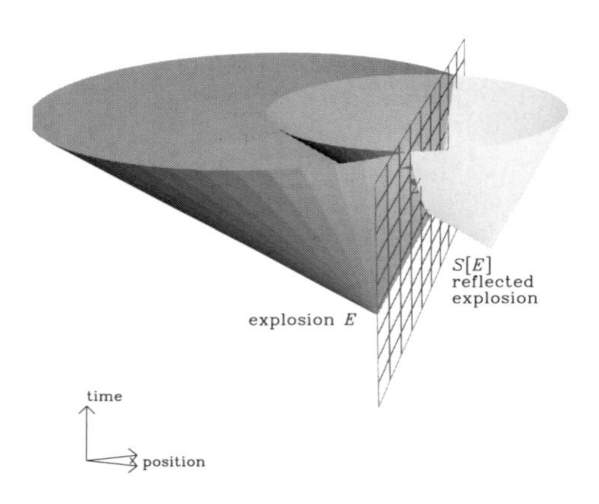

Figure 4.2: *Sound cone in head wind.*

If in Figure 4.1 we subtract the velocity of the mirror by the additive rule, a picture results that can be applied to the propagation of sound, but not to that of light. To get an applicable picture for light, we must proceed as in Figure 3.14. The propagation of light is isotropic even after combination with the velocity of the mirror, in fact with the velocity of any observer. This is in clear contrast to sound propagation, for which the velocity of the mirror must be combined with that of the medium. In the latter case, our figure can be applied. It describes a mirror at rest with a medium streaming from the right.

Figure 4.1, moves after subtraction of the velocity of the mirror. We can state that the scenes of Figures 4.1 and 4.2 are congruent in the Galilean geometry. Corresponding to our operation, the velocity of the wave with respect to the medium is simply added to the velocity of the medium. As long as we experience a material and *tangible* medium as in sound propagation, there is no problem with the relativity principle. With such a medium, we have the necessary external reference for determining a velocity. Without such a medium, the only relativistically invariant isotropic propagation is that with infinite speed. In retrospect, Newton's concept of gravitation that depends only on spatial distance and not on time can be interpreted as propagation with infinite speed.

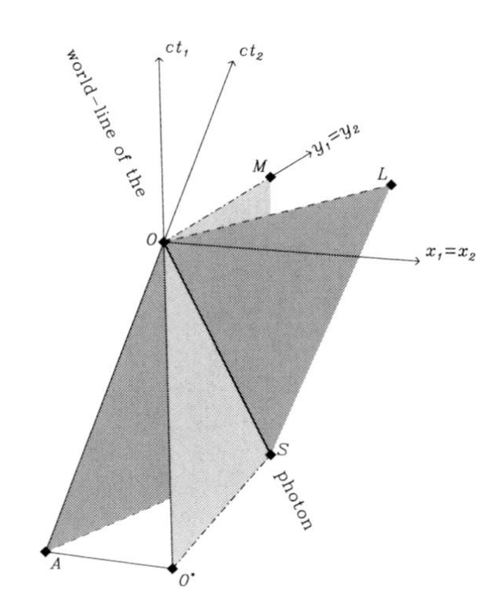

Figure 4.3: *Aberration.*

A photon (particle, signal) comes from S to O. An observer obtains its spatial direction by projection of this world-line onto the space $t = 0$ (just as in Figures 2.3 and 2.4). The observer with the world-line O^*O sees OM; the other with the world-line AO sees the direction OL. The difference of the two is the aberration. The triangle OLM represents the composition of velocities: $\vec{ML} + \vec{LO} = \vec{MO}$.

The propagation of light does not fit to such an explanation. First, light propagation is not instantaneous. It needs time just like sound propagation. Geometrically, this is observed by the Doppler effect (Figure 2.14). In addition, the effect of aberration indicates that the light velocity must be combined with that of the observer (Figure 4.3) and changes at least its direction (when the directions differ). For example, aberration explains certain changes in the apparent position of stars during the year. This position is already influenced by the change of the position occupied by the earth on its orbit (giving rise to parallax, Figure 2.15, which depends on the star's distance) but it is also influenced by the changing velocity, and this gives rise to an aberration ellipse independent of the distance of the star.[6] Consequently, it is to be expected that a particular rest frame of the propagation exists in which, for instance, the propagation is isotropic. The observed velocity of light should be found as the difference of the propagation velocity and the velocity of the observer in this rest frame.

The interference phenomena show that light propagation is wave propagation too. Newton's model of streaming particles cannot explain interference. In this respect, light has properties that we also meet in sound propagation. There is, however, an essential difference. Sound is always transported by a material medium that can be felt, manipulated, and excluded: Sound does *not* propagate through void space. However, light apparently *does*. If a medium existed in spite of this impression, it should be some aether that pervades all space, and that cannot be excluded or extracted from any volume.[7] We have arrived at a curious situation : If the velocity of light must be combined with other velocities by the additive rule, its propagation can be observed as isotropic only for one state of motion of the measuring apparatus. When no material medium is present, this is a state that must be interpreted as absolute rest. It has

[6]For a star in the pole of the ecliptic, its size is $v_{\text{orbit}}/c \approx 10^{-4}$, which results in $20.47''$. The aberration of starlight was discovered by J. Bradley in 1728. At that time, light was interpreted as particles emanating from the source, so nobody was worried by the identification of the direction with the velocity of a stream of particles.

[7]Today, there are many theoretical constructions that pervade space like the zero-point energies of all the quantum fields that have been invented up to now, as well as of those that have not yet been invented. Maybe the aether is something totally new. However, there is no evidence that the conjectured pervading entities have any effect on light propagation.

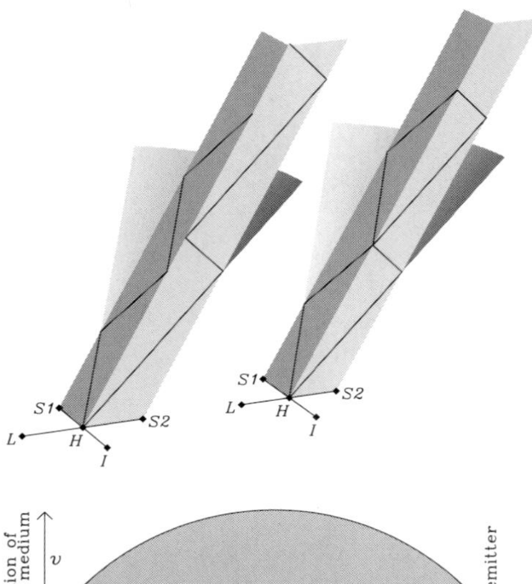

Figure 4.4: *Michelson's aether drift experiment.*

We see the interferometer (upper left) and two of its timetables. The light from a source L is split by the half-transparent mirror H and combined again after the reflection at S_1 and S_2. The difference in elapsed time is measured by the interference pattern. When the arms HS_1 and HS_2 are equally long and the apparatus moves with respect to the reference system of isotropic propagation, a difference is found (left part). It can be eliminated only by accordingly contracting the arm that points into the direction of motion (right part).

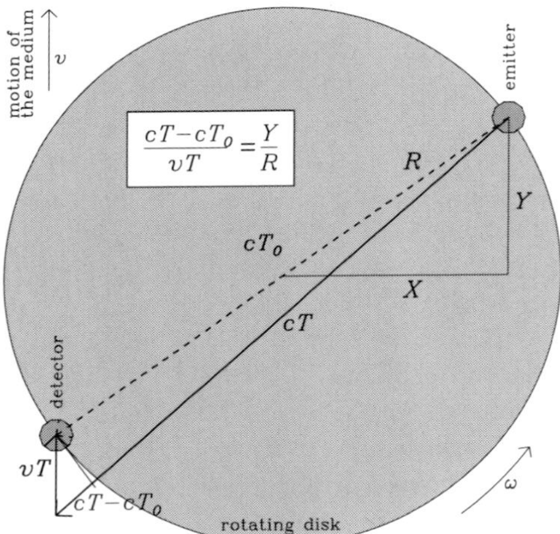

Figure 4.5: *Isaak's one-directional aether drift experiment.*

A rotating disk carries a γ-ray source and an absorber, both of equal frequency. If the carrier medium moves with respect to the device, and if the velocities are combined additively, the arrival times are $t_a \approx t_s + T_0(1 + \frac{v}{c}\frac{Y[t_s]}{R})$, and we obtain the periods $\tau_a \approx \tau_s(1 + 2\frac{v}{c^2}\omega X[t_s])$. The frequencies go out of tune and the absorption is suppressed. Using γ lines, for which the frequency can be determined to a precision of 10^{-15} by the Mössbauer effect, medium velocities should be detectable down to 1 cm/s. No effect has been found [58].

distinguished properties that can be found in an experiment shielded from outside reference objects, for instance, in the ship's interior of Galileo's argument. We must find the anisotropy of the propagation of the light of a light bulb there at rest. Moving with respect to the state of isotropic propagation, the velocity of light should be smaller in the direction of motion, and larger in the opposite direction. This anisotropy should yield an absolute velocity, i.e., a velocity that does not refer to external objects. All this can be expected to be observable in a laboratory shielded against the environment. This would be a contradiction to the relativity principle.[8] However, if we could find this very tenuous or virtual medium, the aether, we shall declare it the cause for the propagation of light. Then this aether serves as an external hallmark just as in acoustics. If we hope to understand light as propagation of some excitation (as it is done successfully for sound), we must expect the existence of such a medium and

[8]Even simpler, this would attribute a definite speed to nothingness [37].

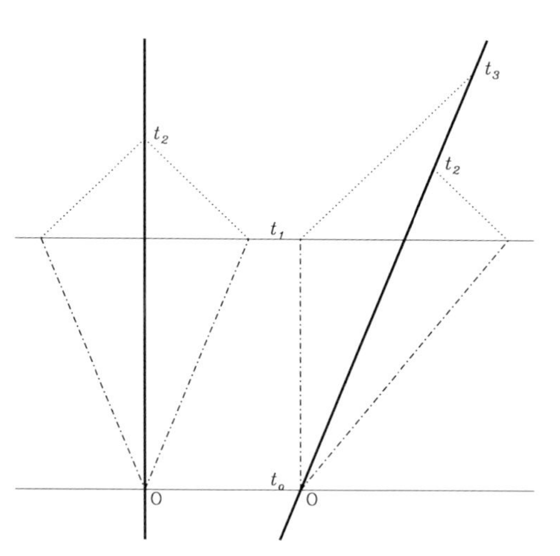

Figure 4.6: *Galilei-invariant mechanics combined with an absolute light velocity.*

When we combine Galilei-invariant mechanics with an absolute light velocity, we get results that depend on the observer's motion. Let us imagine a mechanism at rest on the bank of Huygens' river that expels two particles symmetrically at O, that after a given time t_1 send a light signal back. For the observer on the bank, the signals meet at the position of the mechanism simultaneously (t_2). For the observer in the boat, the return times t_2 and t_3 are different (right-hand side). We find a clear contradiction to the relativity of velocity.

the additive combination of its velocity with other velocities. Surprisingly, all the attempts to establish this additivity failed [57–59]. Michelson tried to measure the difference in time between differently oriented interferometer arms (Figure 4.4). It changed never (although its relative amplitude on the earth's orbit was expected to be 10^{-8}). It looked as if the interferometer arm would be contracted when its direction coincides with the direction of motion against the aether.[9] Einstein interpreted this result as nonadditive composition of the velocity of light with the velocity of the observer, strictly speaking as *universal* isotropy of the propagation of light. More recently, Isaak could perform an experiment able to find effects of first order (relative amplitude on the earth's orbit: 10^{-4}) and directly test this isotropy (Figure 4.5). Contrary to the expectation to find isotropy in one frame of reference only, we see isotropy in any frame of reference. We may suppose the existence of a medium, the aether, but it remains imperceptible. Testing for the anisotropy of light propagation fails to give any indication of contingent motion in this aether.[10]

> **The propagation of light is always found to be isotropic independently of a conceivable motion of the measuring apparatus.[11] The velocity of light combined with other velocities always yields the speed of light.**

The speed of light is not changed in combination with the relative velocities of observers. This expresses the lack of any aether for the mediation of light propagation. However, this alone does not save the relativity principle. We get the impression that our notion of congruence breaks apart. Considering light propagation separately, the scenes in Figures 4.1 and 3.14 are

[9]The Michelson experiment shows that a light signal reflected back and forth along a given arm defines a clock that does not depend on the orientation of the arm. We shall use this light clock again (Figures 5.14 and 5.16).

[10]For the moment, the aberration of starlight contradicts this isotropy. It is the theory of relativity that solves the puzzle (Figures 5.20–5.23). The light *velocity* is isotropic as seen from the moving earth. In contrast to this, the *direction* can change. Deducing the motion of the observer from the appearance of the sky means referring it to distant objects, that is to the stars, not to an aether.

[11]In the process of measurement, the apparatus must not rotate with respect to the observed light ray. Any rotation of the apparatus can be seen in the Sagnac effect just as Foucault's pendulum [12] reveals the rotation of the earth without reference to the stars. Rotating frames are not inertial and require deeper analysis.

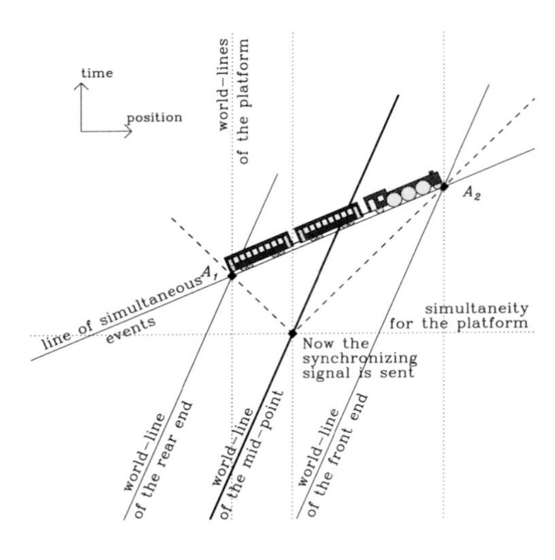

Figure 4.7: *Simultaneity in practice and Einstein's synchronization procedure.*

We draw the world-lines of a train (familiar as Einstein train through many popular presentations) moving through a background at rest. The clocks at the front end and at the rear end are set by a light signal sent from the middle of the train. As evaluated by the ticket collector in the middle of the train, the events A_1 and A_2 are simultaneous. In fact, the light had to pass equal distances with equal velocities after the flash event. As evaluated by the post on the platform, these distances are not equal. The rear end comes to meet the light signal and shortens the distance for the observer on the platform, while the front end runs away from the light signal and makes it pass a greater distance. So the observer standing on the platform concludes that the two events are *not* simultaneous. Simultaneity in the train is *not* the same as simultaneity on the platform.

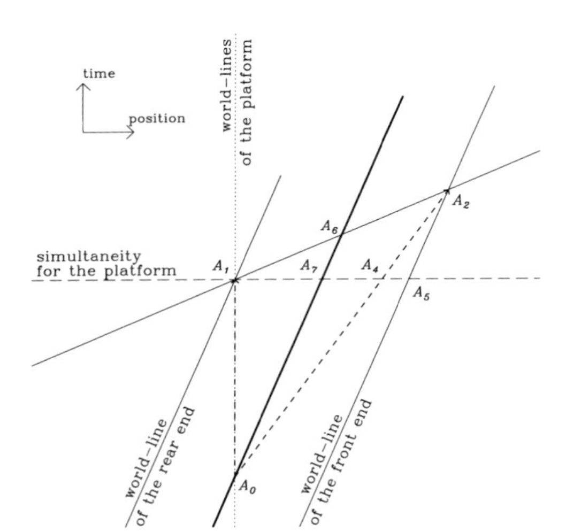

Figure 4.8: *Simultaneity in mechanics.*

We construct the synchronization as in Figure 4.7, but we use ordinary mechanics. At the event A_0, we blast the connection between to equal balls in order that they move away symmetrically. For the train, they arrive at both the ends simultaneously. We let the train move so that the backward ball is at rest of the external observer. The backward ball arrives at the rear end at the event A_1. If the masses do not vary with velocity, the relative velocities must be equal. The forward ball arrives at the front end at the event A_5. The events A_1 and A_5 are simultaneous for both the platform and the train. However, if the mass of the forward ball is larger than the mass of the other because of its motion, the forward ball arrives later at the front end (at A_2). Now the events A_1 and A_2 are simultaneous for the train but not for the platform. Relative simultaneity is equivalent to a variation of mass with velocity.

congruent. But we can have either this kind of congruence or the congruence of Figures 3.15 and 3.14. In the rest frame of any charge, the reflected light cone and the reflected explosion cone are both isotropic. This coincidence is an observable fact that is *not* invariant (Figure 4.6). The relativity principle appears not to be valid.

It was Einstein's idea to modify mechanics in such a way that it conforms to the isotropy of light propagation, i.e., in our language that the reflection of propagating light defines the same image as the reflection of propagating particles. The fact that the speed of light does not depend on the propagation direction (constancy of the speed of light) becomes the first principle for all other constructions (Table 4.1). Before explaining some of them now, we note again that constancy of the speed of light means independence of the direction of propagation and unchangeability by composition. Constancy in space and time is another question. With no other measure available for comparison, it is a convention to take the speed of light as the velocity unit. Then the speed of light is a constant by definition. This is done today in the International System, and the length unit is derived from the atomic time unit. Before the recent achievement of high reproducibility of the speed of light, the length unit was also provided by a spectral line (whose wavelength could be used to calibrate the rigid standards). With such a unit, the speed of light could be conceived as a variable quantity.[12] The ratio of the speed of light to the atomic unit of velocity is determined by Sommerfeld's fine-structure constant. Any variation of the speed of light with position and time must be seen in the variation of this constant. However, there is no variation in the structure of the spectra even of quasars whose light reaches us from a distance of many billions of light years from an equally deep past. No changes in the Sommerfeld constant have ever been confirmed. Nevertheless, the question is the subject of recent work [60, 61].

First of all, the universal isotropy of light propagation has consequences for the physical synchronization of clocks, i.e., for the physical identifiability of *simultaneity* (Figures 4.1 and 4.7). It is precisely these consequences that we meet in the new reflection procedure in the space–time plane.

Simultaneity becomes relative,

i.e., it depends on the state of motion of the observer. Obviously, mechanics is affected by this relativity. After all, we can define simultaneity through mechanical experiments alone, and this definition has to coincide with that of our new procedure. How can we proceed? We consider two objects (balls) of equal mass that are at rest in the middle of the train. The total spatial momentum is zero in the train. If the connection is blasted, they move (in the train) with equal masses and velocities in opposite directions and arrive at the ends at events that must be considered as simultaneous (Figure 4.8). Exactly in the case when the masses vary with velocity the (so defined) simultaneity depends on the motion of the center of mass. Relativity of simultaneity and variation of mass with velocity ($m = m[v]$) are related. However, we must ensure that the different constructions all yield the same result: Only special functions $m = m[v]$ will allow this consistency. We shall return to this question later.

[12] Indeed, it is conceptually simpler to refer to the speed of light that is constant by definition of the unit (Figures 5.14 and 5.16). From this position, it is complicated to define the length unit independently, by using rods. These rods are then cumbersome entities, whose very rigidity is in question, and whose precision and range of applicability is less compared to the one achieved by the atomic clock and light signal method [37].

Table 4.1: Relativity and absolute reference

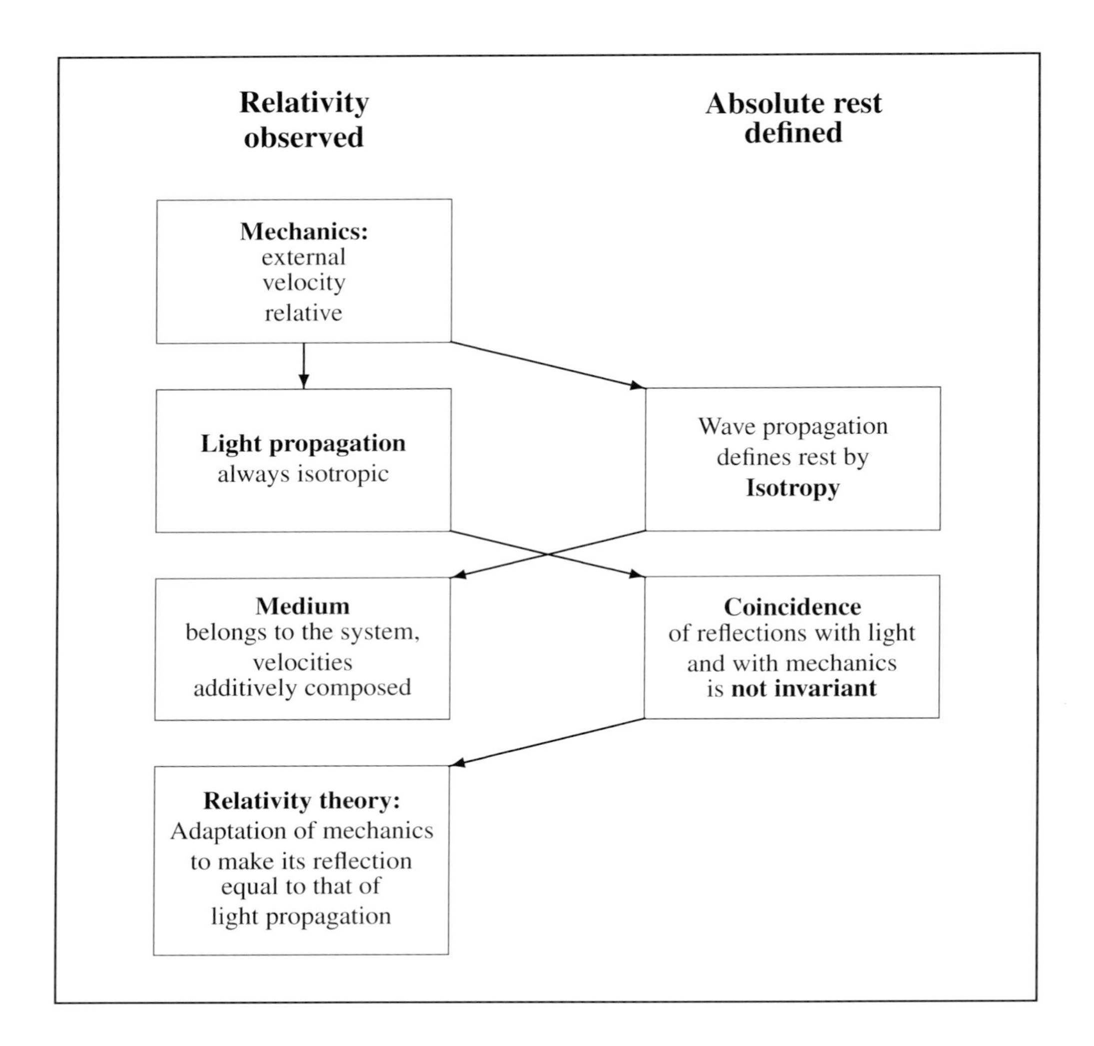

We meet other far-reaching consequences too: If simultaneity is relative, already the measurement of length ceases to give a unique result (Figure 2.12). If simultaneity depends on the motion of the observer, so too does the apparent length of a moving object. Motion is, as we already learned, an orientation in space–time. The apparent length of an object will now depend on it (Chapter 5). It is a bit simpler to calculate the arrival time of a radar echo because no velocity through a virtual medium has to be taken into account (Figure 4.9). The radar echo admits at least in principle the determination of the coordinates of any event. It can be applied in the general theory of relativity too [41].

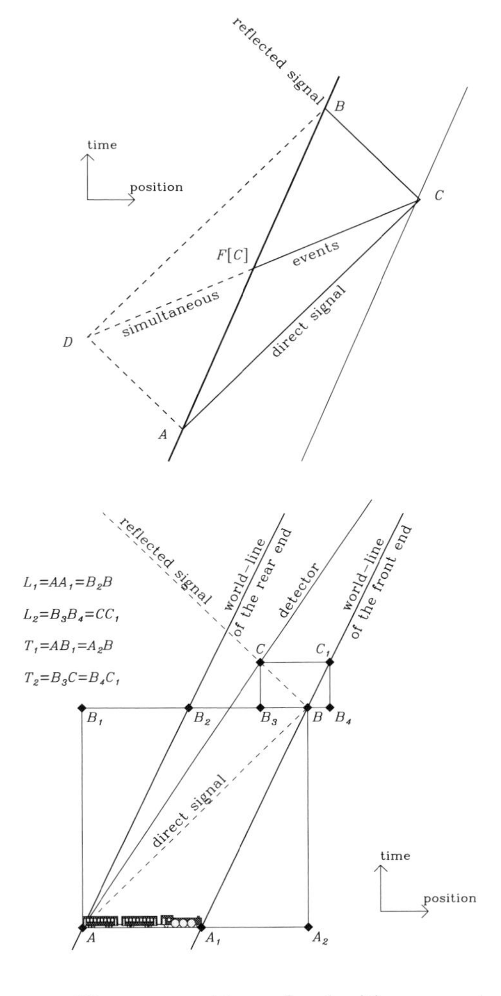

Figure 4.9: *The echo-sounder in the Minkowski world.*

The sketch shows the world-line of an observer in straight and uniform motion. At the event A, he or she sends a signal that is reflected at C and returns at B. If the observer knows about the isotropy of the propagation velocity, he or she can say that the event C is simultaneous with $F[C]$ and that the distance to C is $d[C, F[C]] = ct[A, F[C]]$. We call the quadrilateral $ACBD$ a light-ray quadrilateral.

Figure 4.10: *The composition of velocities.*

We draw the parallel world-lines A_1B and AB_2 of the front and rear ends of a moving train and the world-line of a light signal that starts from the rear end at A, is reflected at B, and meets at C some detector that starts at A too and reaches the front end of the train at some time. If we compare the passed distances and elapsed times, we obtain $c(T_1 - T_2) = w_{\mathrm{D}}(T_1 + T_2), cT_1 = L_1 + v_{\mathrm{Z}}T_1$, $cT_2 = L_2 - v_{\mathrm{Z}}T_2$, that is, $\frac{L_2}{L_1} = \frac{c+v_{\mathrm{Z}}}{c-v_{\mathrm{Z}}}\frac{c-w_{\mathrm{D}}}{c+w_{\mathrm{D}}}$. The same formula is valid in the rest system of the train, in which we must put $v_{\mathrm{Z}} = 0$ and $w_{\mathrm{D}} = w_{\mathrm{DZ}}$. The length ratios are equal in the two cases, and hence we obtain Eq. (4.1).

We now draw attention to a third consequence. The composition of velocities cannot be additive any more because the speed of light does not change in compositions. We can derive the composition law of velocities by a simple *gedanken* experiment (Figure 4.10, after Mermin [42]). The formula found is

$$\frac{c - w_2}{c + w_2} = \frac{c - w_{21}}{c + w_{21}}\frac{c - v_1}{c + v_1} \quad \longrightarrow \quad w_2 = \frac{v_1 + w_{12}}{1 + \frac{v_1 w_{21}}{c^2}}, \tag{4.1}$$

where v_1 is the velocity of the first object, w_2 is the velocity of the second object, and w_{21} is the velocity of the second object with respect to the first. Formula (4.1) is called Einstein's addition theorem of velocities, although it describes no additive composition at all. In everyday language a composition is called an addition in spite of the fact that the mathematical operation of addition may not be applied.[13]

[13]We sometimes joke about the question whether four minus one always equals three: Of course *not*. If we subtract one vertex of a sheet of paper (by cutting it off), we do not get three vertices, but five. The mathematical operation of

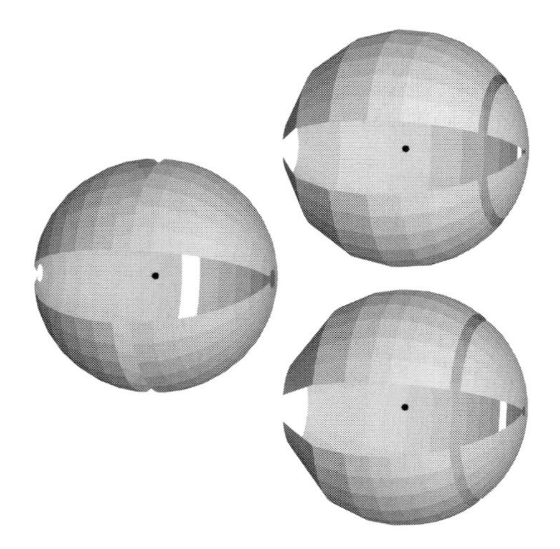

Figure 4.11: *The conformal map of the sky through aberration.*

We show the map of a sphere (at left) around the observer. The sphere is left incomplete to improve the spatial impression. In the upper right, its conformal image is shown for an observer moving with 0.7 c to the right (see Figure D.5). For comparison, in the lower right the result is shown for nonrelativistic aberration. There, the map is not conformal.

The composition described by Einstein's addition theorem of velocities is multiplicative in the quantities $(c - v)/(c + v)$: A central role of the double ratio is signaled (Chapter 8, Figure 8.7). The new addition theorem modifies the magnitude of the aberration too [11]. It now defines a conformal map of the apparent sphere (Figure 4.11).

We observe a strict equivalence:

Velocities are composed by addition.
$\updownarrow$
Simultaneity is absolute.
$\updownarrow$
Masses do not vary with velocities.

If one of the statements is not valid (the relativity principle presumed) the other two cannot be valid either. Only in the case when velocities are combined additively do we obtain absolute simultaneity. Consequently, *independence* of velocity for the masses is only consistent with *additive* composition of velocities. We see how everything is intertwined and how one obtains a consistent picture that necessarily has properties that are surprising as long as one considers them without their interrelation.

The simplest point of view supposes the validity of geometric relations that allow one to consider position, orientation, and velocity of objects as being independent of their other properties. We believe this to be natural to such a degree that it seems to be valid a priori, that is, before considering the dynamics of physical systems. Therefore, the question arises: Can Nature outwit this geometrical point of view? Is this point of view really necessary, i.e., a priori, or rather contingent, i.e., in need of being tested? The latter is correct.[14] We have to

subtraction is not applicable to each physical separation, just as not each composition is an ordinary addition.

[14] In addition, we test only appropriateness. As we already mentioned, Poincaré went so far as to state that the border between a priori geometry and embedded dynamics can be shifted in both directions. Nevertheless, there are more and less appropriate locations of this border.

investigate this question through measurement because we can conceive different geometrical systems, all concerning the same complex of configurational properties. In addition, we know from history arguments *against* the geometric point of view: Prior to the discovery of the laws of gravitation, everything pointed to a distinguished role of the third dimension. If the different quantities that we compare in the measurement react differently under motion, repositioning, or reorientation, we can determine motion, position, and orientation just by these differences without referring to external objects. We know that the different definitions of a sphere could yield different forms. We must always remember that here too we are considering the behavior of the different configurations in our realm relative to each other, never against some absolute space. In our understanding, space is simply the abstraction of a behavior in which all quantities react covariantly to motion, repositioning, and reorientation. Physical relativity demarcates the regime of applicability of geometrical systems. This discovery we owe to relativity theory. In this sense, geometry has become physics.

Before the invention of relativity theory, the nonadditive composition of velocities was the motive for the sophistication of the mechanical model for the hypothetical medium of light propagation, the aether. The mechanical model was believed to be indispensable because mechanics was and is an unrivaled paradigm of conceptual clarity and harmony between mathematics and physics. However, the mechanical model was made obsolete by relativity theory and the new geometry of the world. In this sense, physics became geometry.

5 Relativity Theory and its Paradoxes

5.1 Pseudo-Euclidean Geometry

Seen from the result, Einstein's theory of relativity [8, 62, 63] flows into the statement:

Mechanics has to conform to light propagation.

In fact, all physics (only gravitation needs additional refinement) has to conform too, but in the beginning of the 20th century the relation between electrodynamics and mechanics was of topical interest. Electrodynamics conformed to light propagation, of course—it was the background theory. It had already provided the formal aspect of the new transformation between reference systems.[1] Light propagation defines kinematics and geometry.[2] The relativity detected in mechanics has to correspond to this geometry. Of course, we will cover mechanics here only so far as we used it before, i.e., the definition of momentum and the measurement of inertial mass. We begin the chapter by noting the basic issue of the relativity of simultaneity and by summarizing the elementary constructional means. We then proceed with the formulation of the analog of the theorem of Pythagoras, which yields the procedure for determining distances. The next step is the derivation of the variation of mass with velocity. This is the central argument. We then add the discussion of the so-called time dilation and length contraction and conclude with the necessary remarks about superluminal velocities.

The new geometry is defined through the properties of the propagation of light. The basic difference can be recognized in Figure 4.1 already: The simultaneity of two events B and C will depend on the motion of the mirror, and only for one particular motion will the event C be the reflected image of B and consequently considered to be simultaneous with B. We call this fact the relativity of simultaneity. In contrast to this observation, the assumption of classical mechanics that simultaneity does not vary with the different reference frames is called absolute simultaneity. We found that absolute simultaneity is deeply connected with the invariability of mass with velocity and with the additive composition of velocities. In a mechanical theory that conforms to the new geometry, the velocities will no longer be combined additively, and the masses will depend on velocity. The first is not too big a surprise because it was our presumption: The composition of the speed of light with another velocity was found to yield

[1] Einstein's original paper covered kinematics (equivalent to the geometry that we are about to explain) and electrodynamics. Mechanics was covered more extensively in his subsequent papers. Planck invoked the new mechanics to argue that Einstein should be elected to the Academy of Sciences in Berlin.

[2] One may conceive the isotropy of light propagation to be a mere convention, in particular when constructing clocks by propagating light signals. The construction of the geometry of space–time would be a convention too. Its applicability to mechanics makes all the difference. The geometry constructed by the laws of collision is conceptually independent. Relativity requires the two geometries to be the same.

The Geometry of Time. Dierck-E. Liebscher

ISBN: 3-527-40567-4

the speed of light again.[3] The second will be illustrated when we will become more acquainted with the new geometry.

Let us first construct the geometry generated by the reflection law sketched in Figure 4.1. It is called the Minkowski geometry,[4] the space–time with this geometry is called the *Minkowski world*. In the timetable, the inclination of light signals is always the same. The cones generated by the light rays passing through a given event are called light cones. Each event carries a light cone. The mantle lines are called lightlike or isotropic lines. If we put a mirror in the path of a light ray, it is reflected into another light ray. A congruence of lightlike lines is invariant with respect to all reflections and to all motions of the group, consequently. This is used to construct the general reflection map (Figure 5.1). The central figure of this construction is the *light-ray quadrilateral*. This is a parallelogram of isotropic straight lines. In accordance with the new geometry, its diagonals are perpendicular. They divide the light-ray quadrilateral into four triangles. The opposite triangles can be shifted to form a rhombus that is a square in the new geometry. Two opposite vertices of the light-ray quadrilateral always lie in positions that are symmetric under reflection on the diagonal between them.[5] Of course, the pseudo-Euclidean area is conserved too. However, the *Euclidean* area of a figure is not changed under the reflection. We can use *Euclidean* theorems about areas.

By the reflection prescription, straight lines are divided into timelike, spacelike, and lightlike ones. The relative position of two events is timelike if the connecting line remains inside the light cones carried by the two events. If the connecting line is a mantle line, the relative position is lightlike. If the connecting line passes outside the light cones, the relative position is called spacelike.[6] No timelike line is ever reflected into a spacelike or lightlike line, and no spacelike line is ever reflected into a timelike or lightlike line.

We now consider the first aspects of the definition of distance in the Minkowski geometry. In the Galilean geometry, the arc length of a world-line is determined solely by the change in the time coordinate. Therefore, the increment of the time is set equal to the increment of the arc length. In the Minkowski geometry, this can be transferred only to timelike lines. A timelike

[3] Of course, with the precision of an experiment. That is, the speed of light is not necessarily *the* absolute velocity. The behavior of light reveals the existence of an absolute velocity whether it is the velocity of real objects or not. We will follow the habit to take the speed of light as synonymous to the absolute velocity.

[4] Minkowski stated and propagated the idea that the space–time relations found by Einstein's considerations are in fact a system of geometrical relations, i.e., a structure existing independently of the physical argument by adopting a convention about corresponding axioms. Hence his name is attributed to the new geometric notions. The geometry is also characterized as *pseudo-Euclidean*. This expresses, on the one hand, the resemblance to the Euclidean geometry with respect to incidence and intersection relations of lines and points and, on the other, the difference in the metric properties that we are about to explain.

[5] In sketching timetables, we are forced to use the means invented for the Euclidean geometry of the drawing plane. Adapting to this situation, the inclination of the lightlike lines usually is chosen in such a way that they reflect the choice of coupling the time unit with the length unit by the speed of light. It has no influence on the geometrical derivations and could be chosen arbitrarily as long as the two lightlike lines do not coincide. The choice only helps us recognize figures in our drawings. For instance, we see light-ray quadrilaterals as rectangles of the ordinary interpretation too. Then two straight lines are perpendicular if their angle is bisected by the lightlike directions in the auxiliary Euclidean sense. This convention can be useful for a fast sketch of a figure, but is not founded by the construction means of the geometry just found. In principle, the two lightlike directions can be chosen arbitrarily. The generation of the new geometry requires only their existence.

[6] In the two-dimensional world, the notions timelike and spacelike are pure convention. However, in a four-dimensional world the position coordinate represents three space coordinates, and we can no longer exchange the inside of the light cone with its outside.

Figure 5.1: *Reflection in Minkowski geometry.*

Given the point A and the mirror line g, we draw the light-ray quadrilateral $A\,A_2\,S[A]\,A_1$ and find the perpendicular $AS[A]$ orthogonal to g. Comparing the lengths, we obtain $d[O,A] = d[O,S[A]]$, comparing the angles $\angle AOA_2 = \angle A_2OS[A]$. The connecting line $AS[A]$ is the perpendicular onto the straight line g through the points A_1 and A_2. This is identical to our conclusion in Figure 4.9. We can interpret this construction analogously to Figure 3.4 because, as we shall see in the next figure, all sides of the light-ray quadrilateral $A\,A_1\,S[A]\,A_2$ have the length zero in the now generated geometry. The two light-ray world-lines through the point A correspond to a circle of radius zero around A. The diagonals of a light-ray quadrilateral are perpendicular. Conversely, if two straight lines are perpendicular they are diagonals of a light-ray quadrilateral.

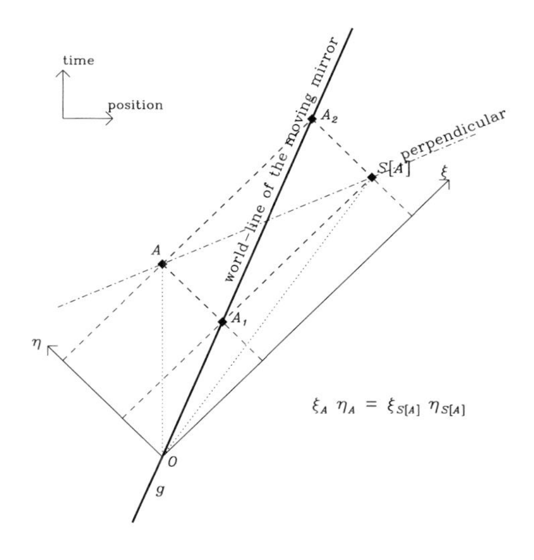

line can be the timetable of an object. Usually, it is assumed to provide a reference frame, that is, it contains the facility of measuring distances in space and intervals of time. In this case, it is usually called an observer. If not accelerated, it moves uniformly and provides an inertial reference[7] (inertial frame). The readout of her or his clock is the only changing coordinate in this frame and tells the length of the world-line. The length of the world-line is the proper time of the object. If the world-line is not straight, we approximate it by an infinitesimal polygon just as we do in the ordinary Euclidean geometry. Each straight infinitesimal edge is part of the time axis of a reference system in which the object (the clock) is momentarily at rest (instantaneous rest frame). At every instant, the increment of the arc length is equal to the increment of the time measured by the clock itself, the proper time. The total measure of the length of a timelike world-line is the time interval measured on the clock with the same world-line. If we now remember the Euclidean geometry, we should expect here too that different world-lines between two events have different lengths, i.e., they show different time lapses: The twin paradox is announced.

We postpone the theorem of perpendicular bisectors (the existence of the circumcenter) to the next chapter and consider the calculation of the proper time (the length) of a world-line segment of a moving clock. This corresponds to the Euclidean problem of determining the length of a segment inclined to the coordinate axes. That is, we now consider the analog of the theorem of Pythagoras in our Minkowski world. We can prove it with elementary means

[7] This is a bit brief. The realization of the necessary means may be intricate. The often used *gedanken* construction is a set of synchronized clocks moving on parallel world-lines (Figure 4.7). The synchronization can be performed by light signals. Nevertheless, the total construction is fixed by the original piece of world-line. The only freedom left is the orientation in space at a given instant. Therefore, we can speak of a reference system defined by the moving object. This does not affect the fact that the reference system does not consist solely of the object in question.

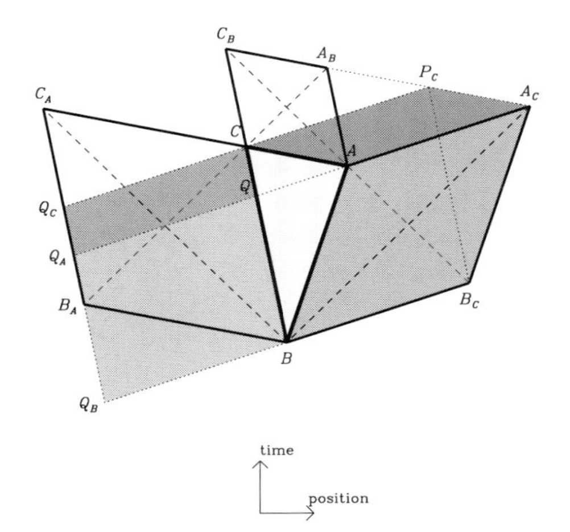

Figure 5.2: *Pythagoras's theorem in the Minkowski geometry.*

We draw a right-angled triangle using the new rule to find right angles. The right angle is at C. Squares are rhombuses with lightlike diagonals (four right angles and symmetry). Then we obtain the squares on the hypotenuse and the sides as shown. Comparing the areas, we find that the square on the hypotenuse is equal to the *difference* of the squares on the opposite sides.

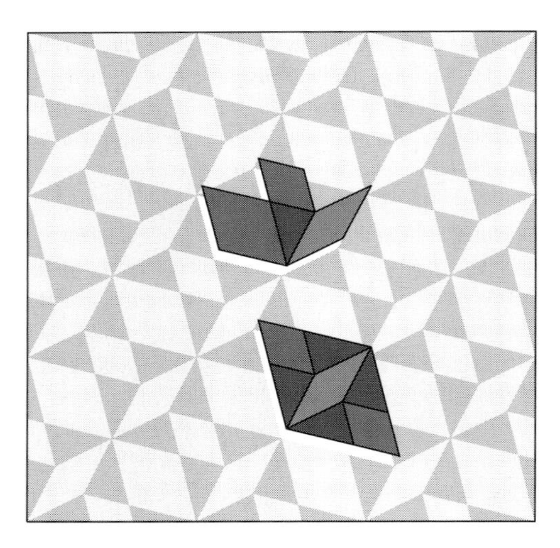

Figure 5.3: *The Pythagoras figure of the Minkowski geometry immersed in a tiling of the plane.*

The triangle and its three attached Minkowski squares can be immersed in a tiling of the plane. Obviously, we can combine four items of the rectangle, the square on the hypotenuse, and two items of the smaller square on the opposite sides to form a rhombus. Up to a shear, the area of the rhombus is equal to $(a + b)^2$, and we conclude $c^2 = (a + b)^2 - 2ab - 2b^2 = a^2 - b^2$.

(Figure 5.2). Through a construction that is completely analogous to that of Euclid, we obtain

$$\begin{aligned} b^2 &= ACC_BA_B = ACP_CA_C = Q_AQ_CCQ \\ c^2 &= BAA_CB_C = Q_BQ_AQB \\ a^2 &= C_AB_ABC = Q_CQ_BBC \\ &\longrightarrow \quad a^2 - b^2 = c^2. \end{aligned}$$

The tiling of Figure 5.3 contains this construction together with a rhombus leading to the same result by the binomial theorem.

The square on the hypotenuse is equal to the *difference* of the squares on the opposite sides.

The minus sign is the characteristic feature of the Minkowski geometry. Consequently, the square on the hypotenuse may be zero or even negative. By convention, we can attribute the

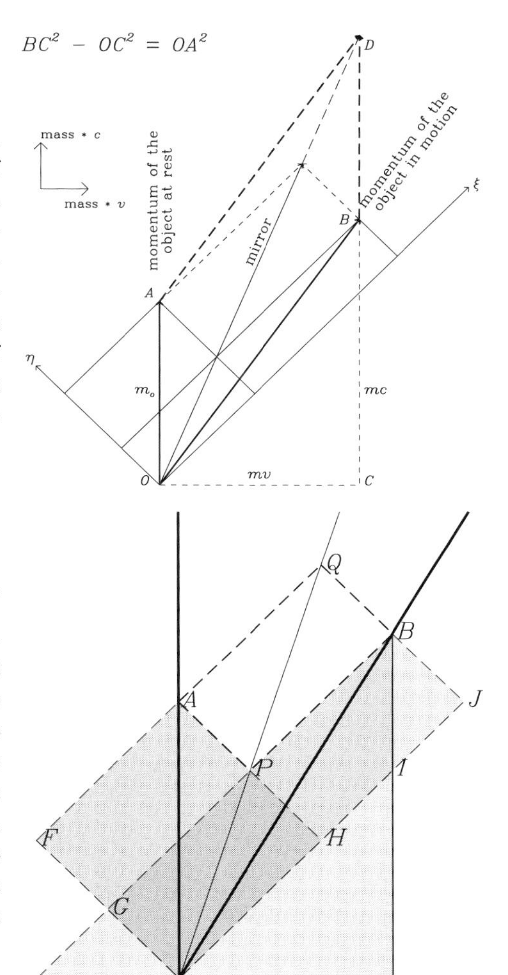

Figure 5.4: *The variation of mass with velocity. I.*

We interpret Figure 5.1 as a momentum diagram. The vector $\vec{OB}$ now represents the momentum of a moving body that has the momentum vector $\vec{OA}$ if at rest. We find it with the help of the reflection prescription of Figure 5.1 as momentum parallelogram to Figure 4.8. In the right-angled triangle OBC, the chord OB is the hypotenuse whose length is equal to that of OA. The theorem of Pythagoras now tells us that we have to calculate with $BC^2 - OC^2 = OA^2$, i.e., $m^2(c^2 - v^2) = (m_0 c)^2$.

Figure 5.5: *The variation of mass with velocity. II.*

Of course, we can evaluate the figure without referring to Pythagoras's theorem. At O, an object decays into two identical fragments. In the frame where one of them is at rest, we can compare the two masses. We draw the auxiliary lightlike lines and the comparison of areas is obvious. P lies on the diagonal of the parallelogram $OJQF$. Hence, the ares of $OHAF$ and $OJBG$ are equal. Because IJB and OGD are of equal area too, we obtain $OHAF = OIBD = DEB - OEI$. This is again $m_0^2 c^2 = m^2 c^2 - m^2 v^2$.

minus sign to the squares on spacelike sides. In Figure 5.2, the square on AC is then to be counted as negative. Now the analog of Pythagoras's theorem in the Minkowski geometry requires addition of squares as in the Euclidean geometry.

The square on a timelike segment is positive. Its square root is the lapse measured in proper time between the two events connected by the segment. The square on a spacelike segment is negative. The square root of its absolute value is the distance of the two events connected by the segment in the frame in which the two events are simultaneous, i.e., the proper distance. The square on a lightlike segment is degenerate, and its area is zero.

If we choose the Cartesian coordinates of the Minkowski plane in such a way that the sides including the right angle are parallel to coordinate lines, we obtain for instance $C = [0, 0]$, $A = [0, x]$, and $B = [ct, 0]$. The square on the hypotenuse is then given by $(AB)^2 = (ct)^2 - x^2$.

5.2 Einstein's Mechanics

We have to reconstruct the momentum diagrams. These must be modified in such a way that mechanically indistinguishable figures become congruent figures of the geometry derived from the rules of the reflection of light. We can apply the new form of the theorem of Pythagoras to the *variation of mass with velocity* (Figures 5.4 and 5.5). The declared scope at the beginning of the chapter was to bring mechanics into agreement with the propagation of light. We find here the first point of the new mechanics that is different from the classical one. Using the prescription of reflection from Figure 5.1, we draw two momentum vectors equally long, i.e., belonging to the same rest mass m_0. We may interpret the figure as representation of a symmetric decay, i.e., the decay of an object into two identical fragments with momenta symmetric with respect to the world-line of the decaying object. The product mc is the time component of the momentum vector measured in collision experiments. In any collision, the sum of the time components is the same before and after the collision just as for the sum of the spatial momentum components. In detail, the time component of the momentum varies with velocity. Only the magnitude of the momentum vector does not vary with motion: It is characteristic for the object in question. For this case, the theorem of Pythagoras yields

$$m_0^2 c^2 = m^2 c^2 - m^2 v^2. \tag{5.1}$$

If the rest mass m_0 is given, the momentum coordinates $[mc, mv]$ describe a hyperbola; in the case of three space dimensions, it describes a hyperboloid shell, which is called the mass shell. The momentum vector of an object of this given rest mass can only end on this shell. We may transform Eq. (5.1) in

$$m = \frac{m_0}{\sqrt{1 - \frac{v^2}{c^2}}}. \tag{5.2}$$

This is the famous formula of the variation of mass with velocity. Now it is time to recall that the inertial mass is defined physically by collision and scattering. So let us consider again the elastic collision (Figures 3.7 and 3.8). The circular cones of the symmetric case are no longer simply shifted to make oblique circular cones, as must be the case for an additive composition of velocities. Instead, we obtain the picture drawn in Figure 5.6. The circular locus of the positions reached by the collision partners is deformed into an ellipse.[8] The center of the circle (i.e., the point that represents the totally inelastic collision) is shifted to an eccentric point. Here we have the intersection of the chords that in the symmetrical case are the diameters of the circle. The ratio of the distances of the intersection point from the right and left periphery of the ellipse is equal to the ratio of the masses of the moving ball and the one at rest. It shows the dependence of mass on velocity in the form that we just determined. However, one does not observe directly the velocities but rather the distribution of directions after the collision; this is the (differential) cross section. In the center-of-mass frame, in which everything goes symmetrically, let us assume that we find the directions uniformly distributed on the circle (the sphere in space). Let us mark for the moment eight equally spaced points on that circle. They are shifted when the figure is sheared into that of the billiard collision. If the velocities

[8] As we shall see, this ellipse is a circle in the Lobachevski geometry of the velocity space (see Section D.3)

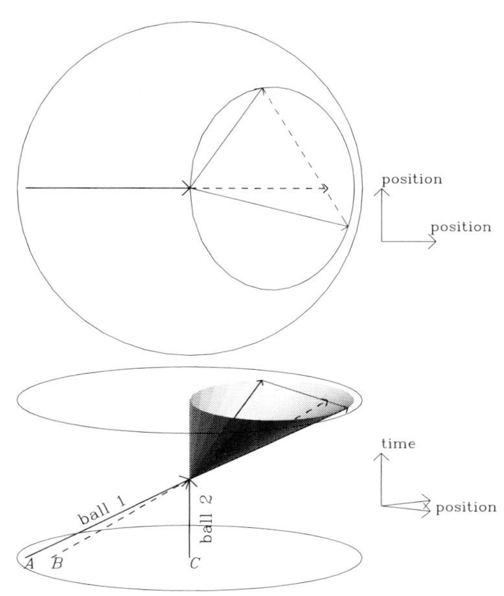

Figure 5.6: *The relativistic billiard collision.*

We draw the setting of Figure 3.8 and note that the speed of light is the maximum speed and that the reduction of momenta to velocities conforms to the variation of mass ($BC/AB = m[v]/m_0$). The Galilean relativity that was the argument for making the transition from Figure 3.7 to Figure 3.8 is no longer valid. If we observe the relativity principle, the figure of the scattered world-lines conforms to another geometry, i.e., that of Minkowski. The points of the upper intersecting plane determine velocities. We will meet them again in hyperbolic geometry (Chapter 8, Section D.3).

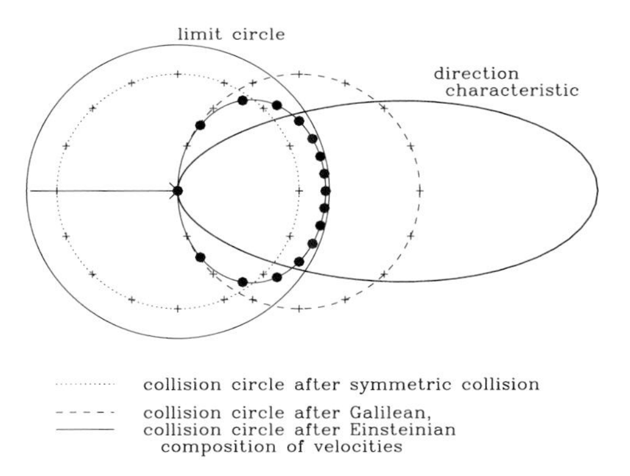

Figure 5.7: *The collision characteristic.*

The ellipse of velocities is marked by points that are distributed uniformly on the circular periphery in the center-of-mass frame (Figure 3.7). They indicate that in the relativistic case the scattering in the forward direction acquires a bigger statistical weight. This is represented by the characteristic, which shows, like an antenna beam characteristic, the probability of scattering into the various directions.

are combined additively, the form of the circle and the relative position of the points remain unchanged. If we compose velocities by our Einstein addition theorem, the circle is deformed into an ellipse. In addition, the points move on the ellipse in the direction of the apex. We can note their density on the sphere and get a characteristic of the collision (the differential cross section) that depends only on the inclination to the apex direction. It is shown in Figure 5.7. In the Galilean case of additive composition of velocities, this characteristic is a sphere with the vertex on the periphery. In accordance with Einstein's addition theorem, it becomes more and more elongated as the collision velocity acquires larger and larger values. The dilation factor increases with $\gamma = 1/\sqrt{1 - \frac{v^2}{c^2}}$ beyond any limit. It is observed and taken into account in all accelerator experiments.[9]

The inertial mass m, Eq. (5.2), increases beyond any limit when the velocity v approaches the speed c of light and the rest mass remains unchanged [64]:

To her friends said the Bright once in chatter:
"I have learned something new about matter:
My speed was so great, much increased was my weight,[10]
Yet I failed to become any fatter!"

5.3 Energy

Einstein's formula for the relation between energy and mass is presumably the most famous formula of all sciences. Its derivation is believed to be complicated; it is simple instead. We need to know Newton's laws, the definition of reflections in space–time through light, and the registration of a symmetric decay like in Figure 5.5. The variation of mass with velocity is an immediate consequence. We now only have to define the notion *energy* to arrive at the famous formula.

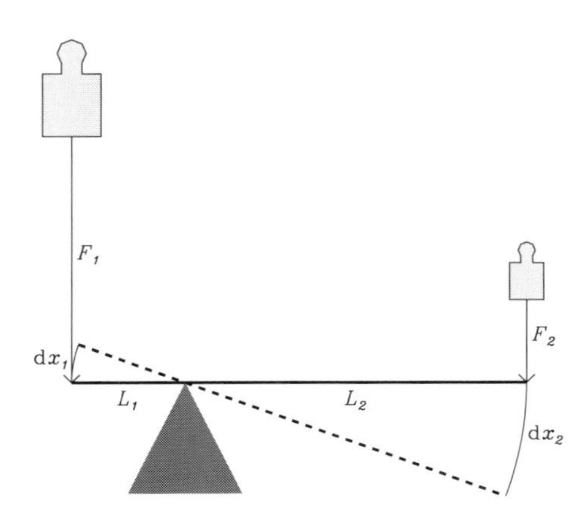

Figure 5.8: *The equilibrium of the lever.*

The formula $F_1 L_1 = F_2 L_2$ is the familiar condition of equilibrium of the lever. It is the requirement of zero net torque on the lever. When we replace the lengths L with the virtual infinitesimal displacement $\mathrm{d}x_i$, we obtain a form that can be applied to any other equilibrium: The sum of all forces weighed with the virtual displacements, $F_i \mathrm{d}x_i$, is zero for equilibrium. This reminds the invariance of the total momentum and indicates a quantity E, the energy, that increases through the action of force, $\mathrm{d}E = \boldsymbol{F}\,\mathrm{d}s$. If such an energy exists, it must increase through the action of force that way.

The energy is the central conserved quantity in all physics. In all canonical classical as well as quantum mechanics, the formal dependence of the total energy on general coordinates and adjoint momenta yields the possible motions of an isolated system. We need only a tiny part of all that which is comprised in the notion energy. This tiny part is the law of the lever of Archimedes. A lever is in equilibrium when the two forces multiplied with the corresponding lever arms are equal, and more general, when the sum of the forces $\boldsymbol{F}_i$, weighed with the

[9]Of course, we need there the complete relativistic mechanics and electrodynamics. The variation of mass with velocity is only one fundamental (i.e., with an elementary geometrical interpretation) part of them.

[10]The word *weight* has to do with gravity here, after all. It anticipates the equivalence of inertial mass and gravitational charge, the latter being the *mass* that is measured in ordinary life on scales.

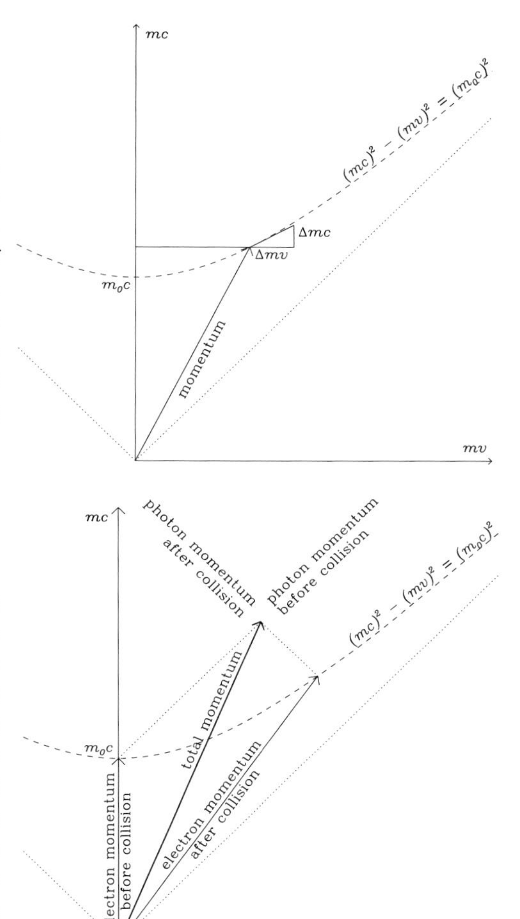

Figure 5.9: *Energy and mass.*

We draw the mass hyperbola and a momentum vector ending on it. The triangle of the increments is analogous to the triangle of the momentum coordinates because the direction of the momentum and its increment are the directions of the diagonals of a light-ray quadrilateral. We obtain $\Delta(mc) = \frac{v}{c}\,\Delta(mv)$. We compare this with Eq. (5.3), and find $\Delta(mc^2) = \Delta E$. The argument can be inverted. Given $E = mc^2$, we obtain $\Delta(mc) = \Delta(mv)\,\frac{v}{c}$, and this defines the mass hyperbola.

Figure 5.10: *The Compton effect.*

We interpret Figure 5.4 as a momentum diagram of a collision between an electron and a photon (dotted lines). The photon is reflected, but its momentum decreases (together with its frequency, as we know from quantum mechanics). This is the Compton effect. Energy and momentum are transferred to the electron. It is repelled, and its mass increases according to the variation of mass with velocity. Inversely, a low-energy photon can be pushed to high frequencies by high-energy electrons. This is inverse Compton effect, which has wide astrophysical applications.

possible changes $\mathrm{d}\boldsymbol{x}_i$ in the position of the acted upon, is zero (Figure 5.8). We interpret the product $\boldsymbol{F}\,\mathrm{d}\boldsymbol{x}$ as energy added by action of the force $\boldsymbol{F}$ through the displacement $\mathrm{d}\boldsymbol{x}$. It is another problem to show that this leads to an integral energy, but the forces that Newton had in mind did it perfectly.

So let us take the increment in energy to be $\mathrm{d}E = \boldsymbol{F}\,\mathrm{d}\boldsymbol{x}$. We then substitute the change in momentum for the force and obtain

$$\mathrm{d}E = \boldsymbol{F}\,\mathrm{d}\boldsymbol{r} = \frac{\mathrm{d}\boldsymbol{p}}{\mathrm{d}t}\mathrm{d}\boldsymbol{r} = \mathrm{d}\boldsymbol{p}\frac{\mathrm{d}\boldsymbol{r}}{\mathrm{d}t} = \boldsymbol{v}\,\mathrm{d}\boldsymbol{p} = \boldsymbol{v}\,\mathrm{d}(m[v]\boldsymbol{v}). \tag{5.3}$$

We can integrate this formula because we know from Figure 5.9 that $v\,\mathrm{d}(mv) = \mathrm{d}(mc^2)$. In our case, Eq. (5.1), we obtain $\mathrm{d}E = \mathrm{d}(mc^2)$ and $E_{\text{kinetic}} = mc^2 - m_0c^2$.

When we accept that photons exist as particles that can undergo collisions with $mc = p = E/c$ as indicated by the Compton effect (Figure 5.10), we draw the following conclusion immediately:[11]

- when we know of a conservation law for the total mass and
- when we know of a conservation law for the total energy, and
- when for a certain part of the energy, this part is proportional to a corresponding part of the mass,
- all energy must be proportional to a corresponding mass. Otherwise, the transitions between the different parts of the energy would hamper either the conservation of energy or the conservation of mass.

The result is the famous formula

$$E = mc^2. \tag{5.4}$$

A supplementary point is the direct transformation for small velocities. Equation (5.1) can be written in the form $m^2c^2 = m_0^2c^2 + m^2v^2$. This is the same as

$$mc^2 = m_0c^2 + \frac{m}{m+m_0}mv^2.$$

For small velocities ($v \ll c$, or $m \approx m_0$), we obtain the approximation

$$mc^2 = m_0c^2 + \frac{m}{m+m_0}mv^2 \approx m_0c^2 + \frac{1}{2}mv^2. \tag{5.5}$$

The second term is the familiar kinetic energy. Consequently, the first one can be named rest energy. For the moment, this is only a label. It acquires interest when we see that this rest energy can really be mobilized. In elastic collisions, the sum of the kinetic energies is conserved, and so is the sum of the rest energies. But in inelastic collisions the sum of the kinetic energies changes and, consequently, the sum of the rest energies too. Kinetic energy and rest energy are transformed into one another.[12] The total of rest energies and kinetic energies is proportional to the time component of the total momentum and is conserved. Consequently, the time component of the momentum of an isolated object is proportional to its total energy. Because it is proportional to the inertial mass too, we obtain the equivalence of mass and energy. This is the content of the formula $E = mc^2$. The total energy of an isolated object can be measured by its inertial mass.

In Newtonian mechanics, energy was defined up to an arbitrary constant. That has now gone because the inertial mass does not contain such arbitrariness. The short statement, Eq. (5.4), requires another consideration yet. For the moment, the total energy of an object

[11] It is reported that Planck draw this conclusion from Einstein's cautious $\mathrm{d}E = \mathrm{d}mc^2$, corresponding to Einstein's conclusion of $E = h\nu$ from Planck's cautious formula $\Delta E = h\nu$.

[12] Collisions in which rest energy is transferred to kinetic energy are known. One calls them superelastic collisions, or collisions of the second kind. The most extreme example is the *annihilation of particle–antiparticle pairs*.

is proportional to its *inertial* mass. In order to measure the inertial mass as ordinary (gravitational) mass by weighing, an additional observation is required. This is the famous strict proportionality (i.e., equivalence) between the inertial mass and the gravitational mass (here the charge[13] in the gravitational field, see also Section 7.2).[14] It would be a mistake to phrase our statement in the form that energy and mass are transformed into one another. *Both* are conserved—the one is the measure of the other. The individual parts only (of the energy and of the mass proportionally) can be transformed.

In a bound state, the total energy is smaller than the sum of the energies of the fragments in a dissociated state. Consequently, the mass of the unbound fragments is larger than the mass of the bound object. The difference is called the mass defect. It is proportional to the binding energy. If the mass defect is negative, then the state is unstable. Such an unstable bound state will decay spontaneously into its fragments. In this case, the mass defect corresponds to the kinetic energy of the fragments after the decay.

5.4 Kinematic Peculiarities

Now we are at the point to reconsider the Doppler effect (Figure 5.11). In contrast to the evaluation in Figure 2.14, we take into account the relativity of simultaneity. The change in period of the reflected signal at the position of the emitter turns out to be a kind of cross-ratio (Figure 8.5). We can use the Doppler effect in combination with the relativity requirement to derive the Minkowski geometry [40]. We show only the first step, that is the transport of units between objects in relative motion. The aim is to get units in which the Doppler effect depends only on the change of relative distance. Then it must be symmetric, i.e., $OB/OA = OC/OB$, if measured in the corresponding units. Evidently, measurement of the intervals with a universal time is inappropriate. Figure 5.11 shows how to construct an interval OH on OB, which is equally long as OA. We assume A to be the unit point on OC and try to determine the unit point H on OB in such a way that $OC/OA = (OB/OH)^2$. By projection onto the particular directions OE and OF, the condition gets the forms $OC_1/OA_1 = (OA_1/OH_1)^2$ and $OB_2/OA_2 = (OH_2/OA_2)^2$. We use the equation $OB_2/OA_2 = OC_1/OA_1$ to obtain $OH_1 \cdot OH_2 = OA_1 \cdot OA_2$. That is, the unit points A and H lie on a common hyperbola that have asymptotes given by the particular directions of the propagation of light. We obtain complete agreement with the findings of Figures 5.1 and 5.4.

When we accept the formula $E = mc^2$ in the case of photons (maybe with the field-theoretical argument that the density of momentum of an electromagnetic wave times the speed of light is equal to its energy density [66], or with the quantum-theoretical argument for $E = h\nu$ and $p = h\nu/c$ [67]), the Doppler effect shows that $\Delta E = \Delta mc^2$ must be valid for any body that can spontaneously emit a symmetric pair of photons. Such a body is at rest before and after the emission when the latter is symmetric and the momenta of the photons

[13]Strictly speaking, one has to distinguish between active (source of the gravitational field) and passive (charge in the gravitational field) gravitational mass. In Newton's gravitation theory, Newton's third law implies for gravitational interaction the equivalence between both. In general relativity, this is a more intricate question [65].

[14]This looks strange at first because one does not usually say: "I weighted myself with the chemist's balance and I discovered that my total energy has increased three MegaJoules." Most of this energy can never be used for real work. In addition, 1 MegaJoule contributes only $\approx 10^{-5}$ g to the weight.

The shortest path to $E = mc^2$

1. Force-free (here horizontal) motion draws *straight* lines on the registration strip. The inclination of the lines with respect to the vertical indicates the velocity (Galileo's law).
2. The *mass* is the factor to weight the velocities in order to obtain a conserved sum (Huygens' law). The product of velocity and mass is called *momentum*.
3. Without external influence, the momentum remains constant. Hence the measure of such an influence, i.e., the *force* K, is proportional to the variation of momentum with time (Newton's second law).
4. *Energy* is the central conserved quantity in a system free of external influence and with constant in time conditions. The increase in energy through acceleration is equal to the *product of force and path* (Figure 5.8).
5. For a decay into two equal fragments, the momentum conservation requires *symmetry*.
6. Symmetry is to be constructed with the *reflection rule for light*. In reflection, the velocity of light changes only its sign. It follows:
7. Mass depends on velocity, $m^2c^2 - m^2v^2 = m_0{}^2c^2$.
8. The increase in energy is proportional to the increase in mass: $mc\,\mathrm{d}(mc) = mv\,\mathrm{d}(mv)$, i.e., $\mathrm{d}E = K\,\mathrm{d}s = \mathrm{d}(mv)\,\mathrm{d}s/\mathrm{d}t = \mathrm{d}mc^2$.
9. When mass is conserved as well as energy, and when one part of the energy is proportional to a corresponding part of the mass, all energy has to be proportional to a corresponding mass, $E = mc^2$.

($p_1 = -h\nu/c,\ p_2 = h\nu/c$) are equal and opposite. In a frame that moves with $-v$ the momentum conservation means

$$m_{\text{before}}v = m_{\text{after}}v + \frac{h\nu}{c}\left(\sqrt{\frac{c+v}{c-v}} - \sqrt{\frac{c-v}{c+v}}\right).$$

Here, the emitted energy is

$$\Delta E = h\nu\left(\sqrt{\frac{c+v}{c-v}} + \sqrt{\frac{c-v}{c+v}}\right).$$

and we obtain

$$m_{\text{before}}c^2 = m_{\text{after}}c^2 + \Delta E.$$

The mass of the body after the emission must be diminished by the already known amount. Today we know that there are particles that can turn completely into radiation. Therefore, we can derive Einstein's formula through reading anew Figure 5.4 (Figure 5.12).

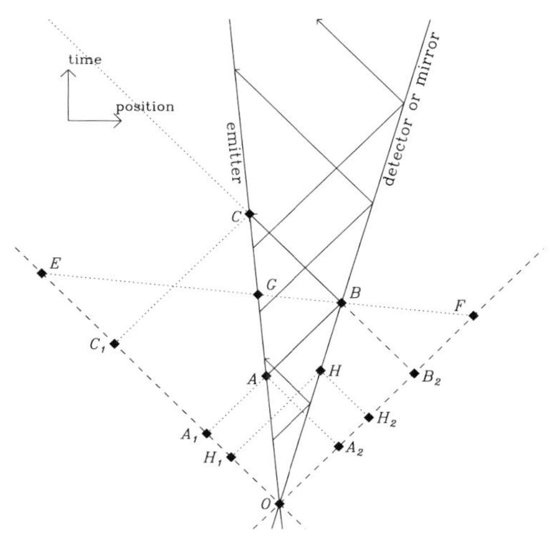

Figure 5.11: *The relativistic Doppler effect.*

We draw the world-lines of an emitter and a mirror in relative motion together with the lightlike directions. The period change depends only on the relative velocity between the emitter and the mirror. To show this, we draw through B the line EF of events simultaneous with B in the frame of the *emitter* ($EG = GF$). GB/OG is now the velocity of the mirror relative to the emitter. The points range CAO is projected first onto CBB_2 and then onto GBF. With the help of the invariance of the cross-ratio in projections (Figure 8.5), we obtain $CO/AO = CB_2/BB_2 = (GF/BF) : (GE/BE) = \frac{c+v}{c-v}$. We can use the diagram to find the point H on OB, which has the same distance from O as A on OC. The argument merely requires the theorems of similitude (see the text).

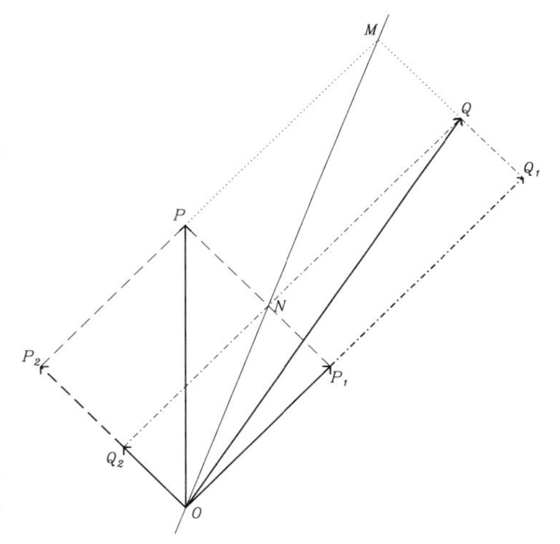

Figure 5.12: *The annihilation of a moving particle*

When a particle at rest (world-line parallel to OP) decays into two photons (world-lines parallel to OP_1 and OP_2), the spatial components of their momenta are opposite and equal. After reflection on the world-line OM, we obtain through the light-ray quadrilateral $PMQN$ the segment OQ as image of OP while the direction of the photon world-lines does not changes. The momentum parallelogram must be similar to OQ_1QQ_2. The spatial coordinates mv of the photons OQ_1 and OQ_2 are determined through the Doppler effect. The velocity that belongs to OQ is known, and we obtain the mass that belongs to OQ as a function of the energy of the photons.

The same pattern that we used to determine the variation of mass with velocity can be employed for the time intervals. We obtain the *time dilation*[15] (Figure 5.13). Let us now consider the time on a moving clock. The time lapse between two events is (in the rest frame of the observer) the projection of the interval of the world-line on the time axis. However, the time intervals $\mathrm{d}\tau$ on the moving clock are modified by the theorem of Pythagoras corre-

[15] This notion is a popular abbreviation of an observation that contains no real dilation of a real object or process but which exemplifies projection on reference frames. In the Euclidean geometry too, projections of line segments do not, in general, have the same size as the projected object.

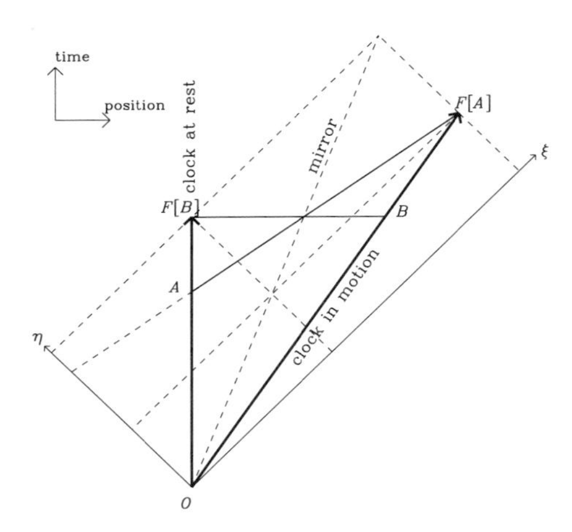

Figure 5.13: *Time dilation and its symmetry.*

We calculate in Figure 5.1 the time coordinates of the individual events. $F[B]$ and B are simultaneous with respect to the observer moving along OA, while $F[A]$ and A are simultaneous with respect to the observer moving along OB. The projection of a chord is always *longer* than the projected chord itself.

$$(OF[A])^2 - (AF[A])^2 = (OA)^2 < (OF[B])^2,$$

$$(OF[B])^2 - (BF[B])^2 = (OB)^2 < (OF[A])^2.$$

This states the time dilation. It is obviously symmetric and homologous to the corresponding statement of the Euclidean geometry. The only difference is that there the projections are being *shorter* than the projected chords. This reflects the changed sign in the analog of Pythagoras' theorem.

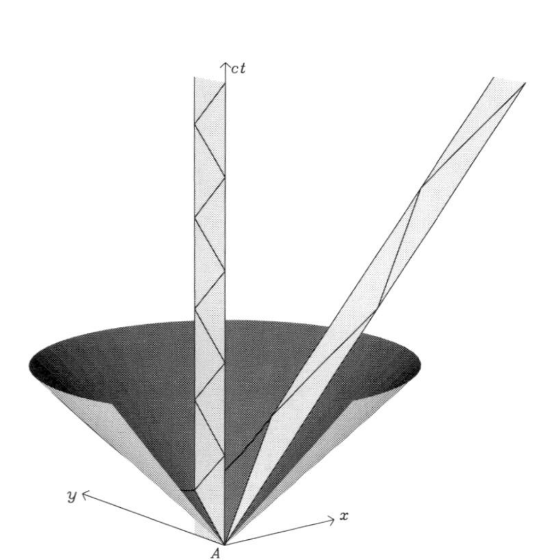

Figure 5.14: *Time dilation and the light clock.*

We show the timetable of two light clocks, one at rest in the reference frame and one moving. The intersection with the light cone at the point A shows the inclination of the world-lines of the photon in the light clock. This inclination determines the clock's pace. Obviously, the moving clock is slower than the clock at rest, if, of course, simultaneous events are chosen by the time of the rest frame.

sponding to the spatial distance covered. Therefore, the projections on the time axis, i.e., the time passed on the clocks of the reference frame, is longer than the time passed on the clock moving with respect to the observer. We have

$$\Delta t = \frac{\Delta \tau}{\sqrt{1 - \frac{v^2}{c^2}}}.$$

The most famous example for this difference in the time interval of the process in its rest frame and its projection onto another reference frame is the muon particle in showers of cosmic radiation. The muon decays by its own clock (determined by its internal physical processes) after about 10^{-8} s. By classical standards, it should use up this time after a flight of no more than 3 m. Nevertheless, for an external observer it can cover more than 100 km in this time span because the projection of its time span onto the time span measured by the observer is much longer if the particle moves fast enough. This time dilation is what makes it possible to observe the muons produced by cosmic radiation as well as to prepare unstable particles in an accelerator.

We must not forget that the time dilation is symmetric. From the point of view of the flying particle, the laboratory is moving, and the time flow of a clock on the table seems to be stretched. However, the time adjustment now concerns completely different events, so it does *not* produce any contradiction (Figure 5.13). The time dilation is not so strange when we compare it with its Euclidean counterpart. In the Euclidean geometry, the two world-lines are replaced by the two legs of an ordinary angle. Segments on the one leg are projected onto the other. There, they obviously appear (i.e., the projections are) shorter than that expected from a measurement on its original leg. This kind of length contraction is perfectly symmetric.

Using the constant propagation velocity of light we can construct an ideal clock using a light signal (photon) continually reflected at the ends of a cavity of constant length. The uniformity of its period depends only on the fact that the speed of light and the length are constant, and does not depend on the subtleties of a complicated inner structure. In addition, we can calculate its beat geometrically (Figure 5.14). We already met this construction in the Michelson experiment (Figure 4.4). The more inclined the strip of its timetable, the slower the clock runs with respect to the time of the reference frame. The unit is determined by the intersection of the strip with the light cone. Strips of different inclination to the direction of the x axis but equal extension in y cut the light cone in a plane $y =$ constant. The curve is a conic section. To be more specific, it is the hyperbola found in Figures 5.1 and 5.4.

The symmetry of the time dilation leads to the formulation of the *twin paradox* or *clock paradox*[16] (Figures 5.15 and 5.16). In order to set it up, we substitute the comparison of *two real* clocks for the comparison of the time coordinates of two reference frames. Two identically constructed clocks are supposed to meet twice (events A and B in Figure 5.15). Only one of them remains at rest (or uniformly moving); the other one moves relative to it, first leaving, then eventually coming back (at C in Figure 5.15) to the first one. The simple addition of the lapses of proper time of the second clock yields a value smaller than that of the first clock. The value is smaller because of the time dilation observed in the frame of the first clock that remains at rest (or in uniform motion). After all, the second clock always was in relative motion to the first, and the time flow on the first clock is composed of the projections of the segments of the world-line of the second.

We obtain a general statement: In a triangle of timelike world-lines, the longest side is *longer* than the sum of the other two. This is the pseudo-Euclidean *triangle inequality*. In the Euclidean geometry, we have a corresponding situation. The only difference is the sign. In the Euclidean geometry, each side of a triangle is *shorter* than the sum of the other two. The straight connection between two points is the *shortest*. In the pseudo-Euclidean geometry, the arc length of a world-line is the time measured by an observer who follows the timetable

[16]The term *paradox* is misleading if one expects something proven wrong by logic or by experience. It merely means something that is unexpected given the opinion or faith that one has with no implication of whether this faith is right or wrong. It denotes *only* the unexpectedness and *no* logical contradiction [37].

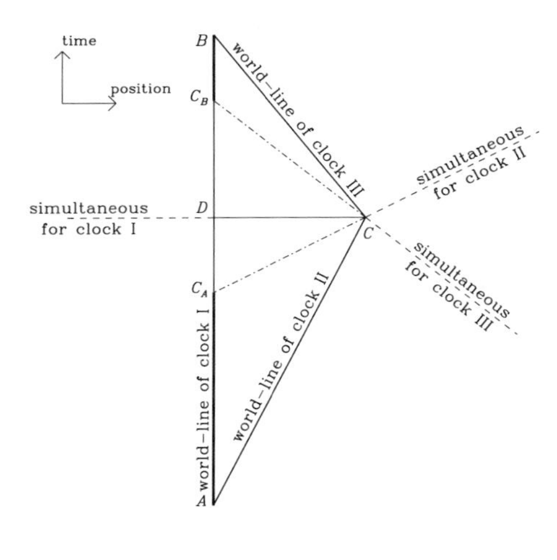

Figure 5.15: *The twin paradox and the triangle inequality.*

We draw a triangle ΔACB with timelike sides and the perpendicular DC to the longest side. Because of the minus sign in the theorem of Pythagoras, the projection AD is longer than AC and DB is longer than CB. Consequently, the side AB is longer than the sum of the other sides. For the voyager going from A to C via B, the projections onto the world-line are longer than the projected segments, too. However, we do not project the total segment AB onto ACB, but only AC_A onto AC and C_BB onto CB. The time dilation yields $d[A, C_A] + d[C_B, B] < d[A, C] + d[C, B]$, and no contradiction arises between the two points of view.

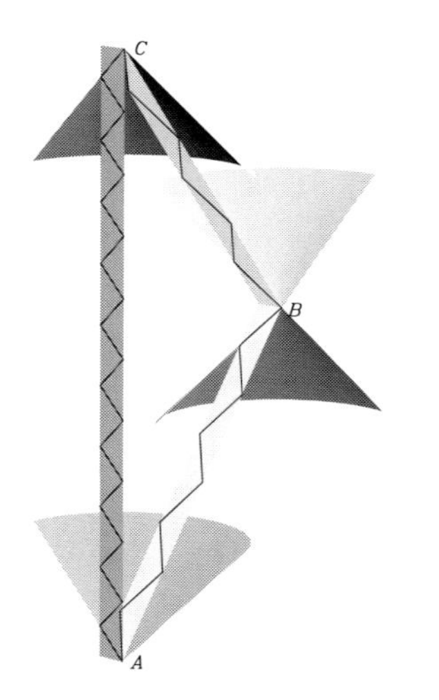

Figure 5.16: *The twin paradox and the light clock.*

The observer at rest and the voyager are furnished with light clocks (Figure 5.14). Their beats are determined by the corresponding light cones. In our figure, we count 7 beats for the voyager and 10 for the observer at rest. Obviously, the intersection pattern of the straight lines that are used to construct the figure cannot be changed by reflections or Minkowski rotations, i.e., by inspection of the diagram from a different reference frame. The result is an expression of the invariant triangle inequality.

represented by this world-line. We found that the proper time that elapses between two events depends on the timetable. The straight world-line between two events is the *longest*.

A paradox is already felt because there are differences between the readings of two clocks that move differently. The comparison with the Euclidean geometry shows that the surprise should be seen in the Galilean geometry or classical mechanics, in which the time lapse between two events (equal to the geometrical length of the world-line between them) does *not* depend on this world-line, i.e., does not depend on the motion of the clock.

Still more curious is the apparent possibility of inverting the statement. With respect to the voyager, the observer at home is in perpetual motion. Should she or he not be subject to time dilation too? Should we do not expect with the same argument that the observer at home experiences a shorter proper time than the voyager? This question constitutes the *twin*

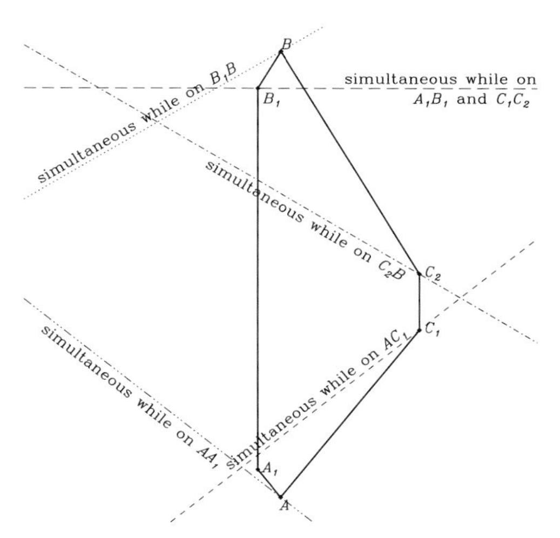

Figure 5.17: *The twin paradox with symmetric acceleration.*

We may conceive the world-lines of the twins in such a way that both experience the same accelerations, differing only in the time when they occur. Figure 5.15 is obtained when the line A_1B_1 approaches AB and C_1C_2 is removed so far that C_1 and C_2 coincide. The lines of simultaneous events are indicated to help the reader in analyzing the outset.

paradox. It seems to show that the reason for the different readings of the clocks when they meet again contains a contradiction. However, there is a difference between the voyager and the observer who stays at home. The world-line of the voyager contains the turning point C, where the voyager's velocity and rest frame change. The observer at rest *is* subject to time dilation, but it has to be calculated for two parts of his or her world-line separately. The analysis shows that only parts of the world-line of the observer at rest are projected onto the world-line of the voyager (Figure 5.15). To avoid the influence of the acceleration of the voyager at C on his or her clock, we can substitute two inertial clocks that meet at the turning event and are compared there for the one that had to be accelerated. In Figure 5.15, a third, oncoming clock takes over at C the reading of the second clock and the following time measurement. The voyager's clocks do not need to be accelerated to state the paradox. Hence, the answer to the paradox is *yes*; the time of the observer at rest seems dilated to the voyager's too, but *not all* of the time that elapses at the position of the observer at rest is to be compared with that of the voyager.

Of course, one can tell something about the effect of acceleration on a clock, although this is an additional refinement and does not alter the statement about the proper times of the two observers. The acceleration can only produce effects that are *not* proportional to the overall size of the triangle. This is the reason why they can always be separated from the geometric effect. The acceleration may produce two kinds of effects. First, it can result in a constant time lag due to the perturbation of the clock during the acceleration. Second, it can permanently change the rate of the clock by changing its mechanism.[17] The first effect would add a contribution that does not depend on the duration of the flight if the acceleration procedure is the same. In contrast to this, the difference of the readings at B due to the triangle inequality is proportional to the size of the triangle, i.e., to the duration of the flight. The second effect would be revealed by comparison of the clocks at B. Consequently, neither effect can shield or mimic the difference of the readings. In addition, there are constructions

[17] A man can get gray hair for instance during the acceleration, but age afterward as before. This illustrates the first effect. He can also get ill and age faster afterward than before. This would be an effect of the second type.

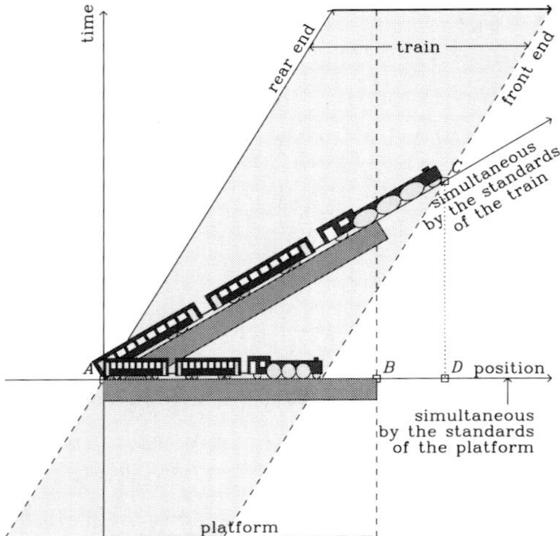

Figure 5.18: *The symmetry of length contraction.*

We draw the world-lines of a moving train referred to a platform at rest. The length of the train (if at rest) is assumed to be equal to the length of the platform. Hence, $AD^2 - CD^2 = AB^2$. The event D lies outside AB. Measured from the moving train, the platform is shorter. However, the event B too lies to the right of the world-line of the front end of the train. Seen from the platform, the train is shorter. So we again have a symmetric effect produced by the relativity of simultaneity (see also Figure 2.12).

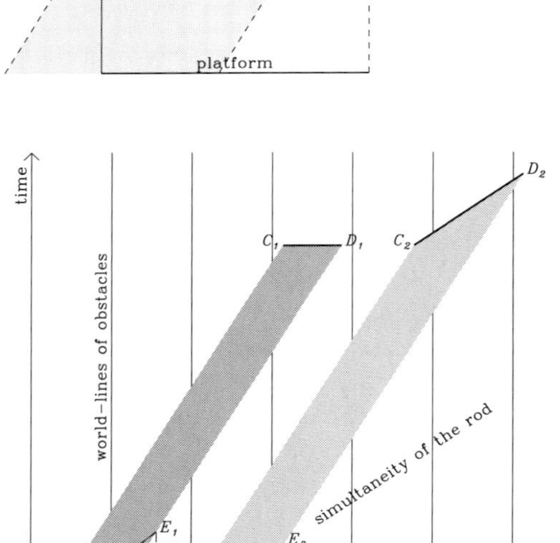

Figure 5.19: *The paradox of length contraction.*

In a (2+1)-dimensional space–time we draw two uniformly moving rods and a row of obstacles (a fence) at rest. The proper length of the rods is assumed to be identical to the proper separation of the obstacles. The left rod is parallel to the fence in the rest frame of the latter, i.e., the space components of the segments A_1B_1 and C_1D_1 are parallel. It seems to be contracted and can pass through the fence. The right rod is parallel to the fence too, but in its own reference frame. Now the segments A_2E_2 and C_2D_2 are parallel. The fence seems to be contracted and passage impossible. In the rest frame of the left rod, the rod appears to be turned into the fence, and it can pass. In the rest frame of the fence at right, the rod appears to be turned off the fence, and it is always stopped.

with world-lines containing equal accelerations (Figure 5.17) [68] and constructions with no accelerations at all on multiply connected worlds [69].

The curious behavior of the time coordinate is met again in the comparison of lengths (Figure 2.12). We determine the corresponding effect for the case of a uniformly moving rod or train. The world-lines of the front and rear ends of a train are parallel. We obtain the *length contraction* (Figure 5.18). The history of an extended, uniformly moving object (the train) is an *oblique* strip in the space–time plane. An object at rest (the platform) is given by a *vertical* strip. The outcome of a comparison of the width of the two strips obviously depends on the definition of simultaneity. If in its rest frame the train is as long as the platform in its rest frame, both appear in the rest frame of the other shorter than expected. When the rear end of the train reaches the platform, the front end is still inside the station if the observation is made at an instant considered to be simultaneous by the observer on the platform. However, if the instant is chosen to be simultaneous for the observer in the train, the front end of the train will have already left the station. The first measurement seems to be made much too early. The moving object always seems to be contracted [64].

A fencing instructor named Fisk
In duels was terribly brisk.
So fast was his action, Fitzgerald contraction
Foreshortened his foil to a disk.[18]

This length contraction is a symmetric projection effect just like the time dilation. Here, one can ask why lengths seem to be *contracted* while the projection of time intervals is *longer* than the intervals themselves. It can be shown that the projections of spacelike segments are also longer than the original ones but the measurement of lengths includes the observation of simultaneity. The relativity of simultaneity produces an effect that overcompensates the projection effect. The result is the apparent contraction.

Again, the symmetry of the length contraction admits the formulation of paradoxes that must be analyzed analogously to the twin paradox. We cite the length contraction paradox presented by Shaw [70] (Figure 5.19). A rod moves uniformly along a row of obstacles, let us say a fence. In addition, it slowly drifts toward the fence. The distance apart of the obstacles is assumed to be equal to the length of the rod (in their own rest frames, respectively). Will the rod pass through the fence or not? Seen from the fence, the rod is moving and should be contracted, so that it can pass. Seen from the rod, the fence moves and should be contracted, so that passing is impossible. Which is right? It turns out that the answer is given by the correct application of what we know about simultaneity. The question whether rod and fence are parallel depends on that simultaneity. Being parallel in the rest frame of the rod is not the same as being parallel in the rest frame of the fence. Being parallel in the rest frame of the rod means no passage because the fence is contracted. In the rest frame of the fence, the rod is contracted too, but it is not parallel to the fence but a little bit skew. Hence, it cannot pass through the fence even though it is contracted. In the other case, when the rod and fence are parallel in the rest frame of the fence, the rod can pass because it is contracted. Seen from the rod, it is not parallel but somewhat rotated to the fence. Thus, it can pass through the fence even though the fence seems to be contracted.

5.5 Aberration and Fresnel's Paradox

The change in orientation that originates in the relativity of simultaneity is the source of aberration (Figure 4.3) in the wave picture. As long as we are justified in imagining a flow of particles or particle-type objects as wave groups, we can use the addition theorem of velocities to obtain the aberration of direction. However, when we observe wavefronts (as happens in adaptive optics, for instance) there is no justification to do that. If we were to observe sound with a device that measures the orientation of wavefronts, no aberration could be found (Figures 5.20 and 5.21). What about light? Fresnel worried greatly about this when he tried

[18] The diameter of the disk is not larger than the diameter of the blade! Button would be more exact than disk.

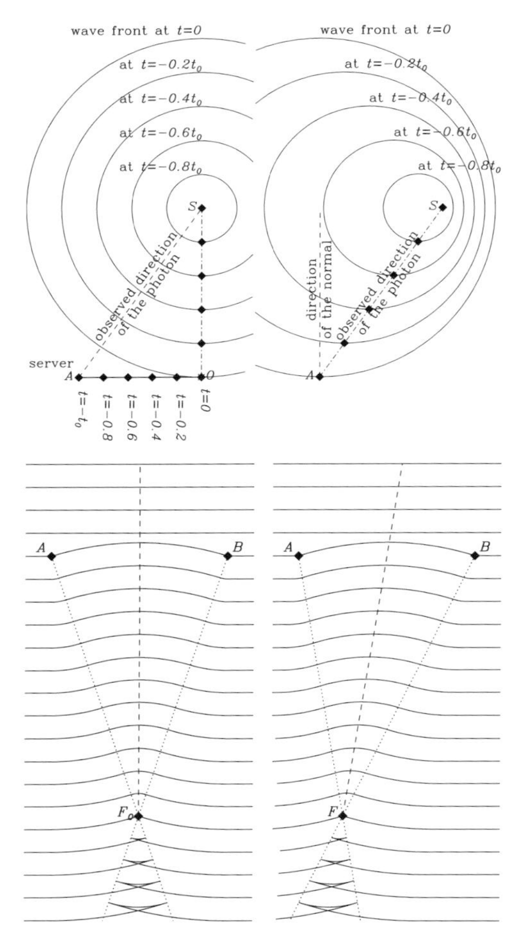

Figure 5.20: *An isotropic wave and an observer in motion.*

We add to Figure 4.3 the fronts of an isotropically expanding wave. On the left, a wave propagates isotropically. On the right, all positions are referred to the position of the observer A. Evidently, the direction of a signal and that of the wavefront normal do not coincide for an observer in motion with respect to the medium. For propagation in a medium, wavefront normals do not show aberration. Figure 5.23 will exhibit the same situation in a space–time diagram.

Figure 5.21: *Fresnel's solution to the aberration problem for waves.*

On the left, the propagation is isotropic. The aperture of the lens AB cuts a piece out of the wavefront that interferes to yield the focus F. On the right, the medium moves from the right as in Figure 5.20. The wavefront normals remain the same, but the interference figure and the focus are shifted to yield the expected aberration.

to recover the known properties of the propagation of light in the wave picture. His difficulty was resolved by the construction of the ordinary telescopes.[19] The aperture of the telescope lens or mirror cuts out a wave group. The motion of this wave group shows aberration as usual. This aberration is perfectly consistent with the lack of aberration for wavefronts if an all-pervading aether is assumed. However, in the theory of relativity, the aether does not exist. It is the *relativity of simultaneity* that produces aberration in the wave picture too (Figure 5.22). Figure 5.23 is the timetable version of Figure 4.3. For the observer B_1, the projection of SO onto the space of simultaneous events is MO, i.e., the intersection with the plane SOO^*. The space of simultaneous events is indicated by its intersection with the light cone. Its front part is covered by the half of a plane disk. For the observer B_2, the plane of simultaneous events is to be intersected with the plane SOA. We see again that this intersection aberrates from MO. However, as long as we do not touch the definition of simultaneity, the wavefront is a

[19] Young discusses the aberration in his Bakerian lectures, and gives this solution, too, but he did not emphasize the paradox as Fresnel did.

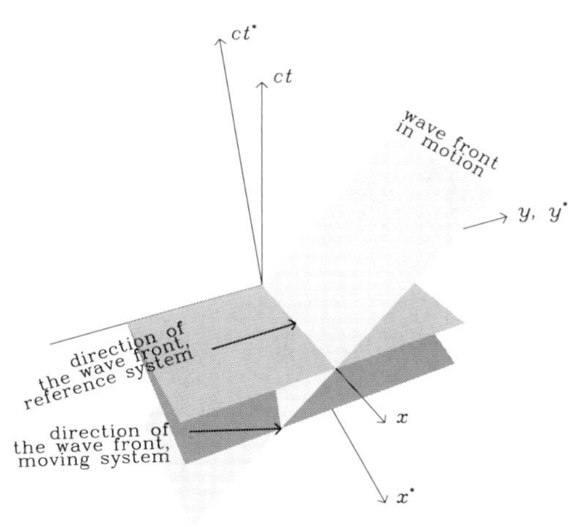

Figure 5.22: *Aberration and relativity of simultaneity. I.*

The normal of a wavefront is a direction in space and does not change between differently moving observers. In Minkowski space–time, simultaneity is relative, and this produces the aberration in the wave picture. Instead of the rod moving slowly toward the fence, we now draw a wavefront moving in the direction of the y axis, together with the planes simultaneous with $t = 0$ (light) and with $t^* = 0$ (dark). The wavefront intersects the plane $ct = 0$ on a line with the normal in the direction of the y axis. It does not change until the simultaneity becomes relative. The two planes must be different; otherwise there is no aberration of phase fronts.

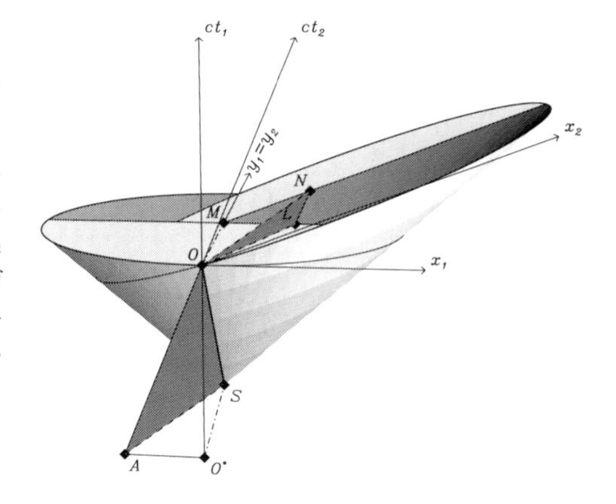

Figure 5.23: *Aberration and relativity of simultaneity. II.*

This is the timetable version of Figure 4.3. The observer determines the direction of the photon coming from S and incident on O [71]. The projection of this rather complex timetable on the spaces of simultaneous events gives us the direction of an incoming photon as well as the position of the wavefront at the moment of observation (see the text).

circle with a normal always pointing to M, even though the direction of the ray is LO. We have to accept relative simultaneity in order to obtain an aberration of wavefront normals too. Proceeding as before, the cut of the plane of events simultaneous with O for B_2 with the light cone has its center at N. We again indicate the plane of simultaneous events by its intersection with the light cone. This time, the rear part is covered. With the appropriate gauge of the coordinates, the curve is a circle, the direction ON is the wavefront normal, and the aberration of the direction of the photon is equal to that of the wavefront normal.

It turns out that in the Minkowski geometry it is equally appropriate, as far as kinematics is concerned, to conceive light propagation as a wave phenomenon or as particle emanation. Because the simultaneity of relativity is the central point, we can even reverse the argument. We can take the observation of particle-type aberration and seek a relativity of simultaneity

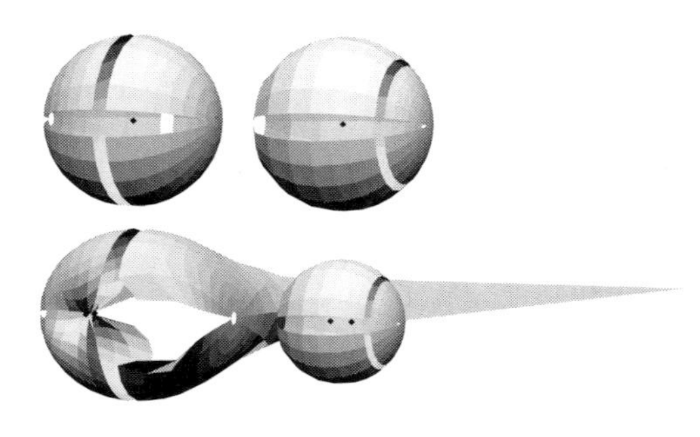

Figure 5.24: *Relativistic aberration.*

We show the map of a sphere (in the upper left) around the observer. The sphere is left incomplete to improve the spatial impression. In the upper right, its conformal image is shown for an observer moving with 0.7 c to the right (see Figure D.5). In the lower right, this map is completed stereoscopically for two eyes lined up in the direction of the motion, i.e., they look sideward. We see an added overall contraction. In the lower left, the eyes look forward. The map becomes singular in the plane spanned by the eyes and the velocity. Here, the relative velocity is 0.4 c.

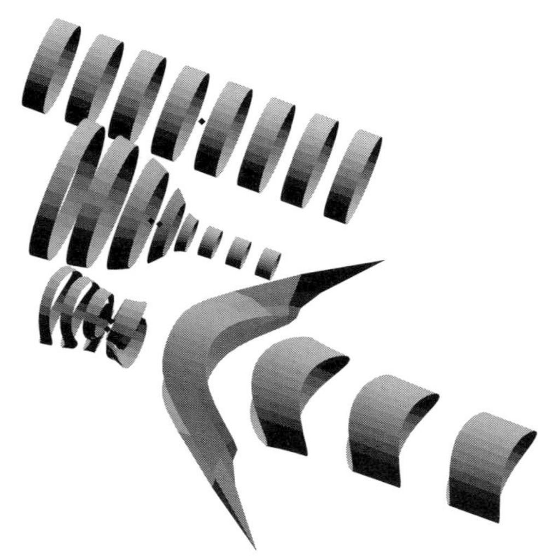

Figure 5.25: *Stereoscopic aberration.*

The momentary image of a cylindrical tunnel is depicted. The first example is the unperturbed image seen by the observer at rest. It is followed by images seen by an observer in motion with 0.7 c to the right. In the middle, the eyes are lined up in the direction of motion, and the observer looks sideward out of a window of a train passing through the tunnel. The tunnel seems narrowed in the direction of motion, and widened in the opposite direction. Some contraction is superposed. The lower image shows the appearance for an observer whose eyes are placed abreast; the observer looks of the front window. There is an obvious dilation in the direction of motion and a contraction in the opposite direction, the tunnel being neither narrowed nor widened.

that reproduces the particle-type aberration for wavefronts.[20] We then obtain the Minkowski geometry [72]. In aberration, it was also observed that the motion of the emitter was not combined with the speed of light. For the wave theory, this was obvious. However, if we require the composition of velocities for particles to include this fact, we arrive at Einstein's composition of velocities too.

At this point, we must add a remark about the visibility of the Lorentz contraction. The apparent image (the photograph) of a moving object not only contains the length contraction but also the much larger effect (because of first order in the velocity) of aberration. Seen with

[20]This was the point where Lorentz and Drude found to be forced to introduce an *effective time* that relaxed absolute simultaneity. However, they did not dare to call it the one and only physical time.

one eye, the aberration compensates the length contraction. An object flying by is seen as if rotated but not contracted in the direction of motion [11].

However, we can embed the conformal map (Figure 4.11) of the apparent sky in a map of the whole space that is constructed by triangulation from two eyes. The two eyes see slightly different positions subject to occasionally different aberrations. With these apparent positions, the true position in space can be calculated as in the case of a parallax (Figure 2.15). The relative orientation of the pair of eyes with respect to the velocity is essential. If we observe with two eyes, one behind the other in the direction of motion, the length contraction of the distance of the eyes is reflected in a length contraction of the stereoscopic image (Figures 5.24 and 5.25).

5.6 The Net

This is the point to note that we have now a net of facts in hand that are all equivalent to the Minkowski geometry. We derived it first through the isotropy of light propagation. Our first and fundamental finding was the relativity of simultaneity. With the exact coefficients, it represents the essential part of the Lorentz group, i.e., a group of motions containing such a relativity of simultaneity can only be the Lorentz group. We then derived the variation of mass with velocity. In addition, we found the relativity of simultaneity to be a consequence of this variation too. The formula $E = mc^2$ was shown to follow, but with the same success we may start with it and derive through the definition of the increment of energy, Eq. (5.3), and its integration the solution, Eq. (5.5). We may even start with Einstein's composition theorem of velocities to see the invariance of the speed of light or the relativity of simultaneity (Chapter 8, Figure 8.7). Bondi [40] started with the symmetry of the Doppler effect. Finally, we can require aberration of light to be the same for wavefronts and photons. This also leads to the relativity of simultaneity. This net is shown in Table 5.1.

5.7 Faster than Light

Is there motion faster than light? Let us ask geometry. We derived it through reflections. The world-line of a particle at rest can only be congruent (i.e., reflection image) to a timelike line, and such a line describes a motion slower than light. It is often argued that the variation of mass with velocity yields the reason for that the velocity of light cannot be attained. In the end, any acceleration will increase only the energy (together with its mass) but leave the world-line timelike. A body at rest cannot attain the speed of light, even if any large amount of energy is applied. The length of the space–time momentum vector is not changed and the velocity is bound to remain smaller than c. Nevertheless, there are particles that move as fast as light. They are not accelerated to that speed, but they are that fast since their formation. The momentum vector of a particle with the speed of light has length zero, i.e., it has no *rest* mass m_0. Nevertheless, it has a mass that can be measured in collision experiments. This is the inertial mass, which is proportional to the energy. The only obstacle is that the particle

Table 5.1: The net of relativistic propositions

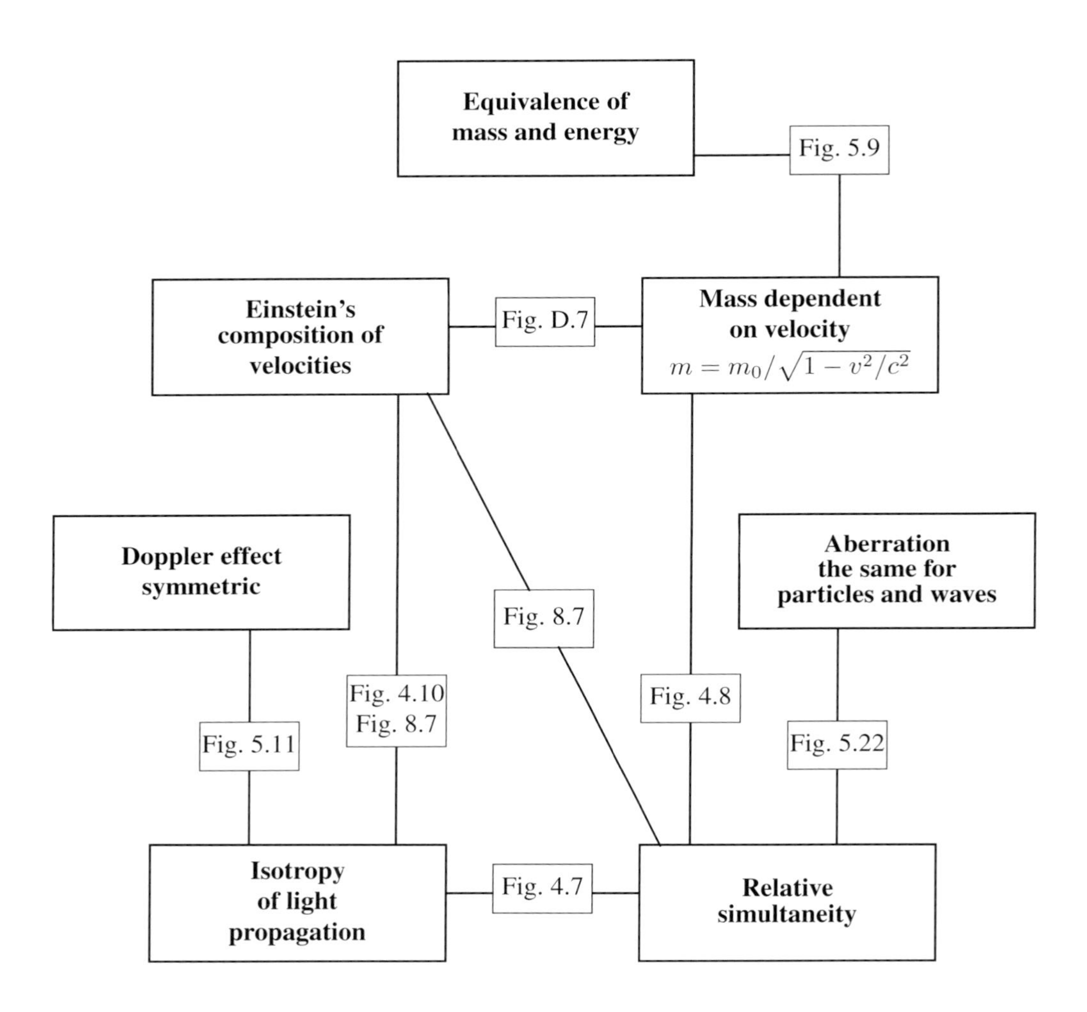

cannot be brought to rest, just as it can never be accelerated.[21] Also for photons, reflections do not change the direction of time. As long as the velocity of real motions does not grow beyond the speed of light ($v \leq c$), a *causal order* is established: Action is transported only in one direction of time, and in every respect the future follows the past.

[21] In a *gedanken* experiment, one can, surprisingly, put a photon on a balance without bringing it to rest. One has to enclose it in a box with perfectly reflecting sides. Its weight due to its mass has the consequence that the collisions with the bottom transfer a little bit more momentum than the collisions with the top. The difference that accumulates in a unit of time is the weight of the photon. At least in principle, the mass of the photon can be determined by ordinary weighing. This was the argument that persuaded Einstein to choose the equivalence principle as the basis of the theory of gravitation.

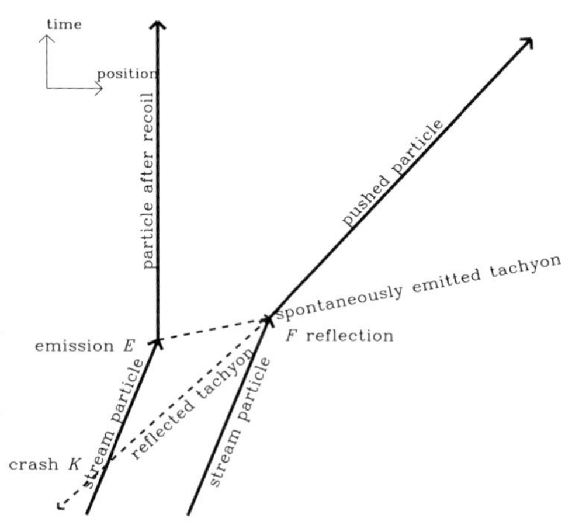

Figure 5.26: *Tachyons are reflected into the past.*

In a stream of particles of given rest mass a tachyon with a given norm of the momentum is emitted and later reflected elastically by another particle of the stream. In this case it can meet the particle that emitted it in an event long before the emission event. In our sketch, the velocities of the tachyon are determined by the methods that we used in our collision analysis: We take the conservation of momentum at emission and reflection into account and use the invariance of the rest mass or the momentum norm, respectively (Figure 5.27).

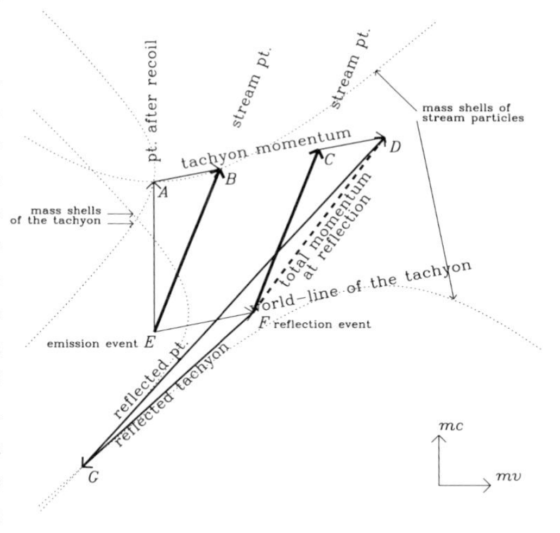

Figure 5.27: *Conservation of momentum in tachyon reflection.*

We supplement Figure 5.26 with the auxiliary lines necessary for the construction. The momentum vector $\vec{EB}$ is decomposed into the momentum vector $\vec{AB}$ of the tachyon (which ends on a hyperbola defining its lengths, i.e., the mass shell of the tachyon) and the momentum vector $\vec{EA}$ (which ends on the same mass shell as $\vec{EB}$). Consequently, we find A as the intersection point of two hyperbolas. At the reflection event, we construct the total momentum $\vec{FD}$ as the sum of $\vec{FC} = \vec{EB}$ and $\vec{CD} = \vec{AB}$. This total momentum is decomposed again into a tachyon momentum $\vec{FG}$ and a particle momentum $\vec{GD}$. The point G is, as the point A, the intersection of two hyperbolas with the central points F and D, respectively.

However, the mere prescription of Minkowski reflections does not prohibit motion faster than light as it does not forbid motion as fast as light. Of course, one may find other arguments for a possible existence of particles faster than light. The only condition: Relativity must not be changed.

Now let us imagine such particles that move faster than light. They are called *tachyons*. Their momentum vector is spacelike, whose norm is negative, and their rest mass is imaginary. Tachyons too cannot be brought to rest. This is because the negative norm of the momentum of a tachyon forces its world-line to be outside the light cone of any event that it passes. Tachyons must be generated with speed greater than light because no acceleration pushes slower object beyond the limit given by the light velocity.

Three important arguments speak against the existence of tachyons. First, the time component of a spacelike momentum vector is not bounded from below. This component is, of course, the energy of the tachyon. A tachyon could therefore release an infinite amount of energy unless we impose conditions that violate relativity. If we analyze the balances of ordinary particles, this would be observed as spontaneous generation of energy. No experiment ever gave evidence for such an effect. Secondly, a particle at rest could start motion by spontaneous emission of a tachyon. This is not possible with photons or particles slower than light. Spontaneous emission of a photon or a particle slower than light necessarily reduces the rest mass of the emitter, i.e., in the ground state of the emitter it does not occur and the motion cannot spontaneously change. In contrast to this, spontaneous emission of tachyons does *not* necessarily reduce the rest mass of the emitter (it may even augment it): It is possible in the ground state of the emitter too. One should be surprised that something like Newton's first law can be obeyed at all if tachyons exist. Finally, reflection can alter the sign of the time component of a spacelike velocity or momentum. A tachyon emitted spontaneously in a stream of particles can collide with its emitter *before* the emission if it is reflected appropriately by another particle of the stream (Figures 5.26 and 5.27). A limerick puts it nicely [64]:

> There was a Young Lady named Bright
> Whose speed was much faster than light.
> She started one day in a relative way
> and returned by the previous night.

Causality seems to be at stake. However, we observe *no* spontaneous emission in a ground state, *no* deviation from the first law of Newton, and *no* breakdown of causality. It is difficult to circumvent these three arguments. In macroscopic experiments tachyons are absent.

6 The Circle Disguised as Hyperbola

In the preceding chapter, we constructed the reflection prescribed by the universal isotropy of light propagation. The light-ray quadrilateral turned out to be the central figure (Figures 4.9 and 5.1). It allows the construction of a geometry that contains homologous statements for most of the theorems of the Euclidean geometry. However, at the first sight all the corresponding propositions look strange. The reason is simple. It is because we must abandon the usual ideas of orthogonality, angle, or circle. But we are on a registration strip, in a timetable, in a space–time, and no longer in the Euclidean plane. However, the basic facts survive: Straight lines play a fundamental role (they represent force-free motions), and reflections can be properly defined. When the configurations are constructed by the *new* reflections and when the *old* names are given the new content, the old theorems are valid again. Sometimes, it seems indeed surprising that one recovers the old theorems at all. As we shall see in Chapters 8 and 9, this indicates a deep relation between different geometries.

The reflections allow us to establish equality between angles and between segments. First, we ascertain that the theorem of perpendicular bisectors is valid, which is a necessary premise for a consistent comparison of lengths (Figure 6.1). The perpendicular bisectors on the three sides of a triangle are found through the light-ray quadrilaterals for each of them. We get a net of light-ray quadrilaterals and apply the area theorems for parallelograms.[1] The intersection of the perpendicular bisectors is, by definition, equally distant from the three points of a triangle. Strictly speaking, distance is to be *defined* for differently oriented intervals in such a way that equal distances result in our construction.

We can reverse the argument. The necessity of a mid-perpendicular theorem defines the geometry also in the case when we have no access to experiments with light (Figure 6.2). When we know the simultaneity (the perpendiculars) with respect to two different straight world-lines, we are able to construct them for any straight world-line. In the case when the mass increases with velocity, we obtain the existence of lines of length zero, and the ability to construct reflections in the way that we just learned of.[2]

Next, we consider the successive reflections of a point Q on a pencil of lines through some point M of the plane. We want to understand reflections as maps that preserve distances. Consequently, the point Q and all its reflection images are equally distant from the point M,

[1] After choosing a diagonal of a parallelogram and a point on it, we draw the parallels to the sides. The parallelogram is cut into four smaller ones. *The two partial parallelograms that are not crossed by the diagonal are always of equal area.* Any point can be used to split the parallelogram in four. The equality of two opposite parallelograms then indicates that the reference point lies on the corresponding diagonal.

[2] When we find the moving mass to be smaller than that at rest, the construction would yield the Euclidean geometry also for space–time. But then, which coordinate would be the time? Invariant wave propagation would not be possible.

The Geometry of Time. Dierck-E. Liebscher

ISBN: 3-527-40567-4

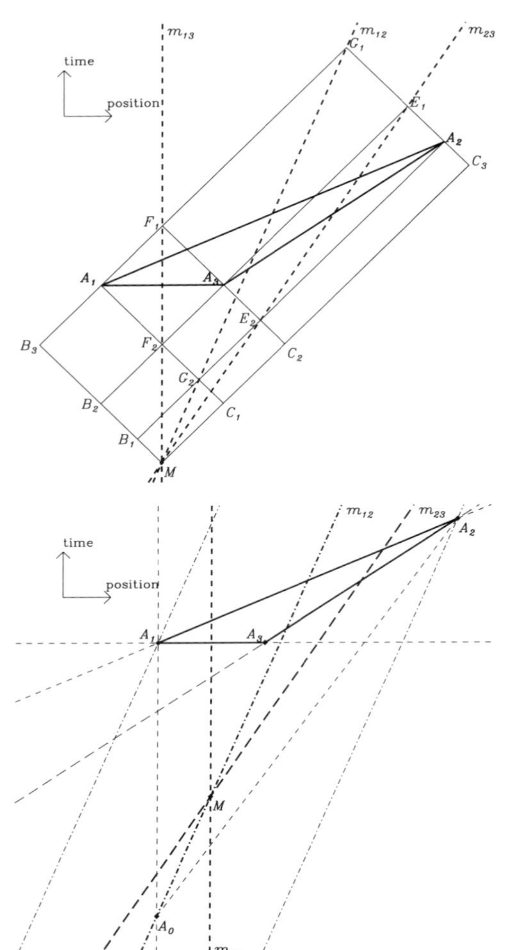

Figure 6.1: *The theorem of perpendicular bisectors in the Minkowski geometry.*

We choose a triangle and construct the perpendicular bisectors with the help of the corresponding light-ray perpendiculars. The proof of intersection is provided by comparison of the individual areas. The perpendicular bisectors onto A_1A_3 and A_2A_3 intersect at some point M and fix $A_1B_3B_2F_2 = F_2C_1C_2A_3$ and $A_3B_2B_1E_2 = E_2C_2C_3A_2$. It follows that $A_1B_3B_1G_2 = G_2C_1C_2A_2$. However, this is the condition for the perpendicular bisector G_1G_2 onto A_1A_2 to pass through M too, which we intended to show.

Figure 6.2: *Simultaneity and relativity.*

This is an extended cut-out of Figure 4.8. There we determined the simultaneity for *one* velocity of the system of reference (i.e., the train). However, when we accept the argument of the circumcenter theorem, we can construct the simultaneity for all velocities. For the event M the events A_1 and A_3 as well as the events A_1 and A_2 are equivalent: In both cases a reference line exists with equal spatial and temporal distance to M (i.e., m_{13} and m_{12}). If we now insist on that A_2 and A_3 are also equivalent in this sense, they have to be simultaneous with respect to the world-line m_{23}.

i.e., they form a circle. In the Euclidean geometry, we see what we expect to an ordinary circle (Figure 3.2). Figure 6.3 shows it with corresponding precision. More precisely, what is to be a circle is *determined* by this construction. Reflection on lines through a point M produces points that are equally distant from M. After choosing a certain kind of reflection, we find by our construction the locus of all points that are equally distant from M and pass through the initial point Q. In the Euclidean geometry, the circle is a particular kind of ellipse. To be specific, it is an ellipse with equal axes, and such an equality is defined only in a metric geometry. The circle of the *new* geometry is the curve obtained by the *new* reflections. Using the light-ray quadrilaterals, we find a hyperbola (Figures 5.1 and 6.4). The name hyperbola means that it is a conic section that passes through infinity at two points. According to the axioms of the new geometry that we are constructing, this is *the* circle, i.e., the locus of constant distance to the center. This is the property of the circle used in geometric constructions. Now the distance is to be calculated by the new theorem of Pythagoras that we discussed in Chap-

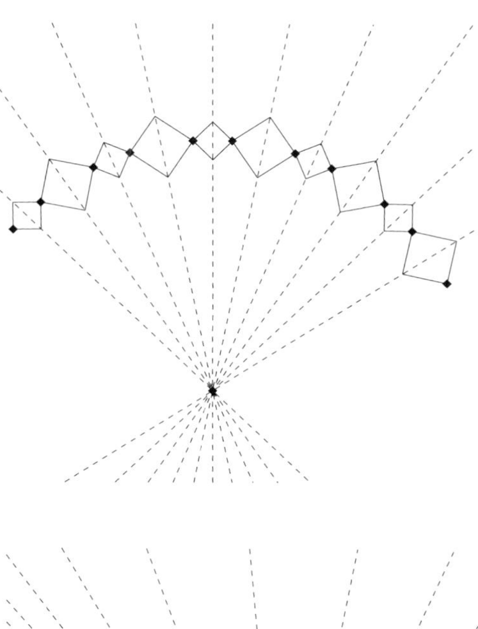

Figure 6.3: *The circle as a result of consecutive Euclidean reflection.*

We reflect a point successively on the rays of a pencil. The reflection images are the vertices of a diamond that has a diagonal, which coincides with the reflecting line. All images lie on a particular ellipse, which is the circle of the Euclidean geometry (cf. Figure 3.2). If the product of three reflections on rays of a pencil did not result again in a reflection on some ray of a pencil, we would obtain on each ray more than one image point, and this would mean that the interpretation of the resulting curve as equidistant from the center cannot be appropriate. We can see that the product property of the reflections on the rays of a pencil is the kernel of the theorem of the intersection of perpendicular bisectors.

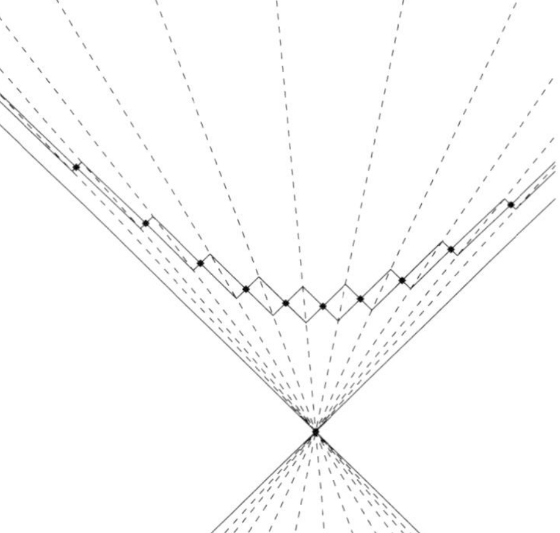

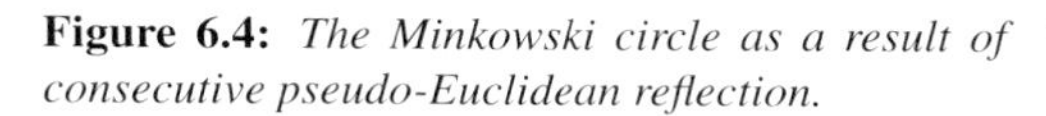

Figure 6.4: *The Minkowski circle as a result of consecutive pseudo-Euclidean reflection.*

We repeat the construction of Figure 6.3, but we adopt the pseudo-Euclidean prescription for successive reflections and obtain a curve that is a particular hyperbola, which plays the role of a circle in the pseudo-Euclidean geometry. Reflection images are vertices of a light-ray quadrilateral that has one diagonal coincident with the reflecting line.

ter 5 (Figure 5.2). The Minkowski circle is not a particular ellipse, but a special hyperbola that is distinguished by the given asymptotic directions.

In everyday language, a circle is usually compared with an ellipse, parabola, and hyperbola. This creates a wrong impression that they all are of the same kind, i.e., particular conic sections. However, on this level a circle does not belong to the same genus. The ordinary circle is a particular kind of an ellipse, and it can be defined only if the plane is endowed with metric properties. Ellipse, parabola, and hyperbola do not presuppose metric properties. The ordinary (Euclidean) circle is a particular ellipse, the Minkowski circle is a particular

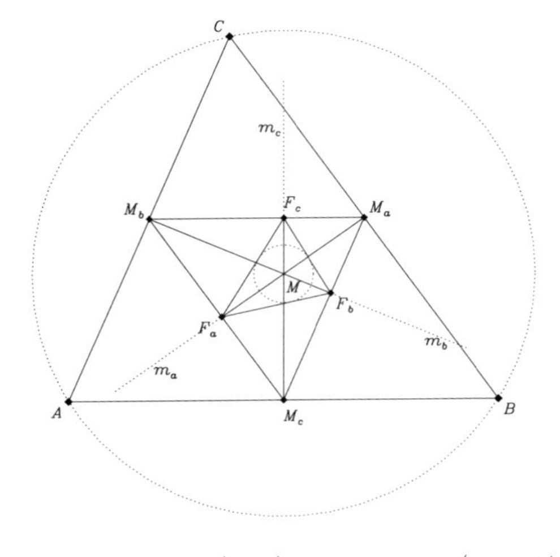

Figure 6.5: *The midpoint triangle and the footpoint triangle.*

In each triangle ΔABC the perpendicular bisectors m_a, m_b, m_c intersect in some common point M. On the perpendicular bisectors lie the altitudes of the triangle whose vertices are the midpoints M_a, M_b, M_c of the sides. The feet F_a, F_b, F_c of these altitudes form a third triangle, whose angle bisectors are m_a, m_b, m_c again.

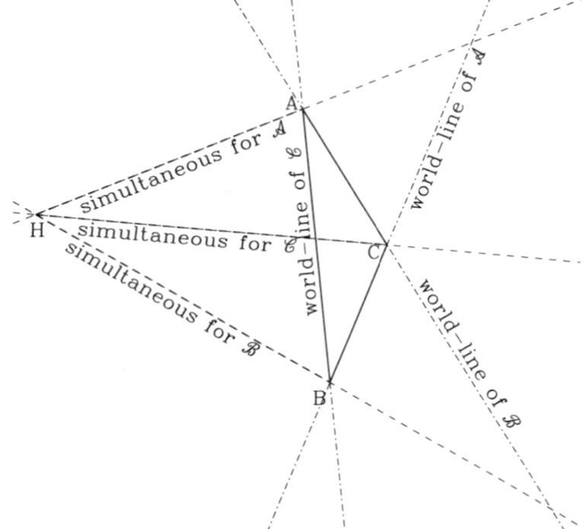

Figure 6.6: *The intersection theorem of altitudes. I.*

The triangle has timelike sides, a, b and c, these sides represent the world-lines of the three mutually moving observers $\mathcal{A}$, $\mathcal{B}$, and $\mathcal{C}$. These meet at the events ab, bc, and ca, respectively. The event H is the intersection of the altitudes of the triangle. It is for $\mathcal{A}$ simultaneous with bc, for $\mathcal{B}$ simultaneous with ca, and for $\mathcal{C}$ simultaneous with ab.

hyperbola. They differ because of different metric properties of the corresponding planes.[3]

As we know from the Euclidean circle, the Minkowski circle is determined by giving three points on the periphery. The construction of the intersection of the perpendicular bisectors yields the center of both kinds of circle. The construction uses a definition of orthogonality, which is not same for the two geometries. The reflection on the perpendicular bisectors reveals that the three points lie on a circle of the corresponding geometry. With reflections on other lines through the intersection, we get all the other points of this circle. We note again that *all* Minkowski circles are hyperbolas with same direction of the asymptotes. The asymptotes are always lightlike. In the projective geometry, which we consider in Chapter 8, the two directions determine two points on the line at infinity. Every Minkowski circle passes through these two points.

The well-known intersection theorems of plane trigonometry are valid in the Minkowski geometry too. We again obtain an intersection point for the three altitudes of a triangle. In

[3] We note that in the Minkowski geometry a circle of zero radius is an instrument for construction (Figure 5.1), whereas it is useless in the Euclidean geometry.

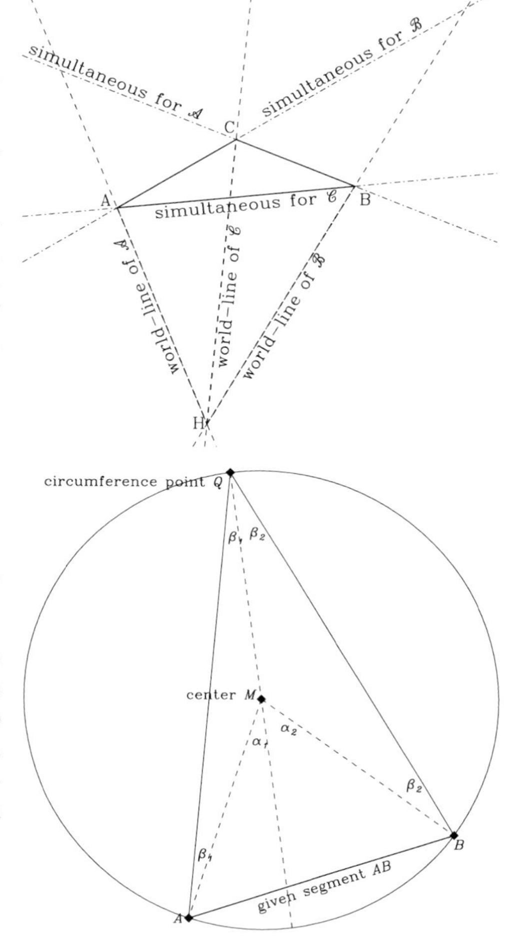

Figure 6.7: *The intersection theorem of altitudes. II.*

$\mathcal{A}$, $\mathcal{B}$, and $\mathcal{C}$ (whose world-lines are in the end of the three altitudes) meet in one single event (H). We choose A on $\mathcal{A}$ and construct B on $\mathcal{B}$ as intersection of $\mathcal{B}$ with the perpendicular from A to $\mathcal{C}$, and C on $\mathcal{C}$ as intersection of $\mathcal{C}$ with the perpendicular from A to $\mathcal{B}$. The intersection theorem of the altitudes yields that the line BC is perpendicular to $\mathcal{A}$.

Figure 6.8: *The theorem of circumference angles in the Euclidean geometry.*

We draw a circle around M and one of its chords AB and choose Q on its circumference. The $\beta_1 + \beta_2$ at the circumference angle is equal to half the angle $\alpha_1 + \alpha_2$ at the center ($\alpha_1 = 2\beta_1$, $\alpha_2 = 2\beta_2$ and $\alpha_1 + \alpha_2 = 2(\beta_1 + \beta_2)$). Consequently, it does not vary with the position of Q on the circumference.

analogy to the Euclidean geometry, we can consider the three bisection points and form a secondary triangle. The perpendicular bisectors of the first triangle are the altitudes of the second. This relation between perpendicular bisectors and altitudes remains valid in the Minkowski geometry because the axiom of parallels still holds. Therefore, we can invoke the equality of alternate angles and corresponding angles on parallels. In geometries that are even more general, this is no longer valid: Two straight lines have in general only one common perpendicular. However, that too suffices to obtain the previously mentioned result (Appendix A). We obtain a dual statement when we consider the foot points of the altitudes in the secondary triangle. They again form a triangle, the third in the sequence. The dual statement is the theorem that the altitudes of the second triangle bisect the angles in the third one (the foot-point triangle, Figure 6.5).

The intersection theorem of the altitudes has a simple physical interpretation [75]. For three mutually moving observers there exists exactly one event E that is for $\mathcal{A}$ simultaneous

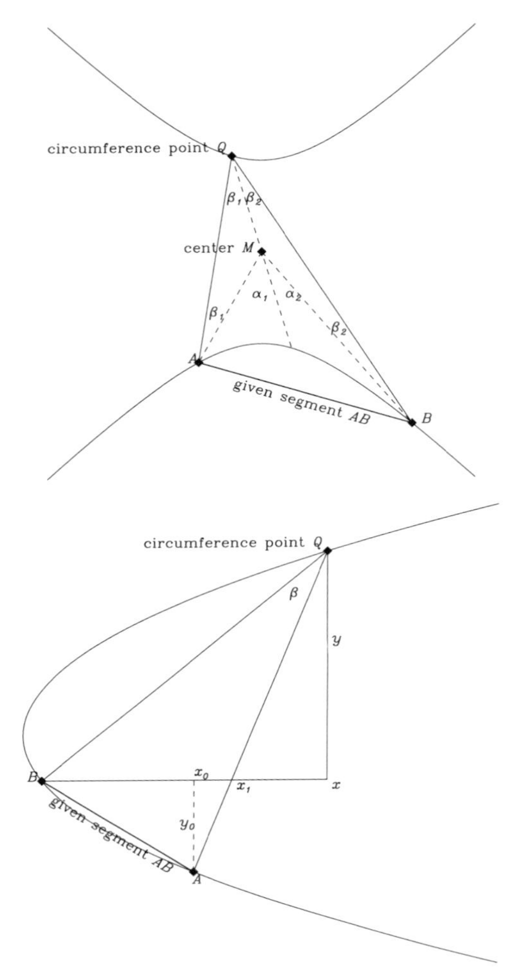

Figure 6.9: *The theorem of circumference angles in the Minkowski geometry.*

For the Minkowski geometry, the proof of the circumference-angle theorem proceeds as in Figure 6.8 if we take into account that isosceles triangles have two equal angles at the base (here invoke the reflection). We then use the theorem of equal corresponding angles as in the Euclidean geometry. The theorem of corresponding angles is equivalent to the axiom of parallels. Hence, we cannot expect any longer to find equal angles on the periphery of a circle for the non-Euclidean geometries.

Figure 6.10: *The theorem of circumference angles in the Galilean geometry.*

In the Galilean geometry, we find a curve different from the circle for the first time. For more general geometries, we find it in Figure 9.13. Let (A, B) be the given segment. The requirement of constant circumference angles is given by $x_1/y = \beta$. The similarity yields $(x - x_1)/y = (x_1 - x_0)/y_0$. If we solve these equations for x_1 in order to eliminate it, we obtain the quadratic equation $x = (\beta y^2 + \beta y y_0 - y x_0)/y_0$. The corresponding curve is a parabola with the axis in the given direction.

with the event where $\mathcal{B}$ and $\mathcal{C}$ meet, for $\mathcal{B}$ simultaneous with the event where $\mathcal{A}$ and $\mathcal{C}$ meet, and for $\mathcal{C}$ simultaneous with the event where $\mathcal{A}$ and $\mathcal{B}$ meet (Figure 6.6). We obtain an alternative version for a triangle with spacelike sides. Let us consider three uniformly moving observers $\mathcal{A}$, $\mathcal{B}$, and $\mathcal{C}$. For any event A on the world-line of $\mathcal{A}$, we can construct B on the world-line of $\mathcal{B}$, which is simultaneous to A for the observer $\mathcal{C}$ and C on the world-line of $\mathcal{C}$, which is simultaneous to A for the observer $\mathcal{B}$. When the three observers meet at one event H, then B and C are simultaneous for the observer A (Figure 6.7).

The points of the triangle and the intersection point at which its altitudes meet form a quadrangle in which each of the four points is the intersection of the altitudes of the triangle of the remaining three points. After all, each line connecting two points of the complete quadrangle $ABCH$ is perpendicular to the connection of the two other points.[4] When we

[4] In the Galilean geometry, the theorem of the intersecting altitudes is trivial: All altitudes are lines t = const,

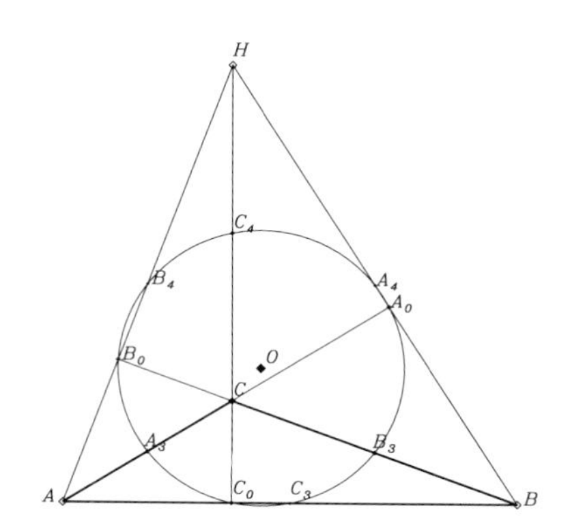

Figure 6.11: *Feuerbach's circle in the Euclidean geometry.*

If we draw a circle through the three feet A_0, B_0, C_0, this circle passes through the midpoints A_3, B_3, C_3 of the sides and the midpoints A_4, B_4, C_4 of the altitudes too. In addition, for any three of the four points A, B, C, H the fourth is the intersection of the altitude of the corresponding triangle and the property of being the midpoint of a side or of an altitude are interchanged.

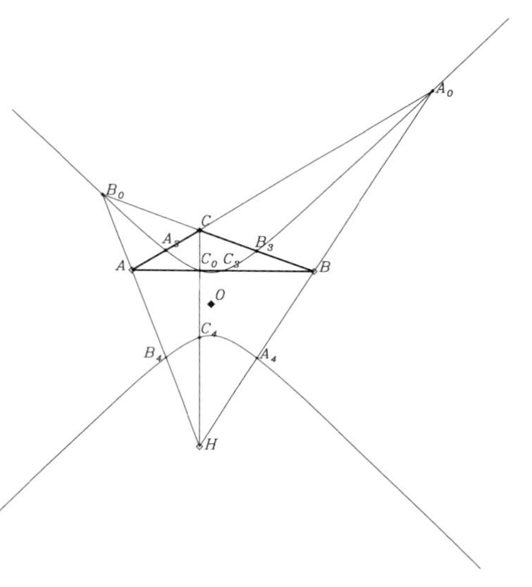

Figure 6.12: *Feuerbach's circle in the Minkowski geometry.*

In the Minkowski geometry, the feet A_0, B_0, C_0 must be constructed by the corresponding rules. The Minkowski circle through the foot points (an equilateral hyperbola with the distinguished asymptotes) again passes through the midpoints A_3, B_3, C_3 of the sides and the midpoints A_4, B_4, C_4 of the altitudes. We can permute the four points A, B, C, and H as in the case of the Euclidean geometry. The projective generalization is given in Figure 8.12.

know the perpendiculars onto two straight lines g_1 and g_2, the intersection of the altitudes can be used to find the altitude from their intersection onto any other straight line. When we define the perpendiculars on three straight lines, we obtain all the geometry, and the intersection of altitudes restricts the possible choice of definition (Appendix D, Figure D.2).

An interesting case is given by the theorem of circumference angles. For comparison, we show its Euclidean form (Figure 6.8), its pseudo-Euclidean form (Figure 6.9), and its Galilean form (Figure 6.10). In the Euclidean geometry, the theorem of circumference angles is exactly valid for the circle, and the circle can be defined as the locus of all points Q which form a given constant angle with a segment AB. It is identical to the locus of all points equidistant from the intersection of the perpendicular bisectors of the triangle ΔABQ. This is valid in the

i.e., parallels, and the intersection point lies at infinity. In general, given three events no other event exists that can be interpreted as above.

Minkowski geometry too. In the Galilean geometry, the two properties separate. The circle defined as locus of equal distance degenerates into a pair of horizontals with equal distance to the midpoint. This point is no longer uniquely defined, since it can lie anywhere on the horizontal in the middle of the other two. The circle defined as the locus of the points at which a given segment subtends a constant angle is instead a parabola (Figure 6.10). Nevertheless, it is still a conic. In the nondegenerate geometries it will be a curve of fourth degree (Figure 9.13).

We end the chapter by considering the Feuerbach (or nine-point) circle. In the Euclidean geometry, we can show that the circle through the feet of the three altitudes bisects the sides of the triangle and the segments between the intersection of the altitudes and the vertices (Figure 6.11). The circle in question passes through nine distinguished points of the triangle. If we now define the circle by the pseudo-Euclidean rule, this property remains unaffected (Figure 6.12).

We could proceed with all the theorems on triangles that use only arguments based on the axiom of parallels and the existence of a metric, but not the Euclidean triangle inequality. The triangle inequality is the point at which the Euclidean and pseudo-Euclidean geometries differ. Up to this point, the Euclidean and pseudo-Euclidean geometries are structurally homologous. For later use, we write the triangle inequality in the form of an existence statement. As in the axiom of parallels, we consider a straight line g and a point P not coincident with the line. In the Euclidean geometry, as well as in the Galilean and pseudo-Euclidean geometries, there exists exactly one line through P that does not intersect g. This is the axiom of parallels. The triangle inequality concerns a dual statement. It is not about lines through the point P but about points on the line g. In the Euclidean geometry, *no* point on g has zero distance to P. We learned that in the Galilean geometry *just one* point on a straight world-line g has zero distance to P. It is the event on g that is simultaneous with P. In the Minkowski geometry *more than one* point on the (timelike) world-line g has zero distance to P. We used these points to construct the light-ray quadrilateral in Figures 4.9 and 5.1, for instance.

7 Curvature

7.1 Spheres and Hyperbolic Shells

If the circle and the hyperbola determine geometries that are so intimately related, what about the sphere and hyperboloids? We know a lot about the geometry of the sphere. After all, the surface of the earth is a sphere to better than one percent. Navigation at sea and in the air has to take the spherical geometry into account. We know that straight lines on the surface of a sphere (strictly speaking the shortest lines or geodesics) are arcs of great circles. In the coordinate net of charts of the earth, the meridians and the equator are such great circles. Now all great circles through one point intersect again at the antipode. The meridians through the North Pole, which meet again at the South Pole, are a well-known example. This property is the reason why two different straight lines on a sphere have two intersection points. No parallels exist on a sphere. The axiom of parallels is not valid in spherical geometry. All the theorems proven with the help of this axiom cannot hold without change. In particular, the sum of the angles of a triangle exceeds the flat angle. The excess of the sum is proportional to an area of the triangle (Figure 7.1). One of the consequences is that if the circumference U of a circle is given, its area F is larger than that expected by the Euclidean calculation (Figure 7.2). Circumference and area depend on the radius ϱ, and we obtain

$$U[\varrho] < \sqrt{4\pi F[\varrho]} < 2\pi\varrho \quad \text{for} \quad \varrho > 0. \tag{7.1}$$

We generally speak of positive curvature at a given point if this is observed in its vicinity. Consequently, the surface of the sphere is homogeneously and positively curved. When we are on another surface where the excess angle is negative, or where the radius of given circle is smaller than expected, we say that the curvature is negative. On a general surface, the curvature can vary, and the excess rules that we just presented are valid only in small enough neighborhoods of the points.

Generalizing these effects, curvature yields a characteristic integral rotation of directions that are maintained as true as possible at each point on a closed path. How is such a *parallel transport* to be defined?[1] Let us suppose that we try to navigate a ship on a straight path to a certain place. If this place is a lighthouse, the task is simple. We follow the line of sight. The natural definition of parallelism along the path would be to consider the directions to the lighthouse at each point to be parallel. Then our path is an autoparallel curve. If the lighthouse is not straight ahead, we could try the same. An autoparallel path then follows a direction that

[1] Here, the notion *parallel* is not used in the global sense that we meet in the axiom of parallels. Instead, it refers solely to the property of equal step angles and alternating angles, which parallels exhibit in the Euclidean geometry.

The Geometry of Time. Dierck-E. Liebscher

ISBN: 3-527-40567-4

keeps a fixed angle to the line of sight to the lighthouse. This results in a logarithmic spiral, or loxodrome, not in a geodesic. The situation will gradually improve if the lighthouse is replaced by the magnetic pole, but the path remains a loxodrome (Figure 7.3). We understand that if we intend to find all geodesics in the form of autoparallel curves, we cannot proceed this way. We have to connect the definition of parallel transport with the geodesics. Two directions at different points of a geodesic are called parallel, if their angle to the tangent of the geodesic is the same. If we have to transport along a general curve, we must approximate the curve by a polygon of geodesics. Any curve can be approximated by a polygon of geodesics: This completes our definition. The parallel transport defined in such a way is called the *geodesic parallel transport*. We illustrate it with a geodesic triangle on the sphere (Figure 7.1). We consider the path from P through A, C and B back to P again. If we look south-east at the starting point P, this direction has an inclination to the intended path of $\pi/4$ to the right. It is this angle that we hold fixed till we arrive at A. Here, our path turns left. With respect to the new direction of the tangent vector, we now keep an inclination of $3\pi/4$ to the right, still to the south-east. After the next turn at C, our chosen direction has an inclination of $3\pi/4$ to the left. For the next part of the path, this is north-east. After the third turn at B the inclination is $\pi/4$ to the left. We arrive at P and find that the chosen direction has been rotated $\pi/2$ to the left, although we did our best to maintain it. This indicates a net rotation of the tangential plane. It is a common property of parallel transport around closed curves. Only in geometries in which the axiom of parallels holds it is zero. The amount of the net rotation is proportional to the curvature and the area enclosed by the path.

The fact that we can invent different physical prescriptions for such a transport indicates that in mathematics it can be chosen freely in the beginning. Afterward, one can try to find a distinguished definition, for instance in relation to metric properties. In an axiomatic approach, the parallel transport is defined first, and the curvature is defined subsequently by its property of rotating the tangential plane in parallel transport along a closed line. As regards parallel transport ruled by the magnetic needle, its curvature is zero up to the poles. The net rotation is zero for a closed path not surrounding a pole, and a multiple of 2π for a path surrounding a pole.

It was an important discovery that the curvature is an intrinsic property of the surface or space in question. Usually, we try to imagine a two-dimensional surface as embedded in the three-dimensional space, and we conceive curvature as inhomogeneity of the direction of the normal vector of the surface. In spite of this intuitive picture, the conception of embedding and of normal vector is nowhere used in defining geodesics, parallel transport, and the derived curvature. All these notions contain only internal properties of the surface, and do not depend on a possible embedding in higher dimensional spaces. This is of extreme importance because it allows to consider the three-dimensional physical space and the four-dimensional physical space–time as curved. Supplementary dimensions are not necessary even though one can pose and consider the problem of which curved spaces or space–times could be interpreted as embedded in higher dimensional Euclidean or pseudo-Euclidean spaces.

Surprisingly, a machinery exists that realizes the geodesic parallel transport of a chosen direction. This is the *South Seeking Chariot* (Zhǐ nán chē) [73], which can sometimes be inspected at exhibitions of as ancient Chinese technology (Figure 7.4). A subtracting differential gear guides the central vertical axle in such a way that the direction of the pointer remains fixed even when the chariot under it is turned around on a plane (Figure 7.5). The rotation of

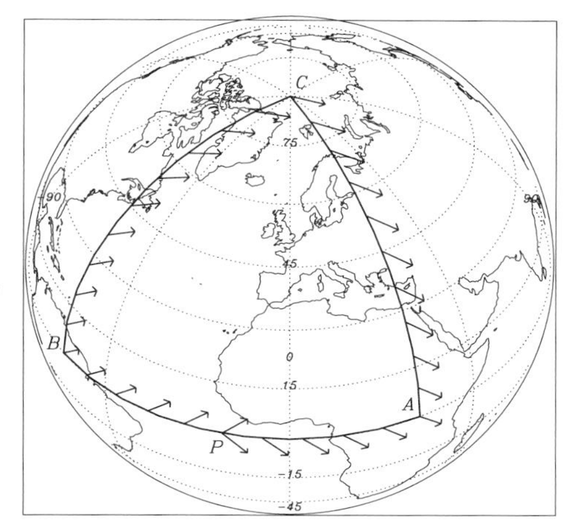

Figure 7.1: *A triangle on a sphere with excess and rotation of the tangents.*

At point P we choose an arbitrary direction (here SE). Along the geodesics to point A, from A to C, from C to B and from B back to P we keep the same inclination relative to the direction of motion. Only at the points A, C, and B do we take into account the fact that the path turns through $\pi/2$ to the left and that this angle is added to the inclination of the sight direction to the direction of motion. After returning to P, the sum of these additions is only $3\pi/2$ to the right. In total, we observe a turn of the tangential plane by $\pi/2$ to the left. This value is identical to the excess of the sum of angles of the triangle ACB. This sum exceeds the Euclidean value (π) by precisely $\pi/2$.

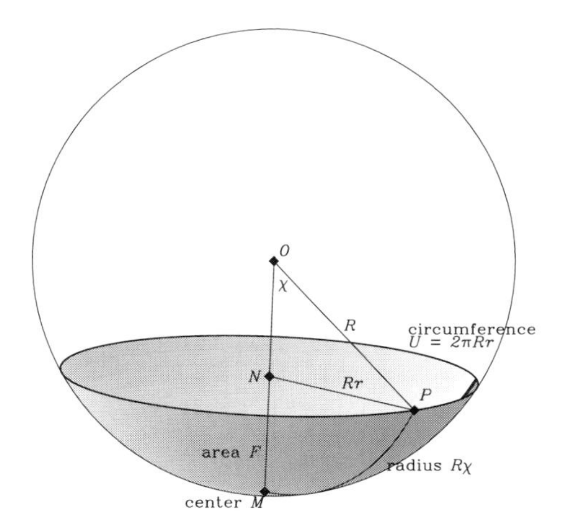

Figure 7.2: *Area and circumference on a sphere.*

A circle on the surface of a sphere is the boundary of a spherical cap. Its radius around the midpoint M is an arc $\varrho = R\chi$ on the surface. The projection Rr of this radius onto the intersection plane is shorter ($NP < MP$). Consequently, the circumference $U = 2\pi Rr$ is smaller than the value $2\pi R\chi$ expected in the Euclidean calculation. The area of the cap is proportional to its height, $F = 2\pi Rh = 2\pi R^2(1 - \cos\chi)$, and is also smaller than the expected value $\pi R^2\chi^2$. In relation to the circumference, the area is larger than the expected value: $4\pi^2\varrho^2 > 4\pi F > U^2$ for $\varrho > 0$.

the indicated direction induced by the curvature of the surface is realized by a pointer too if the chariot is tracked along the curve $PACBP$.

We already noticed that a *magnetic needle* does *not* accomplish geodesic parallel transport. Such a needle keeps the direction to the (magnetic) North Pole independently of the particular great circle on which we are moving. The curves of constant inclination to the magnetic needle are loxodromes, not geodesics. We capture the difference in stating that the angle of this direction to the direction of the geodesic is continually changed.

Geodesics are intuitively defined to be the shortest connections. Strictly speaking, geodesics are *extremal* connections. Depending on the local type of the metric properties (Euclidean or pseudo-Euclidean), they are the shortest or the longest connections. In our example, on the surface of a sphere in a Euclidean space, i.e., any path will get only longer if

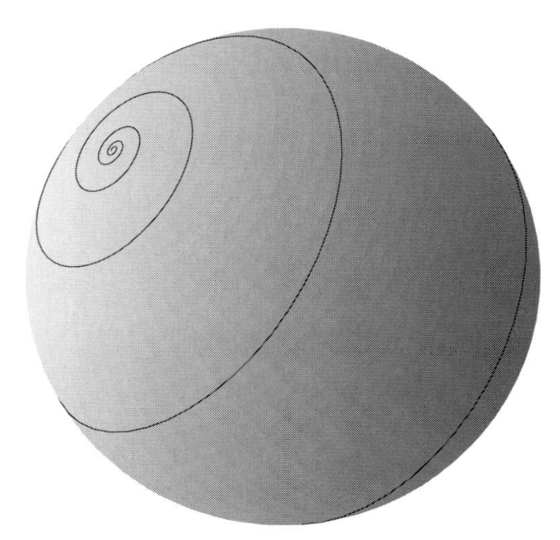

Figure 7.3: *A loxodrome.*

A loxodrome is a curve of fixed inclination to a given line congruence, in our figure the meridians of a sphere. It can be steered by the magnetic needle. In our figure, the loxodrome has a constant deviation from the meridian of approximately $\delta = 0.35\pi$. It approaches the pole like a logarithmic spiral without ever reaching it (the loxodromes of the plane are merely logarithmic spirals). In spherical coordinates (longitude λ, latitude ϕ), the equation of a loxodrome is given by $\cos\phi\ \mathrm{d}\lambda = \tan\delta\ \mathrm{d}\phi$. For $\delta = 0.5\pi$, we obtain circles of constant latitude.

small deviations are included. Consequently, an extremal path can only be the shortest.[2] This property depends on the validity of the ordinary triangle inequality. In a Minkowski world, timelike geodesics are the *longest* connections. We already noted this in connection with the twin paradox.

For the *geodesic* parallel transport, geodesics are *autoparallel* too. Furthermore, they are the only autoparallel curves. We can illustrate this again with the South Seeking Chariot (Figure 7.6). If we always take the direction of the pointer of the South Seeking Chariot, we can be sure that the paths of both wheels have the same length. On the other hand, if the length of the path of the chariot could be shortened by shifting the path sideways, the paths of the two wheels could not be equal and the pointer would indicate this by turning relative to the chariot. Consequently, the path chosen by the pointer is the shortest possible. As we already noted, one can define parallel transport independently of the metric properties and differently from the geodesic parallel transport. In this case it turns out that the curvature is a property of the parallel transport.[3]

We add a short note about the *gyro compass*. In navigation on the surface of the earth, it renders good service. It keeps the direction to the rotation pole of the earth and is to be compared with the magnetic needle. However, a rotating top is a three-dimensional device. Can we make it formally two-dimensional by forcing its axis into the tangent plane? Remarkably, a gyroscope with its axis *constrained* to be tangent to a surface is not appropriate at all to fix a direction. That is because it not only rotates uniformly around its axis, but the axis also rotates around the surface normal. The latter rotation can be made equal to zero initially. However, if the direction of the normal is changed along the path on the curved surface, the axis inevitably starts to turn by precession. In this case, the result depends, for instance, on

[2] On the sphere, geodesics can be extended past the antipode of the initial point. Then they loose the extremal property in the large. Nevertheless, they remain still the shortest lines in the small. The antipodes are an example for focal points in general.

[3] For instance, the parallel transport by the magnetic needle is free of curvature by definition and does not relate to the metric properties of the surface: The length of the path enters nowhere. In general, autoparallel curves differ from geodesics. These subtleties are the background for the unified field theories searched for by A. Einstein, H. Weyl, and E. Schrödinger, among others.

Figure 7.4: *A model of a South Seeking Chariot.*

Tales claim that the mythical emperor Huang Di and his army found their way through fog to defeat a dangerous enemy by using a chariot that always pointed to the direction of the enemy strongholds. In order to show the feasibility of such a chariot, a model chariot was built in the Song dynasty. Nothing has survived except for a description of the construction and the drawing of a jade model of the pointing figure. The model shown in the photo here stands in front of the Taipeh National Museum. Yinan Chin helped me to get the photo.

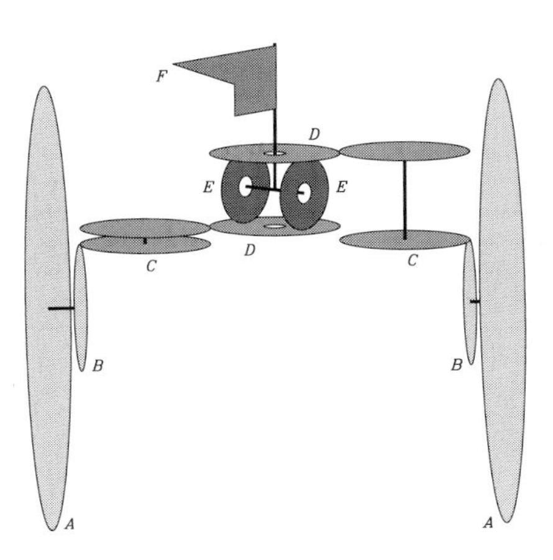

Figure 7.5: *The mechanics of the model.*

A turn of the South Seeking Chariot produces unequal rotation of the two wheels A. Both are clamped to the wheels B, which transfer the rotation to the wheels C, which obtain a different sign of rotation if the wheels B have the same. This rotation is transferred again to the wheels D, which form together with the cursor wheels E a proper differential gear as is used, for instance, in cars. The rotation of the axle of the flag F is half the sum of the rotations of the wheels D. Hence it is half the *difference* of the rotation of the wheels A. The gearing ensures that the rotation of the flag compensates for the turning of a car. Then on a plane the pointer F always indicates the same direction. If the chariot moves on a curved surface, the pointer maintains the direction locally as well as it did in Figure 7.1.

the velocity of the gyroscope in the surface. In contrast to this useless *constrained* gyroscope, the gyro compass works by virtue of the *free* mobility of its axis, the free support in the earth's gravitational field, and the rotation of the earth. At last, we leave the surface of the earth. In three-dimensional navigation through space, freely falling in the gravitational field of the earth or the planetary system, the motion of a *free* gyroscope is subject to the conservation laws of the angular momentum. In curved spaces, the transport of the vector of angular momentum is nearly autoparallel, but not precisely. Instead, it is subject to small changes through a characteristic spin–orbit interaction mediated by the curvature. Gyroscope experiments in orbit around the earth can measure the integral effect of parallel transport.

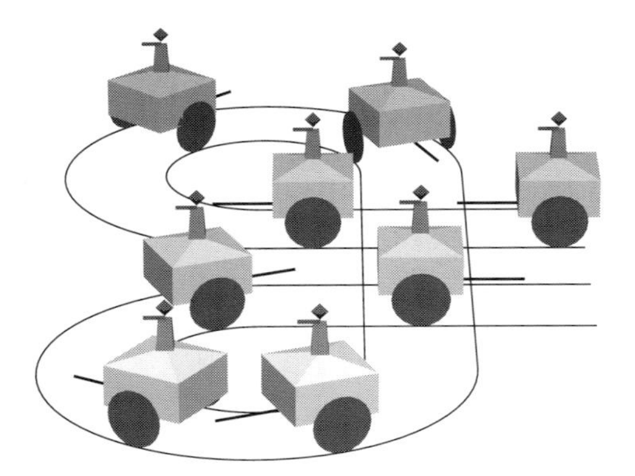

Figure 7.6: *The motion of a South Seeking Chariot.*

The South Seeking Chariot realizes parallel transport of a direction. Irrespective of how the chariot is drawn along its path, the gears inside ensure that the direction in which the figure on the top points is kept the same.

The integral change of orientation considered here appears in parallel transport along closed curves. It also defines the curvature in spaces with any dimensions. The curvature can vary from point to point. In this case, we enter the realm of differential geometry. In the following, we meet only spaces and worlds of constant curvature.

Conventionally, a geometry is called non-Euclidean if the axiom of parallels does not hold. It is then that we have to take a curvature into account [74]. The simplest case of a non-Euclidean geometry seems to be provided by the sphere. After all, on a sphere parallels do not exist, and two great circles always intersect in two points. In addition, we can map the sphere onto the plane in such a way that all great circles become straight lines (Figure 7.9). This is done by projecting the sphere from its center. A great circle being the intersection with a plane through the center, its projection is the intersection of this plane with the plane of projection, i.e., a usual straight line (Figure 7.7). We obtain the *elliptic geometry*. Surprisingly, the geometry of the sphere was not considered in the historical debate surrounding the axiom of parallels because central projection does not map the sphere one-to-one onto the plane. On the plane, the straight lines do not intersect twice. Both intersection points on the sphere are mapped onto the same point in the plane. Hence, the clear non-Euclidean character of the geometry on the sphere is masked in the projection plane. As we shall see in the next chapter, the spheres in the Euclidean space leave many relations unseen because some of the decisive constructions are not real.

Here we are helped by the three-dimensional Minkowski world.[4] The counterparts to the spheres in the three-dimensional Euclidean geometry (which are particular ellipsoids) are again surfaces of constant distance to a center. The distance is now pseudo-Euclidean, and we call these surfaces Minkowski spheres. They appear as particular hyperboloids to a given as-

[4]We simply add a second spacelike dimension but keep the only one plus sign in the analog of Pythagoras's theorem, that is, the square of the distance between two points $P = [t, x, y]$ and $P + \mathrm{d}P = [t + \Delta t, x + \Delta x, y + \Delta y]$ is given by $\Delta s^2 = \Delta t^2 - \Delta x^2 - \Delta y^2$. Just as in the transition from the Euclidean plane to the Euclidean space the circle is replaced by the sphere, in the transition from the two-dimensional to the three-dimensional Minkowski world the hyperbola is replaced by a (two-shell) hyperboloid. The two lightlike lines through an event are replaced by a double cone, the light cone.

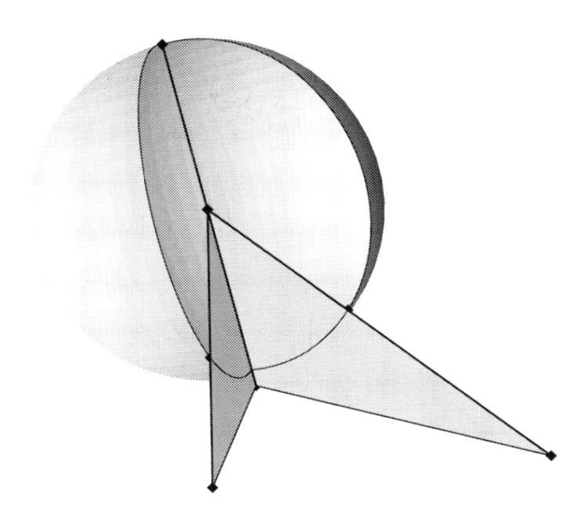

Figure 7.7: *Azimuthal projection of the sphere.*

The figure shows how the sphere is projected from its center onto a plane. A great circle being an intersection with a plane, its projection is a straight line. Two great circles intersect at two points which are mapped onto one single point P of the plane.

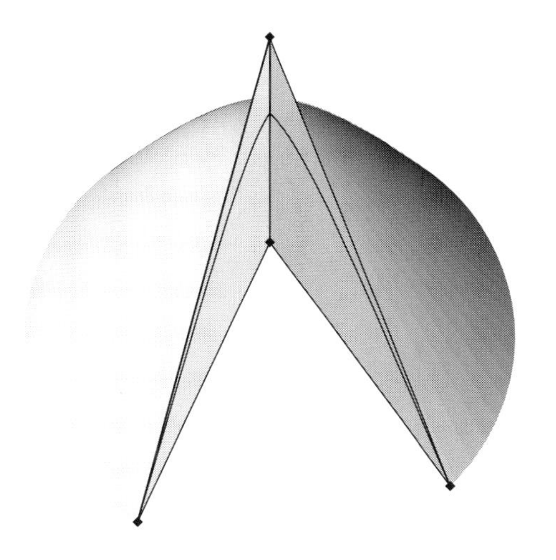

Figure 7.8: *Azimuthal projection of a hyperboloid shell.*

In analogy to the previous figure, the hyperboloid shell is projected from its center onto a plane. In contrast to the sphere, which covers the full plane, the image of the shell only covers the inner part of a circle, which is the image of the asymptotic cone. As in the case of the sphere, the inner part of a circle is covered twice: the second time by the image of the other shell of our hyperboloid, which is left out of the figure. The planes through the midpoint of the hyperboloid cut it in curves that are geodesics in the geometry induced on the shell by the Minkowski world. The projections of these geodesics are again straight lines. The intersection line of two planes through the center defines a point in the projection plane, which is the intersection point of the straight lines corresponding to the two planes. If this point lies outside the absolute circle, the geodesics do not intersect.

ymptotic cone, i.e., the light cone of the event at the center.[5] In the case of timelike separation, the hyperboloids have two shells. We know the physical interpretation of the distance to be

[5]We must emphasize repeatedly that the hyperboloid is always the counterpart of an ellipsoid. The Euclidean sphere is (by virtue of its metric symmetry) a particular ellipsoid, the Minkowski sphere is (again by virtue of its metric symmetry) a particular hyperboloid. *Both* are loci of constant distance to the center. Their difference consists in the fact that the former do not intersect infinity while the latter do. This too is lost when we adopt the projective point of view, where infinity no longer exists.

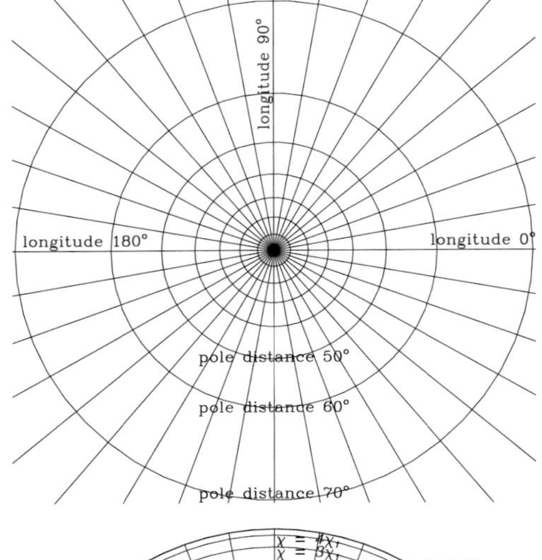

Figure 7.9: *Lines of equal pole distance and meridians in the azimuthal projection of the sphere.*

Maps in azimuthal projection are known from plots of the polar regions of the earth. They seriously distort the area. The equator is shifted to infinity. Despite these distortions, geodesics are projected as straight lines. The map is appropriate for considering the relations to projective geometry.

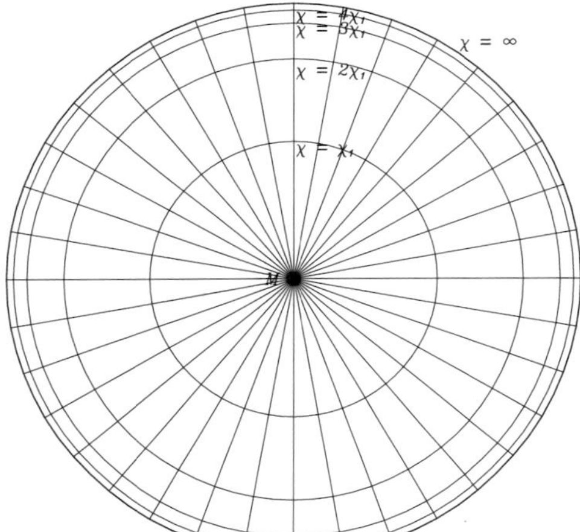

Figure 7.10: *Lines of equal pole distance and meridians in the projection of the hyperboloid.*

As in the case of the sphere, the azimuthal projection preserves straight lines. The distortion of area goes in the opposite direction. The closer we approach the bounding circle, the more (by the Euclidean calculation) equally distant points move closer together. The reader should compare this figure with the evaluation of the collision in Figure 5.6. The plane of the endpoints of the velocities is such an azimuthal projection of the mass shell. The circle shifted from the central position now looks like an ellipse.

the proper time. Hence, we call the shells of these hyperboloids *time shells*. In the analogous momentum space, such a shell contains all end points of momentum vectors of a given rest mass and it is called the mass shell. In Figure 5.27, we already used this construction. In the case of spacelike separation, the hyperboloids consist of only one shell. The tangential planes of such shells contain timelike as well as spacelike directions. They represent a curved drawing surface for locally pseudo-Euclidean geometries with curvature.

Now the geodesic lines (corresponding to the great circles) are the intersections with planes through the center. The reflection on such a plane leaves the hyperboloid unchanged. That is, a plane through the center is a symmetry plane of the hyperboloid. Hence, each intersection of the pseudosphere with a plane through the center is a geodesic. In addition, all the geodesics are such intersections with planes through the center. We note that these intersections have two disconnected branches on the two shells of the hyperboloid. In the projection from the center, they are identified.

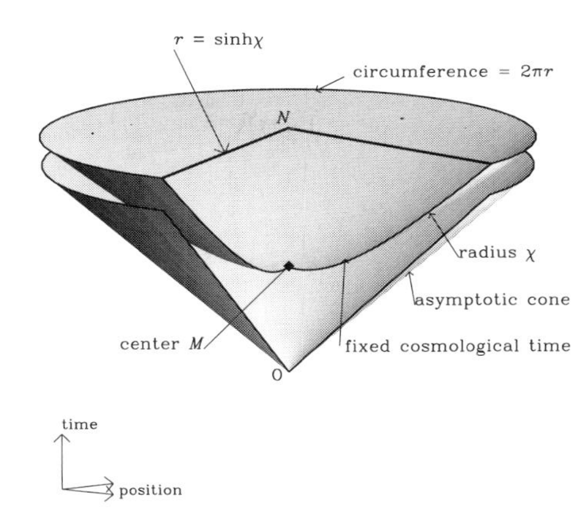

Figure 7.11: *Area and circumference on a time shell.*

In analogy with Figure 7.2, we draw a cap, which now has to be calculated by the rules of pseudo-Euclidean geometry. The radius of the cap (along its surface) is now smaller than the radius of the plane cut. Correspondingly, the circumference is larger than expected from the radius measurement: $4\pi^2\varrho^2 < 4\pi F < U^2$ for $\varrho > 0$. The shell is negatively curved.

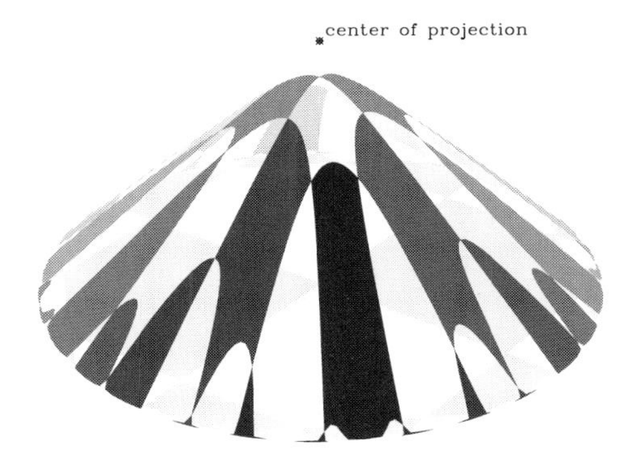

Figure 7.12: *Regular pentagonal tiling of the hyperboloid shell.*

On a surface of constant negative curvature, there exists a regular pentagon with five right angles. It can be used to tile the surface, as sketched on the shell. In the base plane, we see the central projection of the shell. It maps the geodesic sides of pentagons into straight lines. The length of the sides of regular pentagons with five right angles is equal to $\phi = \operatorname{Arsh}\sqrt{(1+\sqrt{5})/2}$. Escher [76] shows in his "limit circles" such tilings with regular hexagons.

When we project these geodesics on a hyperboloid from the center onto a plane, we again obtain the straight lines of ordinary geometry (Figure 7.8). The unexpected new feature consists in the observation that now *not all* points and lines belong to the geometry: The points outside the absolute circle (i.e., the intersection with the asymptotic cone) are excluded. The metrical infinity is mapped onto the finite absolute circle (Figure 7.9). Two straight lines that do not intersect inside this circle are now to be called parallel. The circle, its inner points, and segments constitute Klein's model of the non-Euclidean plane, its geometry is called the *hyperbolical geometry* and was found by Gauss, and Reichardt [74].

At this point, we remember again the relativistic billiard collision (Figures 5.6 and 7.10). The curve of the velocities observed after a relativistic collision is a circle. However, it is a circle in the non-Euclidean measure on the mass shell. We obtain it explicitly if we shift a central circle so far that the center of the absolute circle comes to lie on its circumference. This shift is a Lorentz transformation in the space–time, and a translation in the hyperbolic geometry of the velocity space.

If we remember the definition of curvature in the first paragraph of this chapter, the curvature of the time shell is negative. This seems to contradict the appearance, but the appearance tacitly presupposes Euclidean distances. However, we have to use the *pseudo-Euclidean* measure, and this tells us that the projection is longer than the projected arc (Figure 7.11). The circumference is longer than that expected from the value of the radius (compare with Eq. (7.1)):

$$U[\varrho] > \sqrt{4\pi F[\varrho]} > 2\pi\varrho \quad \text{for} \quad \varrho > 0. \tag{7.2}$$

We illustrate the curvature by a tiling of regular pentagons (Figure 7.12). On a sphere, such a tiling is known. It corresponds to a regular polyhedron, the dodecahedron. Three pentagons meet in a vertex there. Consequently, the angle between the two sides is $2\pi/3$, and the sum for such a pentagon is $10\pi/3$. This exceeds the Euclidean value of 3π, indicating the positive curvature of the sphere. Now we return to the pseudosphere. At each vertex, four polygons meet, the angles are all equal to $\pi/2$, and the sum in such a pentagon is $5\pi/2$. This is smaller than the Euclidean value 3π, indicating negative curvature. The parallel transport around our pentagons can be reexamined in Figure 7.12 analogously to Figure 7.1. If we tour such a pentagon keeping a constant direction as in Figure 7.1, we again obtain a rotation of the tangential plane with $\pi/2$, this time in the *clockwise* direction. The excess of the sum of angles is negative.[6] Tilings of the three-dimensional space are analyzed in connection with the topological structure of the universe [78, 79].

7.2 The Universe

It is the equivalence of inertial and gravitational mass that forces us to recognize the curvature of space–time. It is one of first discoveries of modern physics that all bodies fall (to some approximation) with the same acceleration. In the equation of motion in a gravitational field, the inertial mass (i.e., the factor of the velocity in the conservation of momentum) cancels against the gravitational mass (i.e., the charge in the gravitational field). This observation has been more and more refined. In the experiments of Dicke et al. [80] and Braginski and Panov [45] an accuracy of 10^{-12} has been achieved.[7] If we assume the equivalence of inertial and gravitational mass as a principle before any calculation, we must conclude that even light rays (i.e., Platon's straight lines) are somehow subject to the gravitational field because the light transports energy and the mass of this energy. The rays of a pencil are sheared and screwed depending on the curvature of the space–time. Wave equations (including the Maxwell equations) with constant coefficients will not do. In analogy to the wave equation for a refracting medium, the coefficients of the wave equation now depend on the position in the gravitational field. It is important that the *varying* coefficients of a wave equation are a consequence of a metric of a *curved* space–time. The rays of a pencil can intersect each other again: focal surfaces and focal points form. The common expression "light is bent" has to be taken cautiously.

[6] The rotation of the tangential plane finds its physical expression in the *Thomas precession* [77].

[7] Of course, this number has to be related to the circumstances of the experiment. The issues relating to the equivalence principle are not all settled. In addition, the motion of a body in a curved space–time depends on its spin. It is an intricate question to extract metric properties and the curvature from the motion of objects if their internal structure is *not* known or also has to be found by observation of their motion [37].

It masks the fact that an individual light ray between two events cannot be bent, because there is no alternative curve between these events which is less bent. In addition, the world-lines of photons are, of course, geodesics. If we consider the light deflection by a massive object in the line of sight, we compare in fact with the case when it is taken away, actually or virtually. That is, we compare two different spaces, and the geodesics of the one are bent relative to the geodesics of the other. On the other hand, the *relative* geometry of the lightlike lines can be used to define "bending," for instance, by the appearance of focal points and lines.

The world is curved, and in general space too. The curvature is not homogeneous in detail because the massive objects generating the gravitational field constitute inhomogeneities themselves and are not homogeneously distributed either. They cause the light deflection that we have just discussed. This deflection can be estimated already in Newtonian mechanics. In general relativity theory, the light deflection is twice as large because not only the world, in general, but space too is curved by the local mass distribution. Light deflection was the first effect predicted by the equivalence of inertial and gravitational mass, and it was later confirmed with its general-relativistic value by observation. After being found by Eddington in 1919, it is now not only well confirmed (the best determination is achieved with the radio position of the quasar 2C379, which is occulted and has its image deflected by the sun every October 8) but is also an instrument to explore the universe. Cosmic objects of large mass are revealed through the light deflection produced by their gravitational field even if their luminosity is too low to be seen by our telescopes. They are called gravitational lenses (Figure 7.13). The gravitational field of a massive deflector perturbs the propagation and folds the light cone (Figure 7.14). In a picture of spatial photon orbits, we see the form of the different rays (Figure 7.15). The images are distorted the better the alignment of source, lense, and observer is (Figure 7.16). Supermassive and superdense mass concentrations not only bend the light, they can prevent light escaping from their gravitational field. They then constitute the so-called black holes. Such black holes are assumed to be at the center of big galaxies and to be the cause for some extremely high luminosities observed in quasars.

If we confine ourselves to the consideration of a universe in which no point can be distinguished from any other (the positions are relative, the big mass concentrations in stars galaxies, clusters of galaxies and other large-scale structures are to be neglected), the world can be considered as temporal sequence of homogeneous spaces which necessarily have constant curvature (i.e., which correspond to spheres, planes or pseudo-spherical shells) and which can contract or expand (Figures 7.17 and 7.18). This is a gross schematization of the actual observation but it is the basis of cosmology.[8] The sound foundation of this average picture of the universe is a complicated question from both the theoretical[9] and the observational point of view.

It is necessary to distinguish between the curvature of *space* and the curvature of the *world*. It is the curvature of the *world* (i.e., space–time), which is determined by Einstein's equations of general relativity theory. And it is the curvature of the world that curves the planetary orbits,

[8] A sicilian proverb says: *Tuttu lu munnu è comu casa nostra* (All over the world, things are going like at home).

[9] If we base the theoretical model on Einstein's equations, which have been proved valid, as we know, at a planetary scale, we must still find their *macroscopic average*, since it is the latter that would correspond to the *averaged* observations. But it is not so easy to show that the averaged equations are Einstein's equations too [37].

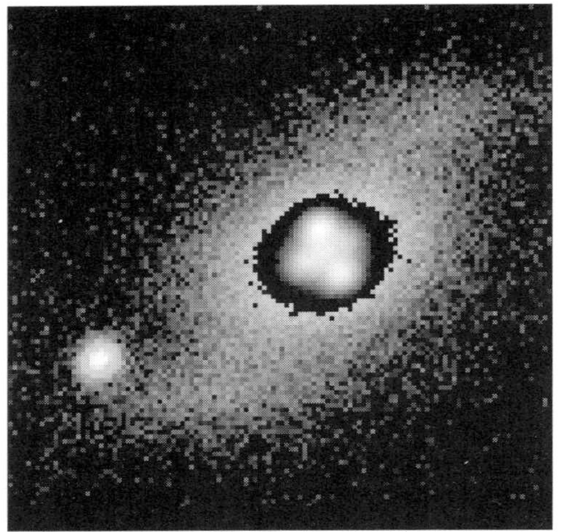

Figure 7.13: *The Einstein cross.*

The central mass condensation of a galaxy in the foreground lenses the background quasar Q2237+0305 (constellation Aquarius), giving rise to a fourfold image known as the Einstein cross. The position of the quasar is nearly central behind the nucleus of the lensing galaxy. The luminosity of the latter is suppressed in this figure to make the images of the quasar visible. At the lower left, a star in the foreground is seen. The picture of the Einstein cross was taken by P. Notni with an earthborn 1.2 m telescope in Maidanak (Uzbekistan) on September 17, 1995. Because of the proper motion of the lensing galaxy and its components, the image has changed distinctly since its discovery in 1988.

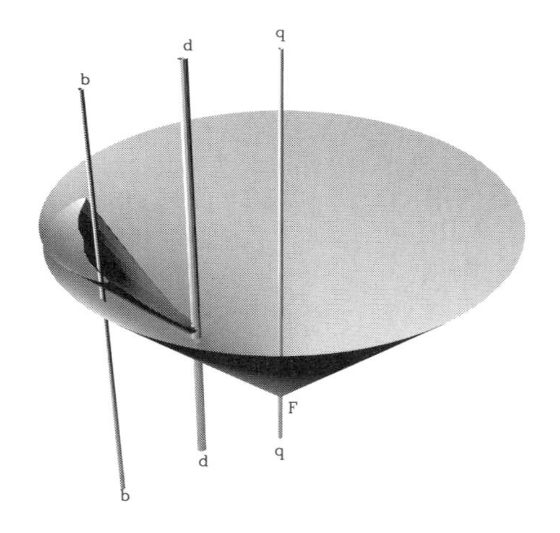

Figure 7.14: *Folding the light cone.*

We see the world-lines of a source q, a deflector d, and an observer b at rest in the space–time reference frame and the light cone of a flash F. In the vicinity of the deflector, the propagation of light is influenced by gravitation. Independently of its exact form, this influence leads to a disturbance that necessarily folds the light cone. In the figure, the simplest kind of folding is shown. The world-line of the observer, here in a suitable position, intersects the light cone of an event E three times. The observer sees the emission event at these three times in three different directions that correspond to the normals of the wavefront at the three events of intersection, here first with a deflection to the right, next to the left, and finally least deflected. The folding always leads to an odd number of images. However, the least deflected is mostly absorbed by the lensing object.

not the curvature of space as one may erroneously believe when one plays with the funnels that can be found in exhibitions to illustrate the planetary motions.

Space is a three-dimensional cut through this four-dimensional world, a (hyper) surface. Its curvature depends on our choice of the cut. In general, the freedom in choosing the slices is not restricted except for the requirement that the chosen space only contains spacelike directions. Here, we restrict this freedom: We allow only homogeneous spaces.

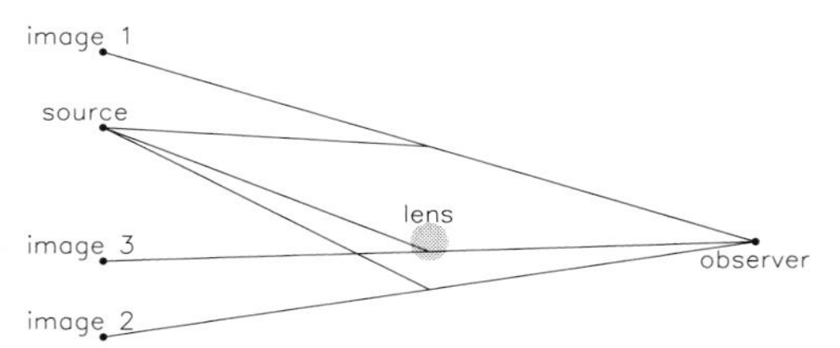

Figure 7.15: *The simple picture of the deflection of light.*

The deflection decreases with decreasing gradient of the potential of the source.

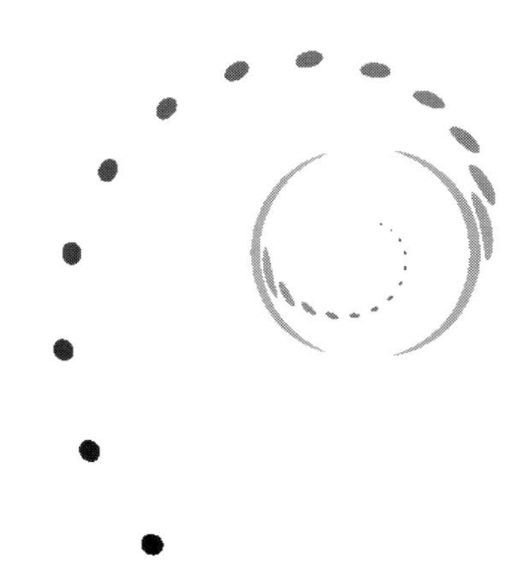

Figure 7.16: *The image of a spiral of sources.*

The first two series of images of sources on a spiral centered behind the lensing object are shown in order to indicate the distortion that increases with closer alignment of observer, lense, and source.

In our examples, the world is meant to be the universe. The homogeneous space sections are labeled by a time coordinate that is usually called the cosmological time.[10] The curvature of the world K_{world} can be decomposed into the curvature K_{space} of the homogeneous space sections and the square of the expansion rate H. Einstein demonstrated that the curvature of the world is determined by its matter density. In our case, the curvature of the world is equal to some basic value $K_{\text{world}0}$ (i.e., the cosmological constant) plus an appropriately normalized mass density.

$$K_{\text{world}} = H^2 + K_{\text{space}} = K_{\text{world}0} + \text{density of mass} \tag{7.3}$$

This is Friedmann's equation. In the limit of negligible[11] mass density, not only the space section but also the space–time itself is homogeneous. These solutions were found by W. de Sitter. We want to consider them now.

We get a first impression in the case of expanding pseudospheres (Figure 7.19). As in the case of positive curvature, where we represent the world as temporal sequence of spheres of increasing size and increasing curvature radius, the world is now represented by a temporal

[10]A time coordinate that is given such a high title should also be given a definition in terms of real physical measurements. This is not so easily done because our observations are restricted to the cone of light rays reaching us from the past. In addition, it will be affected too by the averaging process. In a homogeneous and isotropic cosmological model filled with matter, the cosmological time is a function of the curvature of the world.

[11]That is, negligible with respect to the spatial curvature or cosmological constant.

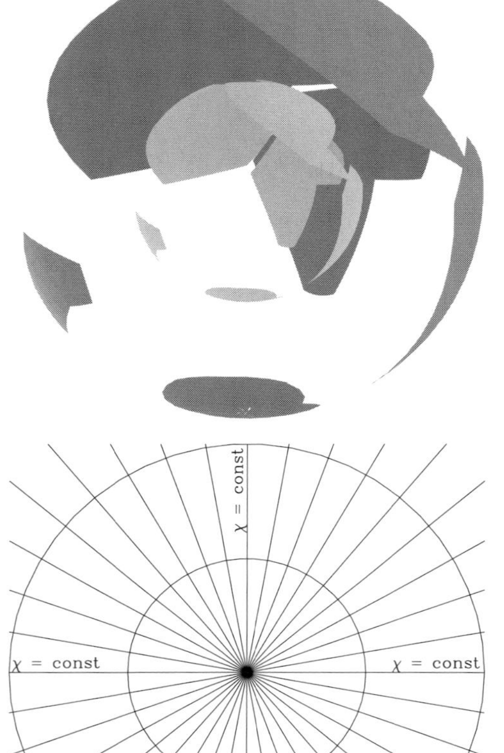

Figure 7.17: *Homogeneous expansion of a sphere.*

If a sphere expands, all distances of points of a given spherical coordinate increase although the points themselves do not move and each figure preserves its shape. The apparent relative velocity of two points is proportional to their distance. The ratio of all distances to the distance of the pole and equator remain constant independently of the size of the globe. The spherical coordinates on the surface correspond to the comoving (expansion-reduced) coordinates of cosmology.

Figure 7.18: *Plane intersection with the expanding sphere.*

We drop the geographic longitude and obtain the plane representation for world-lines in an expanding closed universe. For a particular expansion law (Figure 7.23), also a closed universe can be conceived as void of matter.

sequence of Minkowski spheres of increasing curvature radius (and always infinite size, of course). The curvature radius is identical to the time elapsed since the singularity. After all, it is still the Minkowski geometry, only with new coordinates. The cosmological time τ is constant on the time shells. On these shells, we use coordinates χ that do not change with expansion as in the case of spherical coordinates. The new coordinates τ and χ are chosen in such a way that τ marks the time lapse from the origin ($c^2\tau^2 = c^2t^2 - r^2$) and that a constant χ describes the expansion ($r = f[\chi]\, a[t]$). The transformation is given by

$$t = \tau \cosh[\chi], \qquad r = \tau \sinh[\chi].$$

The distances of events that are indexed by a constant coordinate χ from $\chi = 0$ are increasing with the same rate, $d[\chi] = a[t]\chi$ (Figure 7.20). We call the function $a[t]$ (which here is simply $a[t] = c\tau$) the expansion parameter. The surface of a sphere with the radius $d[\chi] = a[t]\chi$ is

Figure 7.19: *Homogeneous expansion of a shell (Milne universe).*

We see the homogeneous expansion of a negatively curved space. The lines with constant pseudo-spherical coordinates are straight lines through the vertex of the asymptotic cone. The vertex is apparently a singularity. This cosmological singularity contains in some sense all the asymptotic cone. The loci of constant proper time are the time shells already considered. Nevertheless, the space–time is our Minkowski world. We can find a sequence of spaces that do not expand and are not curved. This possibility of choosing foliations with spaces of different curvatures is found because the back-reaction of the substratum on the curvature is neglected or suppressed. The space foliations are only determined by the kinematics of the expanding substratum.

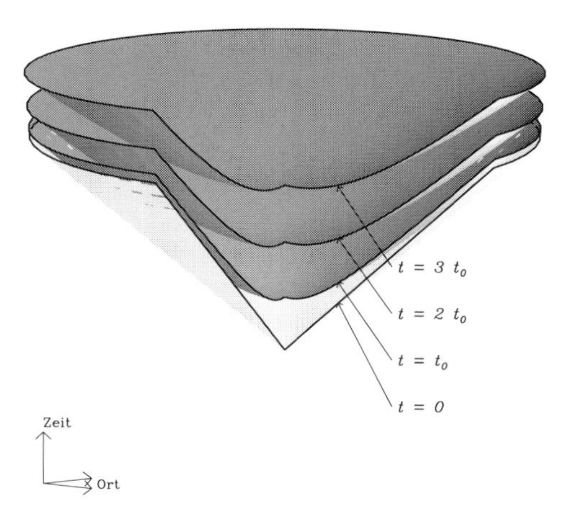

Figure 7.20: *The Milne universe in two dimensions.*

In a plane containing only the radial distance and the time, the world-lines of the substratum are timelike straight lines through an event O. The isotropic lines are the same as in the Minkowski geometry. The substratum does not change the Minkowski geometry of this plane.

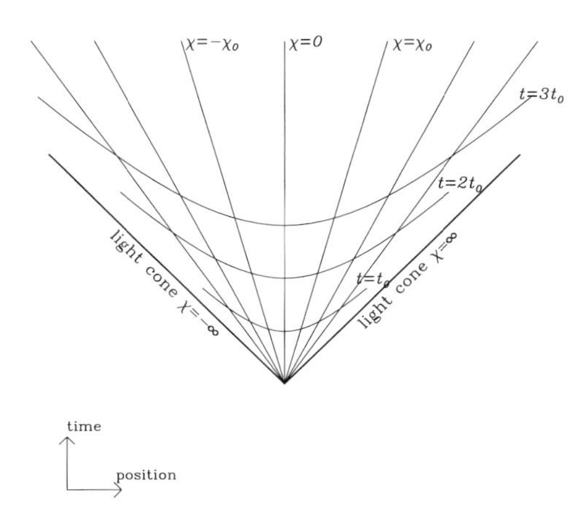

given by

$$O[\chi] = 4\pi f^2[\chi]\, a^2[t]. \tag{7.4}$$

It is larger than expected by the Euclidean calculation, indicating the negative curvature of the spaces $\tau = \mathrm{const}$. The peculiar universe which we constructed here is the *Milne universe*. It is a sequence of homogeneous spaces with negative curvature that expand linearly in cosmological time. Locally, it does not differ from the flat Minkowski world, which in its turn is a nonexpanding sequence of uncurved spaces. Globally, a difference exists: The Milne universe is only a part of the Minkowski world. It has a boundary (defined by the isotropic lines crossing the asymptotic cone) and a singularity, the vertex.

The Milne universe (Figure 7.20) is drawn on the flat Minkowski world straight away. We now consider the universes that can be drawn on the surface of a (higher dimensional) *one-shell* Minkowski sphere. This is the counterpart of the two-shell hyperboloids considered above. While the time shells are the loci of constant timelike distance from the center, the spacelike Minkowski sphere is the locus of the events of constant spacelike distance to the center. It is symmetric like the time shells. Consequently, the geodesics on the spacelike Minkowski spheres are again intersections with planes through the center. In contrast to the time shells on which all geodesics are spacelike, the geodesics on a spacelike Minkowski sphere can be either spacelike or timelike. That is the reason why we can try to draw a two-dimensional universe on a spacelike Minkowski sphere.

On such a pseudosphere, we choose a congruence of timelike geodesics with the intention to later take them as the lines $\chi = \mathrm{const}$. We again obtain the geodesics from intersections with planes through the center of the pseudospheres. We do not consider the general case, but restrict the planes to form a pencil, i.e., all planes are to intersect in one straight line, and through the center too, of course. This corresponds to the Milne universe, in which all the lines $\chi = \mathrm{const}$ have a common point, the vertex. We begin with a pseudosphere of the form

$$T^2 - W^2 - X^2 - Y^2 - Z^2 + 1 = 0. \tag{7.5}$$

We use polar coordinates with $R^2 = X^2 + Y^2 + Z^2$ and illustrate it in the three-dimensional T–W–R space. The other two spatial dimensions are suppressed, but we keep them in mind. We now draw timelike geodesics on the pseudosphere $T^2 - W^2 - R^2 + 1 = 0$ by intersecting it with central planes. In order to get a one-parameter pencil of geodesics, the central planes should form a pencil around some axis (the carrier line) in the T–W–R space. The lines cut out by the planes in the hyperboloid are taken as world-lines of objects with a proper time indicating the cosmological time for the universe, just as was done in the Milne universe. There are three different possible arrangements of the pencil of world-lines $\chi = \mathrm{const}$. In the first case, the carrier line is light-like and just touches the hyperboloid. We choose the coordinate axes $[T, W, R]$ in such a way that the carrier line has the direction $[1, -1, 0]$. The normals of the planes of constant χ can be parametrized in the form $[\chi, \chi, -1]$ (Figure 7.21). We then get the formula

$$T + W = \mathrm{e}^t \;\rightarrow\; R = \chi \mathrm{e}^t .$$

When we now account for being on the quadric (7.5), we obtain

$$T - W = \chi^2 \mathrm{e}^t - \mathrm{e}^{-t} .$$

The formulas for R and the surface O, Eq. (7.4), show that the world-lines $\chi = \mathrm{const}$ describe a universe with spaces expanding exponentially in the cosmological time. Its individual space sections $t = \mathrm{const}$ have the Euclidean geometry. The central projection of the three-dimensional hyperboloid into the plane shows the pencil of world-lines as pencil of straight lines that all intersect a hyperbola (Figure 7.22). The coordinate mesh $[t, \chi]$ fills only the outer part of this hyperbola. The hyperbola represents the metrical infinity.

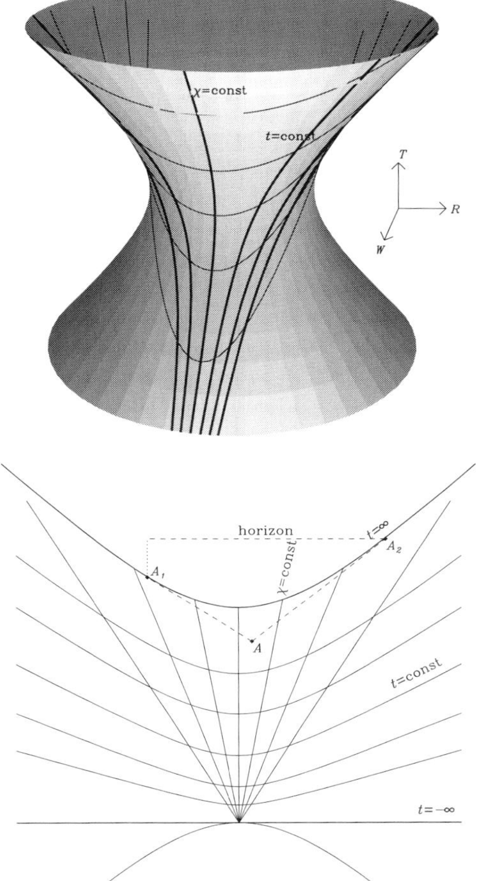

Figure 7.21: *The de Sitter universe. I.*

On the five-dimensional pseudosphere $T^2 - W^2 - R^2 + 1 = 0$ $R^2 = X^2 + Y^2 + Z^2$, the coordinates $T + W = e^t,\ T - W = \chi^2 e^t - e^{-t},\ R = \chi e^t$ are introduced. For our surface formula, Eq. (7.4), we obtain the functions $a[t] = e^t,\ r[\chi] = \chi$. In these coordinates, the spatial sections are flat and expand exponentially. They are represented by the lines of constant cosmological time t, which are cuts with the planes $W + T = \mathrm{const}$. The picture is analogous to Figure 7.20. We only draw on the hyperboloid and not on the plane.

Figure 7.22: *The de Sitter universe in projection. I.*

We project the hyperboloid of Figure 7.21 onto the plane $W = \mathrm{const}$. The infinite is a hyperbola. The substratum is a pencil of rays carried by a vertex at metrical infinity. The timelike lines intersect the hyperbola, but lie in its outside region. For our choice of the substratum, the spaces of constant cosmological time are flat (i.e., Euclidean). The equation of the lines of constant cosmological time in the plane projection is $y^2 - 1 = x^2 - (y+1)^2 e^{-2t}$, and that of the lines of constant χ is given by $y = -1 + x/\chi$.

Through each point of the filled part of the plane, two tangents to the hyperbola can be drawn. They are lightlike directions and represent the light cone at the point in question. Metrical infinity is reached by the light rays at finite values of the spatial coordinate, that is, before all the space has been traversed. We obtained a universe with a horizon of motion:[12] Even if we start from A with the speed of light, we cannot reach every position in the substratum.

If the carrier line of the pencil of planes does not intersect or touch the hyperboloid, we can choose it as the T-axis.[13] The carrier line then has the direction $[1, 0, 0]$. The normals

[12]The technical expression is *event horizon*.

[13]This is possible through an appropriate choice of coordinates that was left free in spite of the conditions we already posed. These free coordinate transformations are the Lorentz transformations in the five-dimensional embedding.

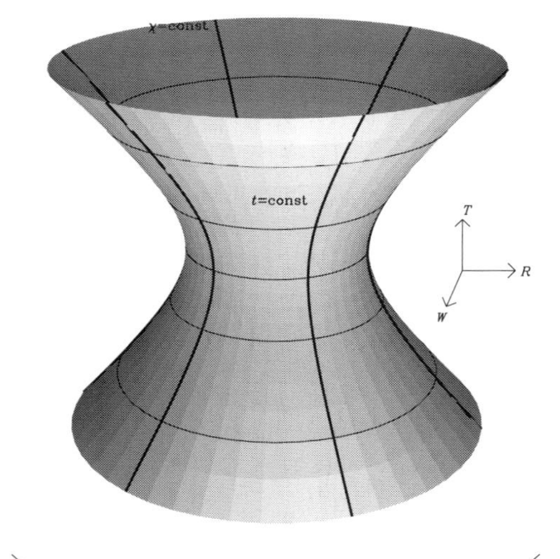

Figure 7.23: *The de Sitter universe. II.*

On the five-dimensional pseudosphere $T^2 - W^2 - R^2 + 1 = 0$, $R^2 = X^2 + Y^2 + Z^2$, the coordinates $T = \sinh[t]$, $W = \cosh[t]\cos[\chi]$, $R = \cosh[t]\sin[\chi]$ are introduced. For our surface formula, Eq. (7.4), we obtain the functions $a[t] = \cosh[t]$, $r[\chi] = \sin[\chi]$. In these coordinates, the space sections are positively curved and the spaces contract down to a certain volume and reexpand afterwards. The space sections are represented by the lines of constant cosmological time t, which are intersections with the planes $T = \text{const.}$.

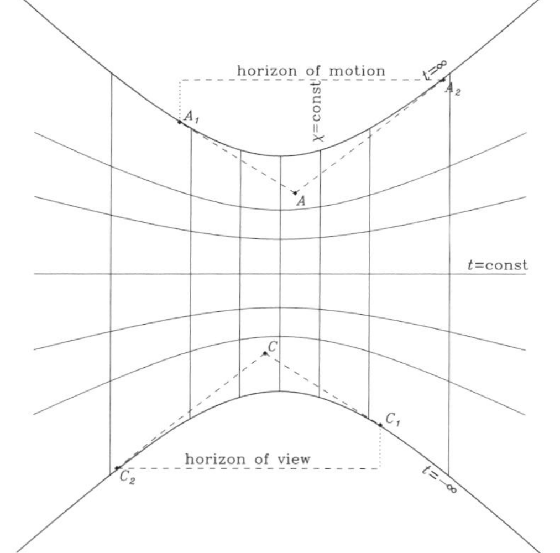

Figure 7.24: *The de Sitter universe in projection. II.*

In this projection of the hyperboloid of Figure 7.23, the substratum is a pencil carried by a point outside the world. The curves of constant cosmological time are hyperbolas of varying opening: $y^2\coth^2[t] - x^2 = 1$. The curvature of the represented spaces is positive.

of the planes of the pencil can be parametrized in the form $[0, -\sin\chi, \cos\chi]$. Everything proceeds as usual (Figure 7.24). Now the hyperbola is both the time $t \to \infty$ (the far future for the upper part) and the time $t \to -\infty$ (the far past for the lower part). Tangents to the latter indicate a finite field of vision, a horizon:[14] The light reaching the event B comes from a finite part of the substratum. No telescope will show the substratum beyond this horizon.

Finally, we put the carrier line of the pencil of planes so that it intersects the quadric. In this case, we can choose the direction $[0, 1, 0]$ for the carrier line and parametrize the normals of the planes in the form $[-\sinh\chi, 0, \cosh\chi]$ (Figure 7.25). After the projection into a plane, we obtain Figure 7.26.

[14] The technical expression is *particle horizon*.

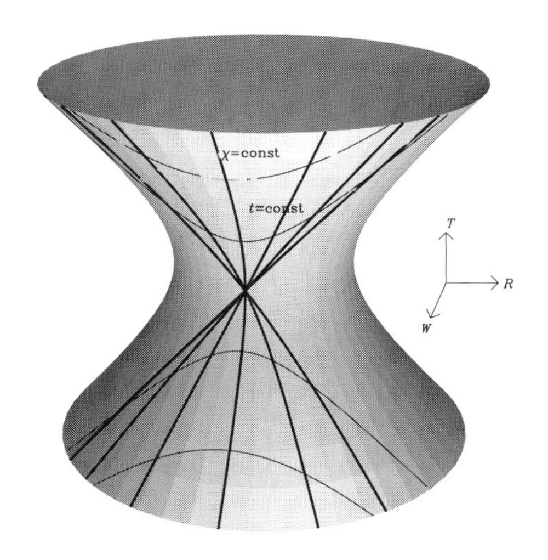

Figure 7.25: *The de Sitter universe. III.*

On the five-dimensional pseudosphere $T^2 - W^2 - R^2 + 1 = 0$, $R^2 = X^2 + Y^2 + Z^2$ the coordinates $T = \sinh[t]\cosh[\chi]$, $W = \cosh[t]$, $R = \sinh[t]\sinh[\chi]$ are introduced. For our surface formula, Eq. (7.4), we obtain the functions $a[t] = \sinh[t]$, $r[\chi] = \sinh[\chi]$. In these coordinates, the space sections are negatively curved and the space expands out of a singularity. The lines of constant cosmological time t are intersections with the planes $W = \text{const}$. This construction is analogous to the Milne universe. The difference is that there the world-lines of the expanding substratum are drawn on a plane but here on a pseudosphere. In both cases, the space sections represented by the lines of constant t are negatively curved.

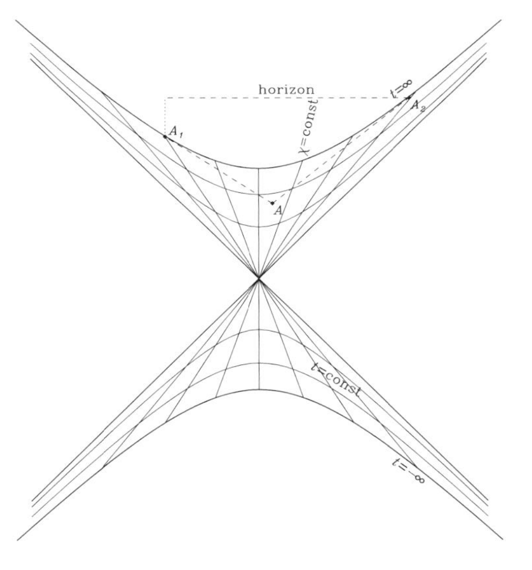

Figure 7.26: *The de Sitter universe in projection. III.*

In this projection of the hyperboloid of Figure 7.25, the substratum is a pencil carried by an internal point of the world. The curves of constant cosmological time are hyperbolas of constant opening: $y^2 - x^2 = \tanh^2[t]$.

Corresponding to the choice of the world-lines of the substratum, we obtain space sections of positive, negative, or no curvature. This freedom has the same origin as the analogous freedom in the Minkowski world: We choose the world-lines of the substratum in a purely kinematical way. A physical effect of the substratum does not exist: When we write down the Einstein equations for the universes considered above, we see that no ordinary matter density is admitted as source of the curvature in the universe.[15] The curvature of the space–time world

[15] The space–time curvature of the de Sitter universe is equal to some constant times the unit tensor. The constant is called the cosmological constant. In the de Sitter universe, it is a pure matter of taste to say that we are considering the Einstein equations without ordinary matter but with a nonvanishing cosmological constant, or the Einstein equations

is a constant. As soon as we make the substratum a *material flow* with a mass density, this freedom of choosing different kinds of homogeneous space sections is eliminated. We have space sections now determined as spaces normal to the timelike world-lines of the matter distribution. In this generic case, the curvature of space is physically measurable.

Equation (7.5) yields only one of the possibilities to obtain a higher dimensional pseudosphere. The other one is obtained by choosing the fifth coordinate W to be timelike:

$$T^2 + W^2 - X^2 - Y^2 - Z^2 - 1 = T^2 + W^2 - R^2 - 1 = 0.$$

The carrier line of the pencil of planes always intersects this quadric. Therefore, we choose its direction as $[T, W, R] = [0, 1, 0]$ and parametrize the normals in the form $[0, -\sinh\chi, \cosh\chi]$ (Figure 7.27). We find the anti-de Sitter universe, which plays a certain role in multidimensional cosmology (Figure 7.28).

The similarity of the geometries induced by hyperboloids or spheres becomes manifest if they are expressed in projective coordinates (which are now coordinates of the three-dimensional world in which the quadric is embedded) and by the polarity (which is now the metric of the three-dimensional world too). Here, we want to illustrate this similarity by presenting the sine theorem for geodesic triangles. Angles and opposite sides stand in a peculiar relation which is given in the Euclidean geometry by

$$\frac{\sin\alpha}{a} = \frac{\sin\beta}{b} = \frac{\sin\gamma}{c}.$$

On the sphere, we obtain correspondingly

$$\frac{\sin\alpha}{\sin a} = \frac{\sin\beta}{\sin b} = \frac{\sin\gamma}{\sin c}.$$

In the Minkowski geometry, in which rotations in the plane have two fixed rays (the light-like directions), we obtain for triangles of timelike sides

$$\frac{\sinh\alpha}{a} = \frac{\sinh\beta}{b} = \frac{\sinh\gamma}{c}.$$

In the Galilean geometry, in which the angles are measured by spatial distances, the formula degenerates into

$$\frac{\alpha}{a} = \frac{\beta}{b} = \frac{\gamma}{c}.$$

On the time shell, we obtain

$$\frac{\sin\alpha}{\sinh a} = \frac{\sin\beta}{\sinh b} = \frac{\sin\gamma}{\sinh c}.$$

On the pseudosphere $T^2 - W^2 - R^2 = 1$, the timelike distance is a hyperbolic angle and the tangential plane is pseudo-Euclidean. Correspondingly, we obtain

$$\frac{\sinh\alpha}{\sinh a} = \frac{\sinh\beta}{\sinh b} = \frac{\sinh\gamma}{\sinh c}.$$

with an exotic sort of matter, i.e., matter with a negative ratio of pressure and density. Such a "matter density" is possible in quantum models for empty space.

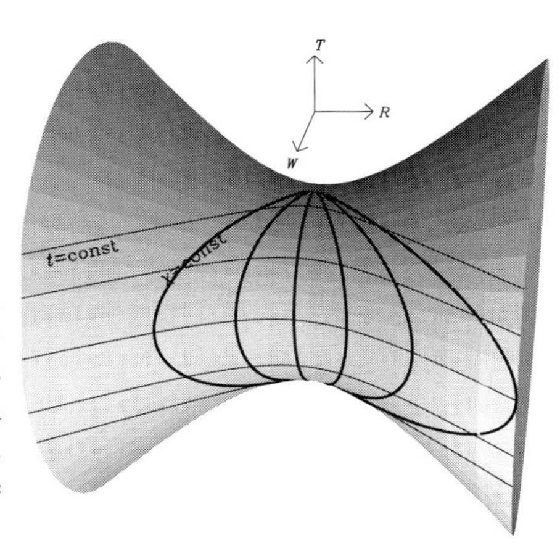

Figure 7.27: *The anti-de Sitter universe.*

On the five-dimensional pseudosphere $T^2 + W^2 - R^2 = 1$, $R^2 = X^2+Y^2+Z^2$, the coordinates $T = \sin[t]$, $W = \cos[t]\cosh[\chi]$, $R = \cos[t]\sinh[\chi]$ are introduced. For our surface formula, Eq. (7.4), we obtain the functions $a[t] = \cos[t]$, $r[\chi] = \sinh[\chi]$. In these coordinates, the space sections are negatively curved. The space expands from a singularity and recontracts again into a singularity. The lines of constant cosmological time t are the intersections with the planes $T = \text{const}$.

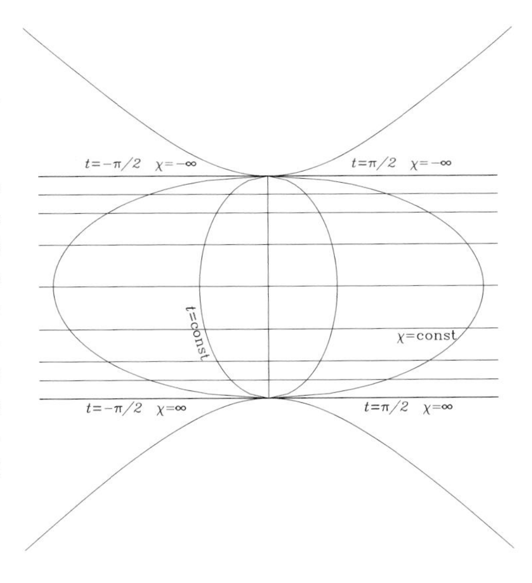

Figure 7.28: *The anti-de Sitter universe in projection.*

In this projection of the hyperboloid of Figure 7.27, the absolute conic section is again a hyperbola, which is not intersected by the timelike straight lines. The substratum is a pencil carried by an internal point of the world (here lying at infinity of the plane). The curves of constant cosmological time are ellipses with a constant major axis: $y^2\cot^2[t] + x^2 = 1$. The lines $\chi = \pm\infty$ coincide with the period boundary $t = \pm\pi/2$. They represent the boundary of the region in which the model and the substratum are defined.

On the pseudosphere $T^2 + W^2 - R^2 = 1$, the timelike distance is a usual angle and it yields

$$\frac{\sinh\alpha}{\sin a} = \frac{\sinh\beta}{\sin b} = \frac{\sinh\gamma}{\sin c}.$$

The proofs are found in Appendix E.

8 The Projective Origin of the Geometries of the Plane

Why do all the geometries described in the previous chapters show so many similarities? They have a common origin, the geometry of perspective, i.e., *projective geometry*. Euclidean and Minkowski geometries, elliptic, hyperbolic, and de Sitter geometry belong to a family called the projective-metric geometries. This chapter is devoted to illustration of its basics. Concerning the formal aspect, the reader is referred to Appendix C. Here, we intend to gain insights that are necessary to obtain a unified view of the geometries of the plane.

The laws of perspective were demonstrated by Brunelleschi in 1425 (Figure 8.1). They were already formed in the hellenistic science, in particular by Euclid in his book about optics [81], but their use in painting was forgotten. Therefore, Brunelleschi's demonstration turned out to be a hallmark in art. We first imagine the Euclidean plane, i.e., our drawing desk. We look at it from the side and project it thereby onto the virtual plane of our field of view, which can be assumed to be perpendicular to some central line of sight. This latter plane replaces the drawing plane (Figures 8.2 and 8.3). We obtain a perspective[1] image of the desk. Infinity in the primary plane is mapped to an ordinary finite straight line, the *vanishing line*, which is also called the *horizontal line*. It represents the image of the *line at infinity*. By projection, we take the infinite by projection to the finite of the drawing plane and regard the points of this line as ordinary points. The result of such a projection of the Pythagorean figure is shown in Figure 8.3. Straight lines that are parallel in the primary plane obviously intersect on the vanishing line. We say that they have a common point on the vanishing line. In addition, the perpendiculars to parallel lines also intersect on the vanishing line. We call the intersection of the perpendiculars to a straight line its pole. The existence of a pole for any straight line is a very important fact. Even if the properties of the vanishing line are dissolved in the geometries with nonzero curvature, the intersection of the perpendiculars to a straight line at its pole remains a central notion.

In perspective and projective maps, angles and lengths lose their elementary comparability. Circles become general conic sections (Figure 8.6). In contrast to these changes, the incidence of points and lines, the collinearity of point triples, and the *cross-ratio* (Figure 8.5) of point quadruples on a straight line (and of ray quadruples of a pencil, respectively) remain unchanged. All theorems concerning the collinearity of three points, the existence of a common vertex for three straight lines, or the separation relation (more specifically the cross-ratio) remain valid even if angles and distances are dissolved.

[1]We call a map a *perspective* map if it can be represented as a projection from a center. Perspective maps do *not* form a group: Two different perspective maps applied successively do not yield another perspective map but a more general *projective map*. The projective maps generated by the perspective maps do, however, form a group.

The Geometry of Time. Dierck-E. Liebscher

ISBN: 3-527-40567-4

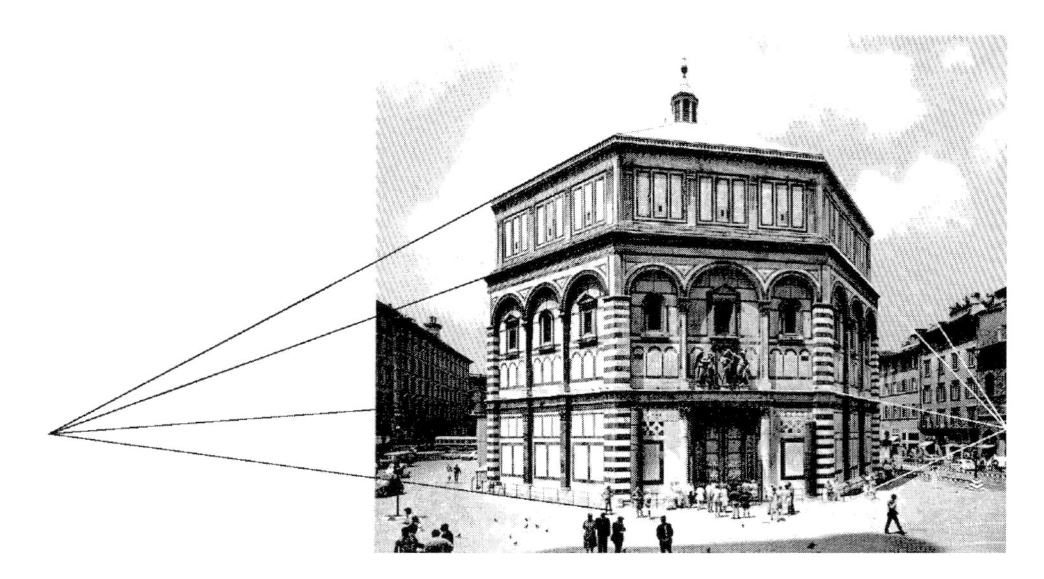

Figure 8.1: *The perspective map. I.*

The laws of perspective (in particular the existence of vanishing points) were found (refound, in fact [81]) in the time between Giotto di Bondone and Filippo Brunelleschi [21]. In 1425, the latter demonstrated these laws by drawings of the baptistery of Florence, which is shown in the figure with two vanishing points.

The geometry that treats two figures as equivalent if they can be projected onto each other (in general in more than one step) is called the projective geometry. In the framework of the projective geometry, all quadrangles (in which no three vertices are collinear) are congruent, and all quadrilaterals (in which no three sides lie in a pencil) are congruent. All the real nondegenerate conic sections are the same figure. The projective geometry provides no means to distinguish between a *hyperbola*, a *parabola*, or an *ellipse*. There is no definition of a circle for the moment, too: The expected form is not invariant. A circle is the derivate of a metric, and no metric of that kind is present at this stage. It is only in the metric geometry that a circle can be distinguished from an ellipse. The distinction between ellipses and hyperbolas does not require the metric in full but the definition of an infinity only (ellipses are conic sections without infinite points). In addition, no conic section is determined by only three points, in contrast to a circle in the metric geometry. In the projective geometry of the plane, a conic section is determined by five points. The construction is considered in Section C.3 and sketched in Figure C.6. From a point of the projective geometry, we need a predefined reference figure in order to distinguish between different kinds of conic sections with respect to this predefined figure. It turns out that this figure is itself a conic section, called the absolute conic section. We can keep it invariant in some particular projective maps. These maps form a subgroup of all projective maps, and they are called the *projective-metric maps*. In fact, any conic section can be elected to be this absolute conic section, even a degenerate or imaginary

Figure 8.2: *The perspective map. II.*

The figure shows the mechanical construction of perspective proposed by Albrecht Dürer [82]. The center of perspective is provided by the ring on the wall. The coordinates of the intersections of the projection rays with the drawing plane (the frame) are determined by the man on the right and marked on the folded page. In this way, the perspective image is constructed point by point.

one (i.e., a conic section that has a real equation but no real points). Under projective maps, they are all equivalent except for degenerate cases. Circles now can be defined in relation to such an absolute conic section. The property of being a circle is invariant to the subgroup of projective-metric maps. It is the required invariance of the absolute conic that reduces the mobility of the projective plane to that of a metric geometry in the plane: The metric geometry implies that if we know the image of a point and of a straight line through this point, the motion in the plane is determined up to a reflection on this line. Ellipses can be distinguished from hyperbolas: Ellipses do not intersect the absolute conic, hyperbolas do. We shall meet this point again in the discussion of the orbits of rotation (Figures 9.16, ff.).

Among projective properties, the cross-ratio of four points on a straight line (Figure 8.5) or of four straight lines in a pencil is the central notion. Pappos already found that it can be transported from line to line by perspective maps: If the lines joining the corresponding points of two four-point ranges are concurrent, then the ranges are equi-cross. Although distances are not defined in projective terms, we can interpret the cross-ratio as a double division ratio of

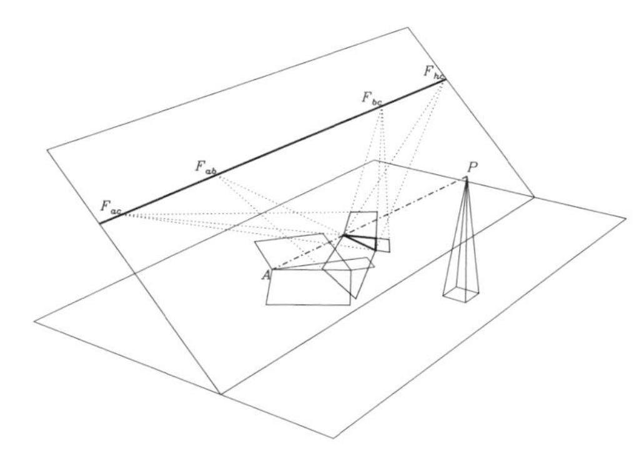

Figure 8.3: *Pythagoras's theorem in the Euclidean geometry under oblique projection.*

The plane α is projected from an elevated point onto a new (field-of-view) plane β. Infinity in the initial plane is mapped to the horizon or vanishing line p, i.e., to an ordinary straight line at a finite distance. Parallel straight lines intersect on this horizon. Perpendicular straight lines yield pairs of points ($[F_{ac}, F_{bc}]$ and $[F_{ab}, F_{h_c}]$) on the horizon. These pairs of points define a map of the horizon onto itself. This map is an involution, i.e., a kind of reflection in the vanishing line. Corresponding sites of the two perspective triangles meet on the intersection line of the two planes. This fundamental fact was found by Desargues. We illustrate it separately in Figure 8.4.

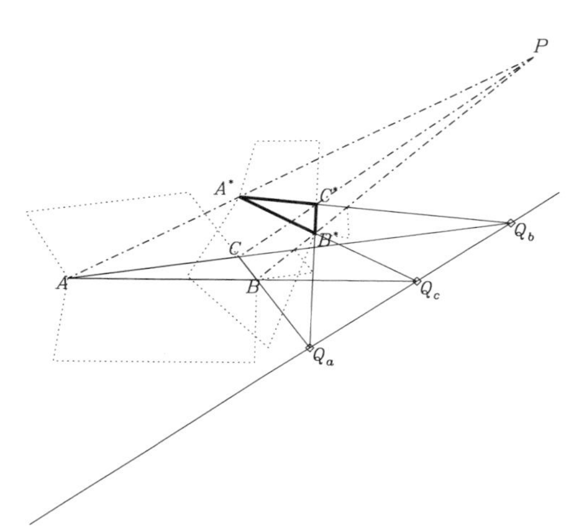

Figure 8.4: *Desargues's theorem.*

If two triangles are perspective, i.e., if the lines connecting corresponding vertices pass through a common point P, the corresponding sides intersect on a common line. We enlarge a part of Figure 8.3. The common line is the intersection of the plane of view with the initial plane. However, our figure is totally plane now, and Figure 8.3 is only a three-dimensional illustration of the theorem.

distances for the moment. As we shall see later, a purely projective definition of the cross-ratio begins with the definition of the harmonic separation. If we want to calculate the cross-ratio in the Euclidean plane, we can put

$$\mathcal{D}[A, B; C, D] = \frac{AC}{AD} : \frac{BC}{BD}. \tag{8.1}$$

The sequence of the four points is essential. If it is changed, the value of the cross-ratio

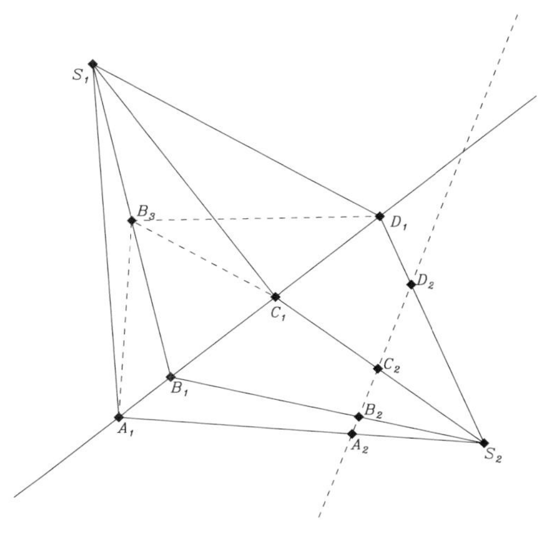

Figure 8.5: *The cross-ratio.*

The cross-ratio of four points on a straight line is the cross-ratio of four areas that are formed by the carrier line with the corresponding rays of some pencil. For instance, $B_1C_1/B_1D_1 = \Delta B_1S_1C_1/\Delta B_1S_1D_1$. However, the cross-ratio of the areas does not depend on the points chosen on the rays if the rays are given. In the figure, $\Delta B_1S_1C_1/\Delta B_1S_1D_1 = \Delta B_3S_1C_1/\Delta B_3S_1D_1$. The points on the rays only determine the lengths of the sides in triangles with given vertex angles. All lengths can be scaled. Consequently, the cross-ratio can be seen as a property of the four rays of a pencil. This fact makes it possible to transport the cross-ratio without change from a line to a pencil and back to another line and again to a second pencil and so on by intersection. Therefore, the cross-ratio is an invariant of projective maps.

Figure 8.6: *Conic sections.*

The plane intersection of a cone can be an ellipse, a hyperbola, or a parabola depending on whether the intersecting plane cuts only one half-cone, both, or is parallel to a mantle line. The intersection on the far left is a circle. We can interpret the pattern as the projections of this circle from the vertex onto the other intersection planes, where it takes the form of all the types of conic section. The mantle lines are the projection rays in this case.

changes too but in a given scheme. It yields

$$\mathcal{D}[B,A;C,D] = \mathcal{D}[A,B;D,C] = \frac{1}{\mathcal{D}[A,B;C,D]}, \tag{8.2}$$

$$\mathcal{D}[A,C;B,D] = 1 - \mathcal{D}[A,B;C,D], \qquad \mathcal{D}[C,D;A,B] = \mathcal{D}[A,B;C,D] \tag{8.3}$$

and the chain rule

$$\mathcal{D}[A,C;E,F] = \mathcal{D}[A,B;E,F]\,\mathcal{D}[B,C;E,F]. \tag{8.4}$$

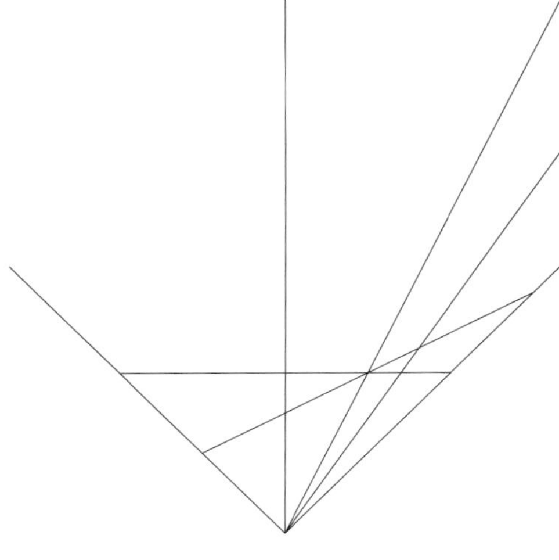

Figure 8.7: *Einstein's composition of velocities.*

We draw the world-lines g_i of three uniformly moving objects and the two lightlike directions, e and f. We draw through the point B on g_2 the (Minkowski-)perpendicular a_1 to g_1 and a_2 to g_2 and obtain the points A, C, and C^*. The composition rule for cross-ratios yields $cD[g_3, g_2; e, f] \cdot \mathcal{D}[g_2, g_1; e, f] = \mathcal{D}[g_3, g_1; e, f]$. This is the composition law of velocities: For instance, $v_{21}/c = AB/OA$ and $\mathcal{D}[g_2, g_1; e, f] = (c + v_{21})/(c - v_{21})$.

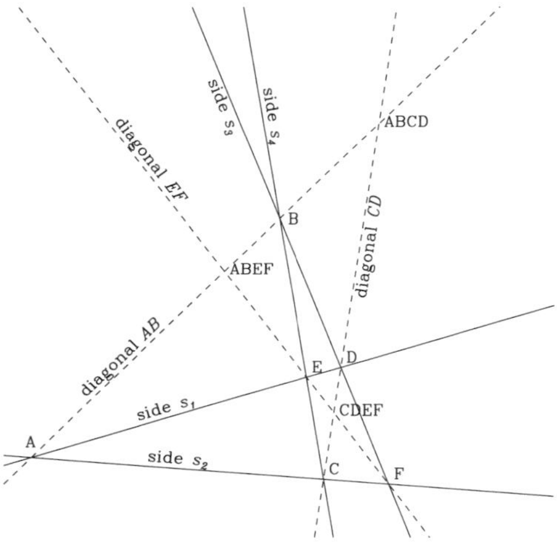

Figure 8.8: *The construction of the fourth harmonic point with the complete quadrilateral.*

In a complete quadrilateral, the sides intersect at six points A, B, C, D, E, F. There are three diagonals AB, CD, EF and three diagonal points $ABCD$, $ABEF$, $CDEF$. Each point lies on one diagonal and can be taken as the center of perspective between the two other diagonals. So any two diagonals can be projected onto each other by two different centers. The difference is the order of the two vertices on the diagonals. The conclusion is that interchanging the two vertices does not alter the cross-ratio: The cross-ratio must be -1. The complete quadrilateral is the standard construction to find the fourth harmonic point to three collinear points or the fourth harmonic ray to three rays of a pencil.

The simplest application of the chain rule is Einstein's composition of velocities (Figure 8.7). To an event in the Minkowski plane, we draw the lightlike directions, e and f, and three uniform world-lines g_i. The line a_1 is orthogonal to g_1. Obviously, the velocities $v_{21}/c = AB/OA$ and $v_{31}/c = AC/OA$ determine the cross-ratios $\mathcal{D}[g_2, g_1; e, f] = (c+v_{21})/(c-v_{21})$ and $\mathcal{D}[g_3, g_1; e, f] = (c+v_{31})/(c-v_{31})$. The essential point is to remember the relativity of simultaneity. The line of simultaneous to B events in the frame moving along the world-line OB is a_2. This line a_2 is orthogonal to g_2. We obtain $v_{32}/c = BC^*/OB$ and $\mathcal{D}[g_3, g_2; e, f] = (c + v_{32})/(c - v_{32})$. The composition rule of cross-ratios,

$$\mathcal{D}[g_3, g_2; e, f] \cdot \mathcal{D}[g_2, g_1; e, f] = \mathcal{D}[g_3, g_1; e, f],$$

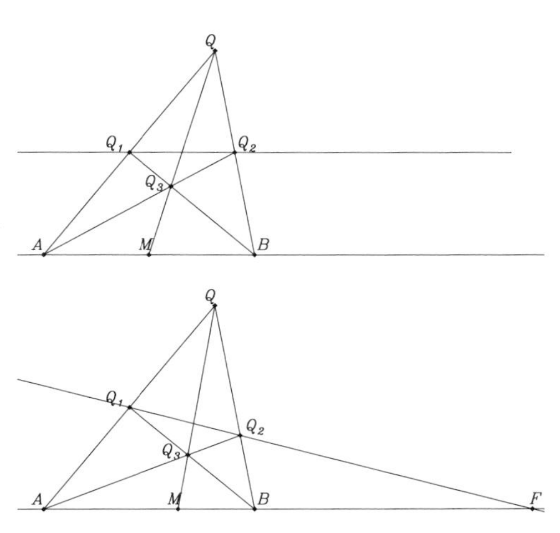

Figure 8.9: *Metric and harmonic bisection.*

Metrical bisection can be represented as harmonic separation because the point at infinity of a straight line can be taken into account. The upper part shows how the bisection in Euclidean geometry uses the parallel to the segment AB. Projectively, this implies designate and retain the "point at infinity" F and the projective construction of the bisection. In the lower part, this is indicated in a projective sketch.

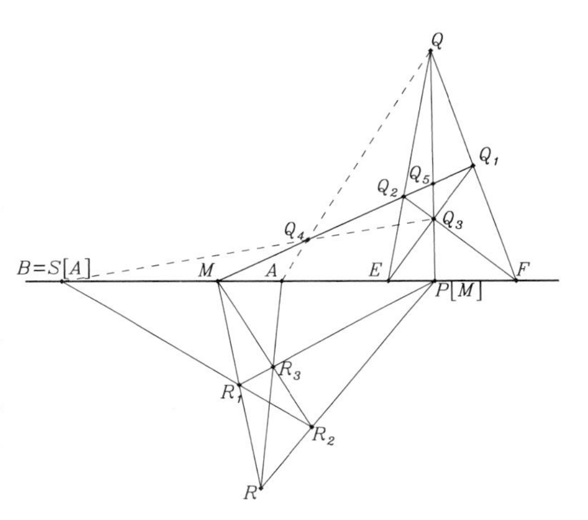

Figure 8.10: *The transfer of chords.*

We give a somewhat more complicated (but more general) construction of the same continuation as in Figure 8.11. We look for a point $B = S[A]$ with the cross-ratio $\mathcal{D}[B, M; E, F] = \mathcal{D}[M, A; E, F]$. To this end, we determine $P[M]$ as fourth harmonic point in the harmonic range $[M, P[M]; E, F]$ and continue determining $S[A]$ as the fourth harmonic point in the harmonic range $[S[A], A; M, P[M]]$. This construction is recovered in the two-dimensional plane (Figure 8.15).

is exactly Einstein's composition of velocities, Eq. (4.1). In contrast, if we intend to use the composition law for velocities, we infer the value of v_{32} from the corresponding cross-ratio. In order to interpret the value v_{32} in an ordinary way as distance–time ratio, we have to introduce the relativity of simultaneity, i.e., we have to declare a_2 orthogonal to g_2. Only by this declaration we obtain $v_{32}/c = G_2G_4/OG_2$.

Evidently, the cross-ratio remains invariant in projective maps (Figure 8.5). The cross-ratio of a range $[A_1, B_1, C_1, D_1]$ is a property of the rays chosen from a pencil carried by S_1, or any other point S_2. If we intersect this pencil with another line, we obtain a new range $[A_2, B_2, C_2, D_2]$. It has the same cross-ratio as $[A_1, B_1, C_1, D_1]$, because of this equality of cross-ratios between pencils and ranges on intersecting lines. The value $\mathcal{D}[A, B; C, D] = -1$ determines a nontrivial particular case. It denotes the *harmonic separation*, which is

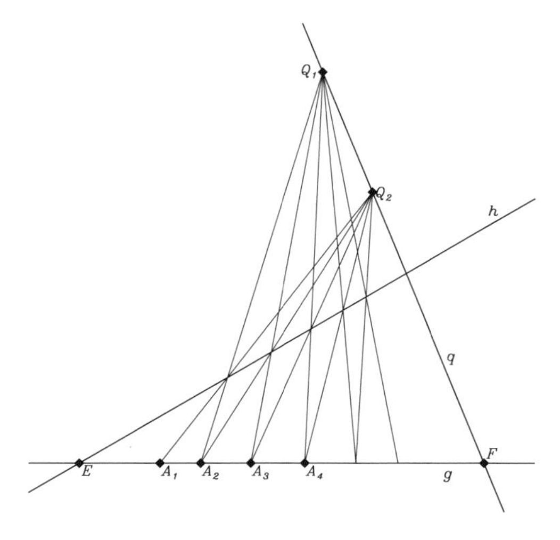

Figure 8.11: *Comparison of lengths through use of two fixed points.*

If two absolute points E and F are given on a straight line, the length of the segment AB is determined by the cross-ratio $\mathcal{D}[A_1, A_2; E, F]$ that the endpoints form with the two absolute points. Through both fixed points we draw a straight line (q and h). We choose Q_1 on q and draw the connecting line from A_1 to the intersection of h with the connection of Q_1A_2. The intersection point of this connection with q is denoted by Q_2. We continue the segment by successive connection with Q_1 and Q_2. We see immediately that $\mathcal{D}[A_1, A_2; E, F] = \mathcal{D}[A_2, A_3; E, F] = \mathcal{D}[A_3, A_4; E, F] = \cdots$.

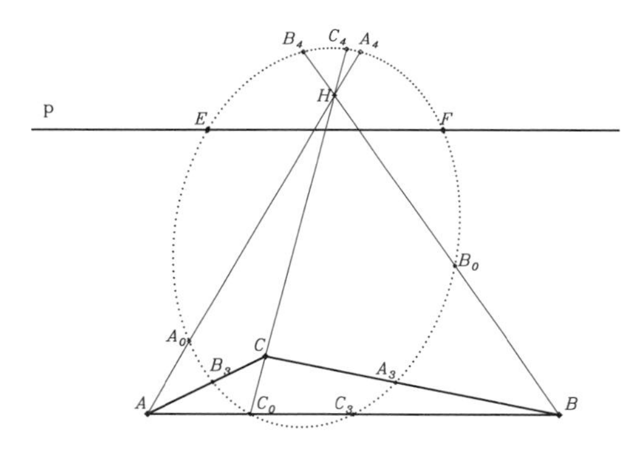

Figure 8.12: *Feuerbach's circle in projective dress.*

We see here a projective representation of Figure 6.12. The line initially at infinity is now a visible line in the plane, the vanishing line p. The asymptotes of the Minkowski circles intersect this vanishing line at two points E and F. Two straight lines are orthogonal if their points of intersection with the vanishing line separate these two points harmonically. By this rule, we get the altitudes of the triangle ΔABC, which meet at H, and their feet A_0, B_0, C_0. Taken with the vanishing point of the line, the midpoint of a segment separates the vertices harmonically. Using this rule, we obtain the midpoints of the sides, A_3, B_3, C_3, and of the altitudes A_4, B_4, C_4. All nine points lie on one circle, i.e., on one conic section that intersects the horizon in E and F.

shown by the construction of a complete quadrilateral[2] in Figure 8.8. Four harmonically separated points on a line are said to form a *harmonic range*. Figure 8.8 shows that the construction of the harmonic separation can be performed without calculating the cross-ratio explicitly: We consider on each diagonal the two vertices of the quadrilateral and the two diagonal points. These ranges are mapped onto each other by projection from other vertices of the quadrilateral. There are two such maps from the diagonal AB to the diagonal CD, for instance, one with E as the center of perspective, the other with F. On the one hand,

[2]That is, four lines together with the six intersection points.

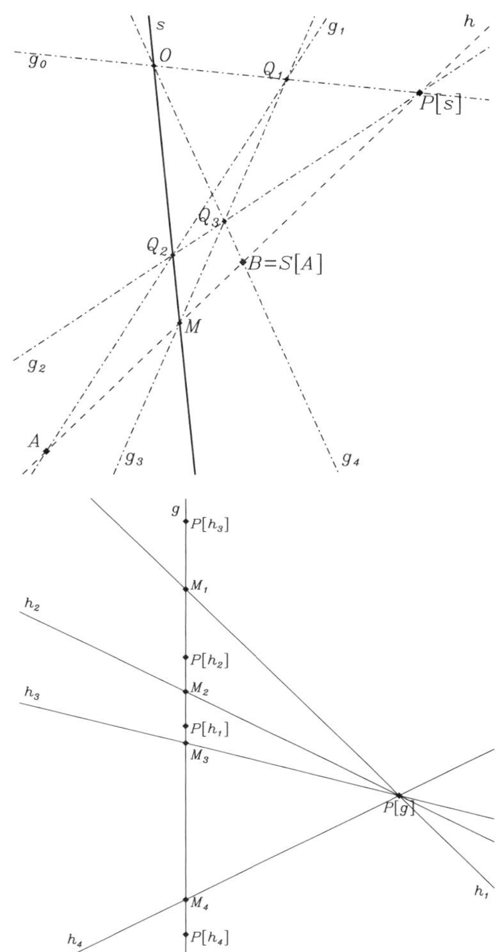

Figure 8.13: *Reflection on a straight line. I.*

Let $P[s]$ be the pole of the line s. We draw the connection with A and obtain the intersection point M. The image $S[A]$ is constructed as the fourth harmonic point: $\mathcal{D}[A, S[A], M, P[s]] = -1$. We choose g_0 passing through P and g_1 passing through A, and draw successively the lines g_2, g_3, g_4 through the intersection points O, Q_1, Q_2, Q_3, B. We see the quadrangle $MP[s]Q_1Q_2$ with the diagonal points A, O, Q_3. Hence A and B separate M and $P[s]$ harmonically.

Figure 8.14: *The poles of a pencil of rays are collinear.*

We draw the line g and its pole $P[g]$. The reflection on g is defined now. This reflection in the plane induces reflections on each line h_i through $P[g]$. The perpendicular in the intersection points M_k must always be g itself because g is invariant in all these reflections. The carrier points of the perpendiculars of any line h_k, i.e., the poles $P[h_k]$, are incident with the line g: $\langle h, P[g]\rangle = 0 \rightarrow \langle P[h], g\rangle = 0$.

$\mathcal{D}[A, B; ABEF, ABCD] = \mathcal{D}[D, C; CDEF, ABCD]$, and $\mathcal{D}[A, B; ABEF, ABCD] = \mathcal{D}[C, D; CDEF, ABCD]$ on the other hand. The cross-ratio $\mathcal{D}[C, D; CDEF, ABCD]$ is equal to its inverse, $\mathcal{D}[D, C; CDEF, ABCD]$. Because it cannot be 1, it is -1, together with the cross-ratios of the corresponding ranges on the other two diagonals.

Beginning with a harmonic range, all other values can be constructed by successive multiplication and inversion and by taking into account the symmetry properties (8.2), and the chain rule (8.4). Consequently, the harmonic ratio can be determined without referring to the Euclidean plane used to write Eq. (8.1).

We now intend to define orthogonality in a projective way and to organize the comparison of angles and segments. We start to analyze bisecting and multiplying of intervals on a line and angles in a pencil when we have chosen two distinguished points on a line, or two distinguished rays in a pencil, respectively. We proceed in three steps. First, we compare segments on a straight line and find out how to multiply and how to bisect them. Secondly, we transfer

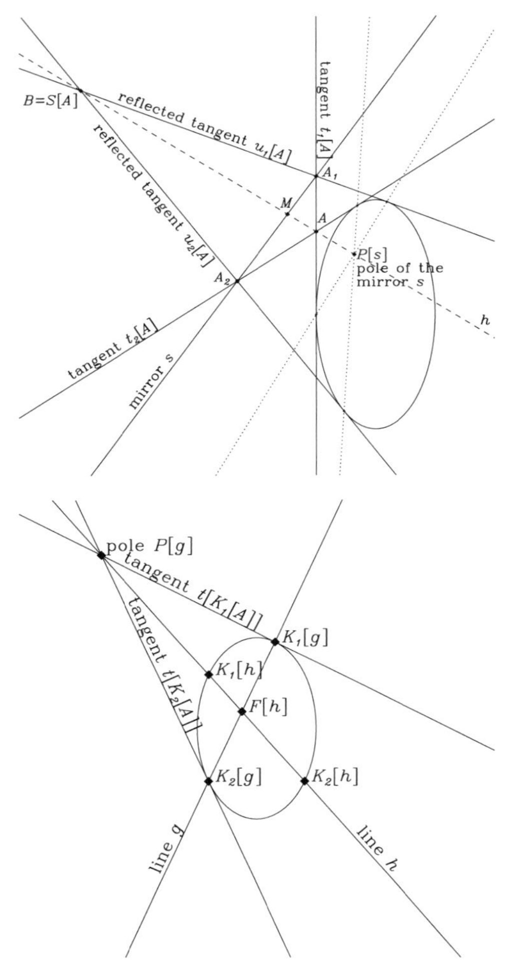

Figure 8.15: *Reflection on a straight line. II.*

Given an absolute conic section, its bundle of tangents is mapped onto themselves in reflections. This determines the reflection. For some point A, we first draw the tangents to the absolute conic. From their intersections A_1 and A_2 with the mirror we again draw tangents to the conic. They are the reflected images of the two former tangents from A and must now intersect in $S[A]$. The pole of s is given as the intersection of the polars to A_1 and A_2 (dotted lines). Alternatively, A_1 and A_2 are situated symmetrically under reflection with respect to the line $AS[A]$. On the line $AS[A]$, we obtain Figure 8.10.

Figure 8.16: *The pole of a straight line.*

The pole of a straight line g is the intersection of the tangents t to the conic section drawn at the points $K[g]$ where it is intersected by the line g. $P[g] = t[K_1[g]] \times t[K_2[g]]$. The intersections $K_1[h], K_2[h]$ of *any* straight line h through the pole separate the pole and the foot point $F_g[h]$ harmonically. This can be used to define the pole if g does not have real intersection points with the conic section.

the method to a pencil of rays and obtain the method of comparing angles of one individual pencil. Finally, we construct the comparison of segments and angles between different lines or pencils, respectively.

To find our bearings and get some practice, we first consider the plane of perspective, which we used for instance in Figure 8.3. Parallels are defined by their intersection on the vanishing line p. When we bisect the segment AB on the line g by the point M, the intersection $F[g]$ on the vanishing line and this midpoint M separate the segment AB harmonically. We obtain the midpoint M by constructing the fourth harmonic point (Figure 8.9). Correspondingly, we continue the segment AB by constructing the point C that (together with A) separates the pair $BF[g]$ harmonically (Figure C.2). The symmetry of the cross-ratio yields $\mathcal{D}[A, B; M, F] = -1$, $\mathcal{D}[A, B; C, F] = 2$. When we now choose the coordinates in such a way that they represent the cross-ratio numerically, we put $(A, B, F) = (0, 1, \infty)$. In this case, the point M acquires the coordinate value $1/2$ and C acquires the value 2. Thus, the

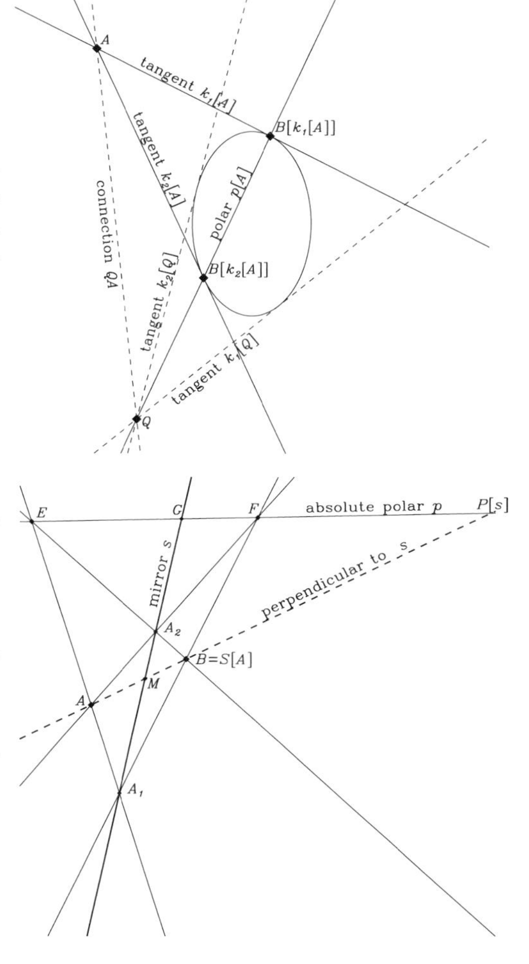

Figure 8.17: *The polar of a point.*

The polar of the point A connects the points of contact B of the tangents $k_1[A], k_2[A]$ to the conic section, i.e., $p[A] = B[k_1[A]] \times B[k_2[A]]$. The tangents $k_1[Q], k_2[Q]$ of *any* point Q on the polar separate the polar and the connection QA harmonically. This can be used for the construction when A lies inside the conic section.

Figure 8.18: *Reflection with the axiom of parallels holding.*

Given two absolute pencils, they are mapped onto each other in reflections. This determines the reflection. We now construct the reflection image of the point A. First, we draw the rays of the two pencils that pass through A. They are reflected into rays of the other pencil, which intersect the mirror at the same points. The intersection of these two reflected rays is the reflection image $S[A]$. The quadrilateral $AA_1S[A]A_2$ is the perspective image of the light-ray quadrilateral $AA_1S[A]A_2$ in Figure 5.1.

harmonic separation with the point at infinity furnishes the length measurement on a straight line (in the Minkowski geometry as well as in the Euclidean or Galilean geometry).

We can understand Figure 8.9 as definition of the reflection on a line: M is a mirror, and B is the image $S[A]$. This reflection has *two* fixed points: Not only M but also $F[g]$, which can here be interpreted as the point at the infinity of a line. However, when we forget the origin of our plane as perspective image of the Euclidean plane, a reflection on a line about the point M is simply defined by a pair of corresponding points E and F (Figure 8.10). We construct the reflection in the form of a projective map that maps E and F onto each other and with M being one of the fixed points. Through construction of a complete quadrangle, we obtain the second fixed point, which we call $P[M]$. The reflection maps any point A on the fourth harmonic point with the pair $M, P[M]$. The figure shows the projection of the harmonic range $[Q, Q_3; Q_5, P[M]]$ from Q_4 onto $[A, S[A]; M, P[M]]$. We obtain $\mathcal{D}[A, S[A]; M, P[M]] = \mathcal{D}[Q, Q_3; Q_5, P[M]] = -1$.

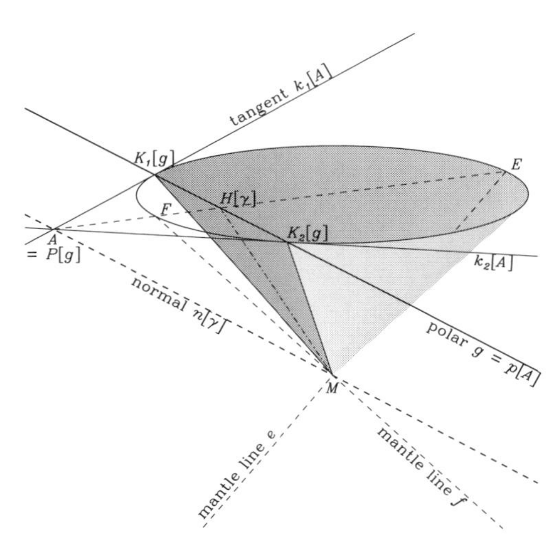

Figure 8.19: *Polarity and orthogonality.*

In the (2+1)-dimensional world, we represent the drawing plane in the form of a plane intersection with the light cone. Each straight line g in the drawing plane is associated with the plane γ in the world that passes through the vertex. The normal $n[\gamma]$ can be constructed with the help of the known light-ray quadrilaterals that are available in the Minkowski world. This normal intersects the drawing plane in the pole $P[g]$ of the line g. In addition, the normal $n[\gamma]$ carries all planes in the world that are perpendicular to the plane γ. The intersection of these perpendicular planes with the drawing plane are rays of the pencil of the pole $P[g]$. It is natural to interpret these rays as perpendiculars to g in the geometry of the plane.

When we recall the projection of the Euclidean plane, $P[M]$ is the point at infinity of the line g and hence for any point M on g the same. The purely projective constructions yield no reason for this property. Instead, the points $P[M]$ are some projective image of the points M. Now for the next step, the projective map of a line onto itself has at most two fixed points as we can see in Figure 8.10. In the real case, we have two points E and F that are mapped onto itself for reflections on *any* point on a line. The fixed points E and F define for any point M a *pole* $P[M]$ as the fourth harmonic point. The map $M \rightarrow P[M]$ is itself an involution on a line.

The simplest construction for transporting a length is achieved by two successive perspective maps (Figure 8.11). For further generalization, we show multiplication with the help of the harmonic range. Again, two points that do not vary with our motions are distinguished on a line. Figure 8.10 shows the construction. With the help of this scheme, we can now transport segments along a line with their length conserved, so that they can be compared. As far as angles are concerned, the construction is the dual or *reciprocal* to the one considered above. Using the same methods as for segments on a line, we can rotate and compare angles of a pencil.

We give here an example for consideration. It is the projective generalization of Feuerbach's circle (Figure 8.12). We obtain it by projection of Figure 6.12 in such a way that the line at infinity is mapped onto an ordinary line (the vanishing line) and the asymptotic directions onto two real points on this line. The story of the projection of the Feuerbach figure in a metric plane can be expressed in purely projective terms: Given some complete quadrangle[3] $ABCH$ and the line p. The line p intersects all the six sides of the complete quadrangle and determines six points $A_3, B_3, C_3, A_4, B_4, C_4$ that separate harmonically the edges on the six sides, respectively. These six points lie on a conic that also passes through the three diagonal points A_0, B_0, C_0 of the complete quadrangle (strictly speaking, it is a circle of the geometry

[3]That is, four points together with their six connecting lines.

in question, an ordinary circle in the Euclidean geometry (Figure 6.11), an ordinary hyperbola in the Minkowski geometry (Figure 6.12), and a parabola in the Galilean geometry). The pairs of points on p at which opposite sides of the quadrangle (for instance, AB and CH) intersect define an involution. The fixed points of this involution (in Figure 8.12, the points E and F) also lie on the conic. The conic is sometimes called the 11-point conic to the quadrangle $ABCH$ and the line p.

If we want to know what happens to the sizes of angles and distances when they are distorted in projective maps, we have to represent them as cross-ratios and recover them in the new geometric forms. We use the intuitive concept that some harmonic separation must generalize metrical bisection of angles and segments. Let us consider distances on a line. The chain rule for the cross-ratio demonstrates which function has to be taken as distance on this carrier line. We have to read the cross-ratio $\mathcal{D}[A, B; E, F]$ as a function of the first two points, i.e., we have to define the second pair as invariantly given. Because the distances should be additive on a line, $d[A, B] + d[B, C] = d[A, C]$, and the cross-ratio is composed multiplicatively, Eq. (8.4), the distance has to be a logarithm of the cross-ratio. The basis of the logarithm determines the unit of length. In our construction, the two points E and F can even coincide. Then we can no longer establish an absolute measure of the distance AB, but we can still compare ratios between two segments on a line (see Section E.2).

What can be expected for the angles between the rays of a pencil? In the Minkowski geometry, two rays of every pencil are distinguished: the isotropic directions. We know that the cross-ratio is conserved in projective maps. Hence, the Minkowski motions map an isotropic direction onto itself. The points at infinity of the isotropic directions are invariants. Any angle acquires a value through the cross-ratio of its sides with two distinguished directions.

There seems to be a difference between angles and lengths. On a straight line, one point is distinguished for reference: The point at infinity or on the horizon, respectively. We can compare lengths, but we have *no absolute* measure. In a pencil, *two* rays are distinguished for reference, and we obtain an *absolute* measure for angles. This is the general (nondegenerate) case. If two elements coincide and become one, we have the degenerate case. If we imagine the two distinguished directions of a pencil to coincide, the cross-ratio becomes indeterminate too. However, the indeterminacy is reduced in the *ratio* of two cross-ratios. In this latter case, we no longer obtain absolute angles, but we can still compare angles. This is realized in the Galilean geometry (Appendix E).

When we intend to define an *absolute* length, we transfer this insight to a line by a dual construction. On a straight line, we need *two* fixed points of motions to get an absolute measure of length. The length of any segment is determined by the cross-ratio with these two fixed points.

We can already read Figure 8.10 as the definition of reflection on a particular straight line. The point C is the reflection of A on B when E and F are given as absolute points. The points E and F are simply the points that remain the same in reflections on arbitrary points of a line in complete correspondence to the absolute directions of the Minkowski world that reflect the isotropy property of light propagation. The two points E and F can not only coincide, as in the Galilei geometry, but can also turn out to be imaginary. The real property that is necessary for constructing is the association of the poles $P[B]$ with the points B of a line. This association is an involution and is called polarity. Involutory projective mappings on a line have two fixed points, which can be real, as the points E and F in Figure 8.10. If they are imaginary, the real constructions are only a little more complicated. We merely need in addition two pairs of real points, which are mapped onto each other. These two pairs compensate for the two fixed points being imaginary.

We now generalize the result to the case of a plane. Here we want to construct the reflection on a line. We expect that the reflecting line is orthogonal to the connection of a point with its image. On the connecting line, we obtain the one-dimensional reflection about the intersection point M, which has the second fixed point P. However, now we have a two-dimensional reflection, and P is the fixed point of *this* reflection too. P carries all lines that connect points A with their images. That is, P carries all perpendiculars to the reflecting line s. We call it the pole of the line s. When we know the pole $P[s]$ of the line s, the reflection on this line can be constructed (Figure 8.13). The points M and $P[M]$ can change their role. The line s is orthogonal to the connection $MP[M]$ and contains the poles of all lines that pass through $P[M]$. Hence we call it *polar* of the point P. The poles of a pencil are collinear, and the polars of the points on a straight line are concurrent.

What kind of curve should be expected when we mark the pairs of absolute points for any straight line on the plane? The simplest possible curves that are intersected by a straight line at just two points are the *conic sections* (Figures 8.6 and C.6). The orthocenter theorem requires that this curve is a conic section. To see this, we first state that the poles of the individual lines cannot be chosen freely. The orthocenter theorem yields a condition already for the poles of three lines (Figure D.1). After the definition of the poles of the three sides of a nondegenerate triangle, we are able to construct the pole of any other line using the orthocenter theorem which now plays the role of an axiom. One can show that this is a projective relation: The rays h_k carried by a given point $P[g]$ intersect the connecting line g of their poles at some points M_k. A reflection on h_k induces an involution on g that leaves M_k and $P[h_k]$ fixed (Figure 8.14). The pairs $[M_k, P[h_k]]$ form the polar involution on g that was already identified as a projective map. Hence, the map $h_k \rightarrow M_k \rightarrow P[h_k]$ is projective. Let us now consider the lines that contain their own pole. If such lines exist, three of them define a conic that contacts the lines in their poles (Figure C.8). Any fourth line containing its pole must be tangent to this conic, the pole being the point of contact. The conic is called the *absolute conic* of the polarity. In any reflection, it is mapped onto itself because the projective property of a line to contain its pole cannot get lost in a projective map. When the line h intersects the absolute conic, the intersection points are the fixed points E and F of the involution that relates the poles $P[M]$ to the points M on h so that the corresponding reflections can be constructed on the line (Figure 8.15). That is, the locus of the absolute points of all lines is a conic, more precisely, the absolute conic of the set of reflections.

There exists an important theorem that states that each conic section defines a polarity in the plane. Given such a conic section, we find to any arbitrary straight line g a *pole* $P[g]$. Let us draw an arbitrary line h that intersects the conic section (and this happens in general in two points E_1 and E_2, as we can see in the case of a circle). The pole of g and the intersection of g with h separate the intersections with the conic section harmonically (Figure 8.16). The dual statement takes the form that with each point Q is associated a straight line $p[Q]$, its *polar* with the corresponding dual property. When we choose on $p[Q]$ some point A and draw the tangents to the conic section, the polar and the line connecting A and Q separate the tangents harmonically (Figure 8.17).

The polarity is a relation between points and straight lines in a plane. If the conic section is degenerate, the invertibility is lost. We can find common poles for all straight lines or common polars for all points and even both together. The infinity of the geometry in the plane is now the absolute conic section. In some sense, the absolute conic section generalizes and replaces the vanishing line. The latter appears in its role for metric relations as a degenerate case of the conic. With the help of its associated polarity, the absolute conic defines first of all the orthogonality of two straight lines:

Two straight lines are orthogonal if one passes through the pole of the other.

We can now construct the reflection on an arbitrary straight line:

A point and its reflection image lie on a perpendicular to the mirror line and separate pole and foot point harmonically.

This is the last step in the definition of the general reflection procedure in a plane. The whole group of motions is found by successive reflections on the lines in a plane. Figure 8.18 shows the construction in the case of degeneracy, for instance, for the Minkowski geometry, and Figure 8.15 shows the generic case.

We illustrate the obtained construction and use the orthogonality in the three-dimensional Minkowski world (Figure 8.19). For any plane γ through the vertex M, we can construct there just one perpendicular $n[\gamma]$ by pseudo-Euclidean rules. By the same rules, all planes that contain this perpendicular are again perpendicular to γ. The projection is the intersection with the drawing plane. By this intersection, planes in space determine lines in the plane and lines in space yield points in the plane. In contrast, each line in the drawing plane defines a plane in space (which contains the line and the vertex) and each point on the drawing plane gives a ray through the vertex in space. The plane γ corresponds to g. The planes orthogonal to γ determine lines in the plane that in their turn are projectively orthogonal to g. The intersection of these lines is the point marked by the normal $n[\gamma]$ in the drawing plane. We obtain the polarity that we are already acquainted with. Namely, the drawing plane intersects the light cone in a conic section $\mathcal{K}$. We can reconstruct directly the result that all perpendiculars to a straight line g pass through a common point $P[g]$, which we called the *pole* of the line g. If g intersects the conic, the pole lies outside the conic. In this case, there exist two tangents that touch the conic at the same points at which it is intersected by the line g itself.

Considering the transport of a segment along a line, we immediately observe that the absolute conic is the metrically infinity. In this sense, the poles to lines that intersect the conic at real points lie outside the conic, i.e., beyond the infinity represented by the conic. In Figure 8.19, the inner region of the absolute conic is the projection of a time shell. The geometry of these time shells embedded in the light cone is non-Euclidean. We already know this. In a geometry in which the axiom of parallels holds, all poles lie on the line at infinity, the vanishing line, which represents in this case a degenerate conic section.

The decisive prerequisite for determining reflection mappings is the choice of a polarity. As soon as we know the poles of all straight lines and the polars of all points in a projectively invariant way, no freedom is left for constructing reflections. The polarity itself defines the absolute conic. It is the locus of all points that coincide with their own polar or, dually, the envelope of all straight lines that contain their own pole. The different ways in which such conic sections can be chosen give rise to different geometries of the plane. The polarity and the projective reflection that is defined by it yield the unifying point of view for all these geometries.

9 The Nine Geometries of the Plane

Now we are in a position to show how the geometries that we considered in the previous chapters are derived from projective geometry. The geometries are simply declared to be parts of the projective plane. In the projective plane, a metric geometry is formed through the definition of polarity (i.e., through the identification of an absolute conic section). Similarly in this case, for any two points two other points are given on the same intersecting line, and for any two straight lines two other rays of the same pencil are defined. The double ratio of four elements becomes a composite function of the first two. This function is the basis for the measure of distance and angle, respectively. We are constructing the metric geometries in a projective plane, with points and straight lines. Points and lines are simply understood as equivalent to reflection maps: The polarity yields for each point a polar and for each line a pole, so that we can make a construction as we did in Figure 8.15.

The induced geometries differ according to the form of an absolute conic section. It assigns to each point two (not necessarily real) tangents and to each straight line two (not necessarily real) intersection points. Any two rays of a pencil define a double ratio with these two tangents of its carrier, just as any two points define a double ratio with the two intersection points of the conic with their connecting line. These double ratios remain invariant in *all* projective transformations. However, we *restrict* the use of projective maps to the maps that keep the absolute conic section unchanged. Therefore, the double ratio becomes a measure for the angle between the two rays and the distance between the two points, respectively. Both remain invariant under the reduced group of motions. We are in a metric geometry in which the allowed motions leave distances and angles invariant, i.e., in which lengths of segments and widths of angles need no reference to other objects. Such a reference is necessary only for embeddings of metric geometry into more general ones, as into projective geometry, in which the reference is provided by a conic. Here, two segments that are equal with respect to one conic can differ with respect to another one. The metric geometry of the plane is that part of projective geometry that defines two figures as congruent if not only the internal projective relations of the figures but also their projective relations to the absolute conic section are the same. The reflections defined projectively by this conic section (Figure 8.15) satisfy the axioms of the metric geometry (Appendices D and E).

Now we want to determine the polarity of the different geometries of the plane by real constructions even in the cases where the absolute conic is not real or contains imaginary elements. In order to do this, we use a three-dimensional picture. This trick enables us to find real representations even in the cases where some elements are imaginary (for instance, in the elliptical geometry). In addition, the three-dimensional representation admits new insight into the unification of the geometries of the plane.

The nondegenerate case in the plane is given by a nondegenerate *ellipsoid* or *hyperboloid* $\mathcal{B}$ (we choose a sphere). The existence of such a nondegenerate quadric guarantees that the

The Geometry of Time. Dierck-E. Liebscher

ISBN: 3-527-40567-4

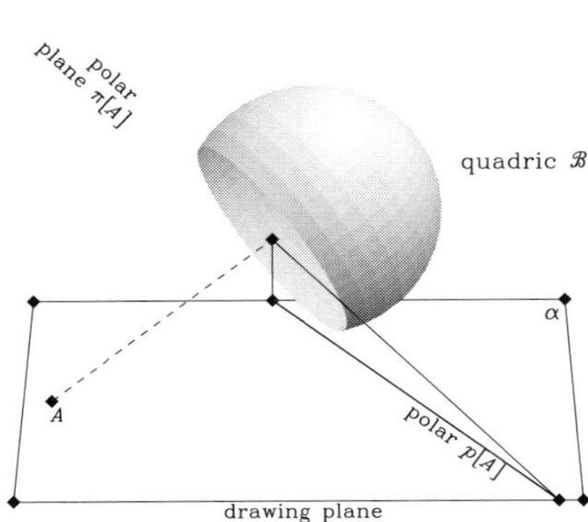

Figure 9.1: *Elliptical geometry.*

In the most important nontrivial case, the auxiliary quadric does not intersect the drawing plane. The absolute conic contains no real points. Nevertheless, for any point A on the plane we obtain a real construction of, first, the polar plane $\pi[A]$ and, second, of its real intersection with the drawing plane. We thereby obtain the polar $p[A]$. Conversely, the points Q on a straight line g in the drawing plane generate a pencil of the polar planes $\pi[Q]$ which is carried by a line that intersects the drawing plane in the pole $P[g]$.

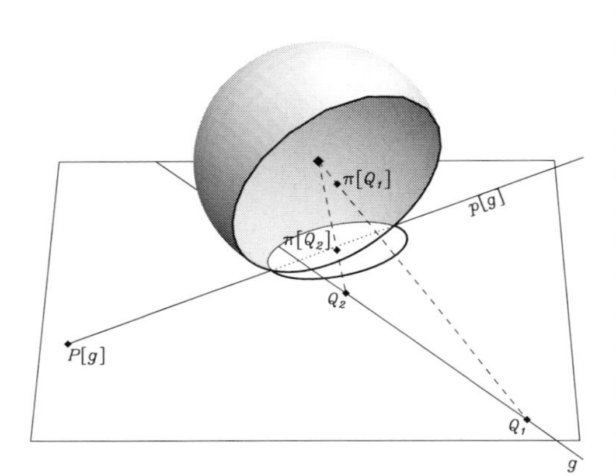

Figure 9.2: *The pole of a line through three dimensions.*

We draw the circles of contact of the tangent cones from two points Q_1 and Q_2 of a line g to the quadric. The two circles define two planes $\pi[Q_1]$ and $\pi[Q_1]$ (in the figure, these designations are written at the centers of the circles). The planes $\pi[Q_1]$ and $\pi[Q_2]$ are the (in the three-dimensional space) polar planes to the points Q_1 and Q_2. The two planes now intersect in the (three-dimensional) polar $p[g]$ of g. The polar plane of any point on g contains this polar. In its turn, the polar meets the drawing plane at a point $P[g]$ that is the (two-dimensional) pole of the line g.

polarity in the space is invertible. We construct the polarity in the plane as follows: To any given point A in the plane α, we construct the cone of tangents to the quadric $\mathcal{B}$. The tangents touch the quadric in a plane conic section and determine (in the three-dimensional space) the polar plane $\pi[A]$ to the given point A. This plane intersects the drawing plane in a straight line that is the polar $p[A]$ to A (Figure 9.1). In contrast, given a straight line g in the drawing plane α we obtain by the same method for all points C of the line g the polar planes $\pi[C]$ in the space (Figure 9.2). Because all points C lie on the straight line g, all polar planes $\pi[C]$ intersect in a straight line[1] $p[g]$ that itself intersects the drawing plane α in the pole $P[g]$. Thus, we find a real construction for the plane polarity even without a real absolute conic section.

[1]In the projective three-dimensional space, the polarity maps points to planes and planes to points but straight lines to straight lines. If a line intersects the quadric, the two tangent planes at the intersection points intersect in the polar line, which then does not intersect the quadric. If the line in questions does not intersect the quadric, there exist two planes of the pencil of planes carried by the line that touch the quadric. The line connecting the two contact points is then the polar. The polarity is an involution in the set of straight lines.

If the quadric $\mathcal{B}$ intersects the drawing plane in a real conic section, the construction of polarity is much simpler: If the conic is real, the polarity needs no auxiliary three-dimensional construction. If the quadric $\mathcal{B}$ does not intersect the drawing plane, the real three-dimensional construction is substituted for the discussion of imaginary elements in the plane. We found the polarity by means of a real figure in space even though the absolute conic of the plane α has no real points.

We draw the quadric $\mathcal{B}$ without loss of generality as a sphere in the three-dimensional space. If it intersects the drawing plane in a nondegenerate curve, it generates a real nondegenerate conic (Figure 9.3). The geometrical constructions that we obtain in such a way do not necessarily contain all the points or all the lines of the embedding projective plane. This we see as follows. Starting from some point, we consider the set of all images that are generated by the reflections we just defined. We call it the transitivity region of the reflections. In the case considered here, it does not cover all the projective plane. This will be divided into more than one transitivity region when a given point or line cannot be moved by reflections into *every* other point or line of the plane. We observe this here where the polarity is defined by a real nondegenerate conic section. Such a conic section divides the points into those in the exterior (which carry two real tangents) and those in the interior (which do not). Correspondingly, the straight lines are divided into those that intersect the conic at two real points and those that do not.

Here, the straight lines are divided into two transitivity regions, one containing the lines that intersect the absolute conic at two real points and the other that miss the conic. In the first case, the points also form two transitivity regions, one containing the points whose polar does not intersect the conic, and the other containing the points whose polar does. We can show that reflection on lines of one transitivity region never produces reflections on lines of the other and that reflection on points of one transitivity region can never be combined to yield a reflection on one of the other. Taken all together, we can reduce the group of motions to three distinct subgroups.[2]

Let us first consider the points in the inner region of the conic, from which tangents cannot be drawn, while the polars of the points do not intersect the absolute conic. Any line connecting two points intersects the absolute conic at two real points. The pole of such a line does not belong to the geometry (i.e., to the points just chosen). Thus, we obtain the *hyperbolic geometry*, the first acknowledged non-Euclidean geometry. The plane is locally Euclidean. That is, the ordinary triangle inequality holds and all segments can be compared. However, at large we find many differences in the Euclidean geometry that are related to the violation of the axiom of parallels. The curvature is negative, and the sum of the angles in a triangle is smaller than π. The excess angle (here negative) is proportional to the area of the triangle. In the limit, the vertices lie on the boundary, where all the angles vanish and their sum too. As we already noted, the pole of a connecting line lies outside the absolute conic, that is, it does not belong to the points of the geometry. Only in the representation of the geometry as part of the projective plane can the pole be constructed as a real point. The absolute conic defines the metric infinity. If the absolute conic is a circle, we speak of Klein's model of the non-Euclidean plane.

[2] This does not imply that we cannot mix all kinds of reflections if we insist on it. The point is that the reduction is possible.

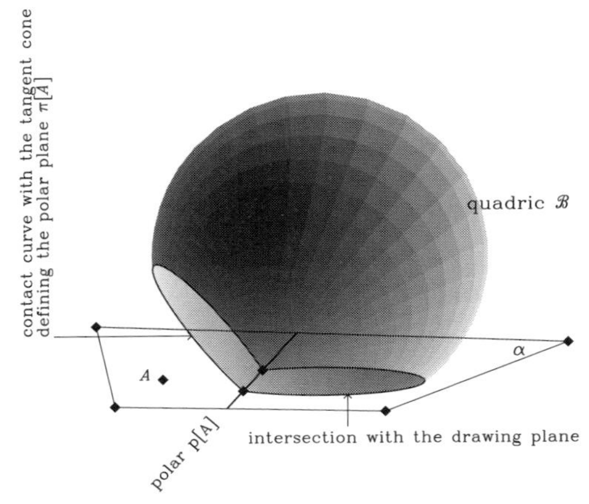

Figure 9.3: *The real cut.*

We look at the drawing plane α from below. The polarity in the plane is mediated by a real absolute conic in space. In the simplest case, a sphere intersects the plane in a real circle. The cone of tangents to the sphere carried by a point A touches the sphere along a circle, where the sphere is intersected by the polar plane $\pi[A]$, which itself intersects the drawing plane in a straight line $p[A]$. Because of the harmonic properties of the poles and polars, it is the polar of A with respect to the circle cutout of the plane.

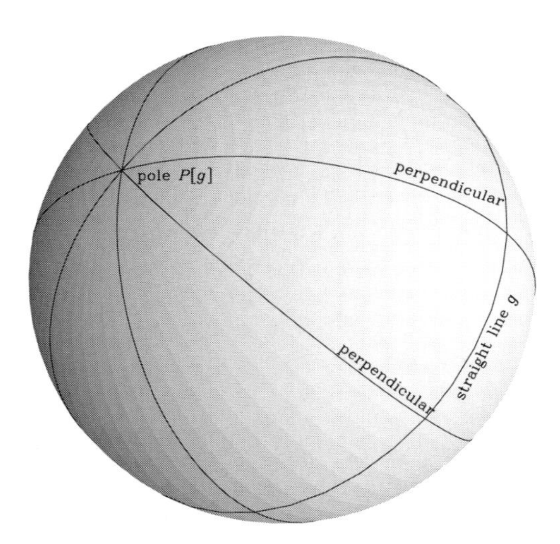

Figure 9.4: *Pole and polar on a sphere.*

Each great circle on a sphere can be interpreted as an equator. For this equator, the assigned poles carry all perpendiculars. They are the meridians.

Let us now consider the points A whose polars intersect the absolute conic twice. We find such points only outside the conic. From each point A, two real tangents can be drawn to the conic. They divide the straight lines through the vertex A into those that intersect the conic and those that do not. There is no transition between these two classes by reflection maps. Hence, we obtain two different cases. If we choose the lines that intersect the absolute conic, we find the *de Sitter geometry* (doubly hyperbolic geometry, Figures 7.22, 7.24, and 7.26). Only the reflections on the lines that intersect the conic are chosen to generate the group of motions. In the physical interpretation, these straight lines turn out to be the timelike lines of force-free motion. Two points (events) are not necessarily connected by such a line (physically speaking, they lie spacelike to each other). The geometrical reflection on a straight line is to be interpreted physically as the reflection seen by an observer in inertial motion along this world-line.

In a triangle whose vertices can be connected, the pseudo-Euclidean triangle inequality holds. The sum of the two sides is shorter than the third when one takes into account the orientation, which can now be defined to the absolute conic. The curvature is again negative. To any straight line g, one can find many parallels h (i.e., allowed lines that intersect g only outside the geometry) through a given point A. The triangle inequality finds an analogous form: To any point A, one can find on a straight line g many points B that cannot be connected (timelike) with A. They are the points now simultaneous with A or at zero distance to A. This is a curious but characteristic similarity between parallelism and simultaneity. From this point of view, the (two-dimensional) de Sitter geometry is doubly hyperbolic. As always, the absolute conic section is the metrical infinity.

The straight lines that do not intersect the absolute conic form the *anti-Lobachevski geometry* (antihyperbolic geometry (Figure 7.28)). In it, the spacelike lines of the de Sitter geometry are declared to be timelike. Again, not all pairs of points can be connected with a timelike line. Again, for a connected triangle the pseudo-Euclidean triangle inequality holds. The curvature, however, is now positive. Two straight (timelike) lines always intersect inside the geometry, in the metrically finite. No parallels exist.

If the drawing plane α is not intersected by the quadric, we obtain a polarity for which no point can lie on its own polar. We find the *elliptic geometry*, which represents the projective image of spherical geometry (Figure 9.1). We already considered this geometry many times. It is the familiar example for our endeavor to uncover unusual interdependences. The sphere and its geometry are particularly transparent because the sphere can be embedded in the three-dimensional Euclidean space and because it is not simplified by degeneracy. We obtain plane elliptical geometry by the projection of the sphere from its center. The great circles turn into straight lines, and the spherical triangles of arcs of great circles turn into ordinary triangles with straight sides. All great circles have *two* poles on the sphere (just as the equator defines the North and South poles), and these become *one* point of the plane. *All* pairs of opposite points on the sphere must be interpreted as *one* point on the projective plane because they lie on the same ray through the center and are mapped onto the same point in the plane. Each point A of the plane is the pole $A = P[g]$ of some straight line g that is its polar $g = p[A]$. The pencil of perpendiculars to this line g is carried by the pole $P[g] = A$ and corresponds to the pencil of meridians to a given equator (Figure 9.4). *Parallels* do not exist on the sphere because there is no infinity and no line at infinity. There exist triangles and circles too (in the plane projection, they become particular conic sections), but there is no infinity. In the elliptic geometry, no straight line exists that contains its own pole. This reflects the fact that the absolute conic section does not contain real points. Two points can always be connected by a line of the geometry. Disjoint sets of points or lines, as in hyperbolic geometry, do not exist. All lines are to be allowed as reflectors in order to generate a geometry.

The elliptical geometry is locally Euclidean. That is, the ordinary triangle inequality holds. The sum of the angles of a triangle exceeds π, indicating a positive curvature. This is no surprise, because we already know about the heritage from the sphere. The polar triangle shown in Figure 7.1 is an example for a sum of $3\pi/2$. The excess (here $\pi/2$) is proportional to the area of the triangle.

In our consideration of projective-metric geometries, we obtain a particular case if the auxiliary quadric $\mathcal{B}$ touches the drawing plane α at only one point P (Figure 9.5). Then all polars $p[A]$ pass through this point P. All straight lines g now have only one common pole, that is P. We call it the absolute pole. This all constitutes the *anti-Euclidean geometry*. Metric distance is now universally referred to this point. Consequently, we find lines of length

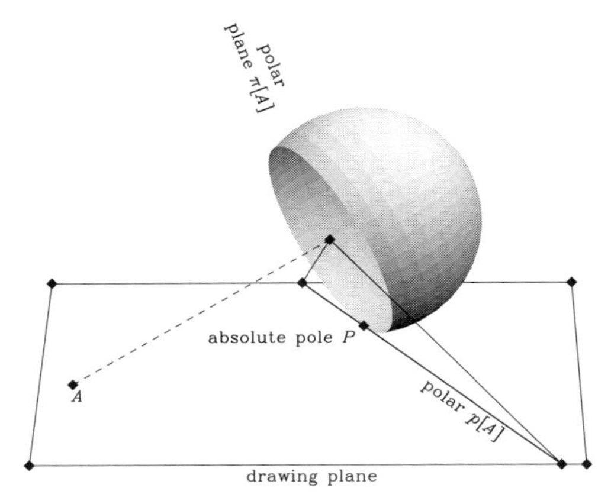

Figure 9.5: *Anti-Euclidean geometry.*

If the drawing plane just touches the auxiliary sphere, the contact point is the absolute pole P. The tangential cone of all points in the plane contains the connecting line to this pole. The pole carries each polar plane of the points in the drawing plane and therefore each polar. The connecting lines AP and the polars $p[A]$ determine an involution in the pencil carried by P. Here, this involution has no real fixed rays.

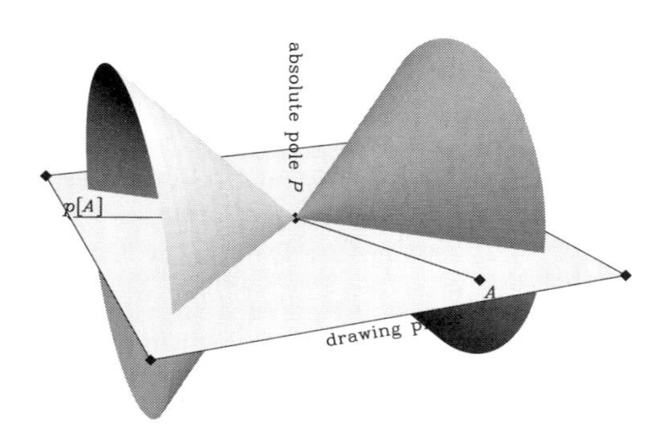

Figure 9.6: *Anti-Minkowski geometry.*

We substitute a hyperboloid for the sphere and let it degenerate into a double cone. Its vertex is assumed to lie on the plane. Then again it is an absolute pole P. Different points A have different polars all passing through P. They are found as the fourth harmonic ray to the two mantle lines in which the double cone intersects the drawing plane. These mantle lines are the fixed rays in the involution between AP and $p[A]$. The result is dual to the Minkowski geometry. We call it the anti-Minkowski geometry.

zero, which are the straight lines through the absolute pole. As in Galilean geometry, there is one side in each triangle that is equal to the sum of the other two. However, the curvature is positive. This *two-dimensional* anti-Euclidean geometry is degenerate although the *three-dimensional* quadric is not. In our representation, only the position of this quadric relative to the plane is special.

Evidently, it is all the same how the quadric looks like. Up to here, it could be a hyperboloid or even a cone. The results are the same and depend only on whether the quadric intersects, touches, or avoids the plane. An intersection is always a conic, of course. If it is not degenerate, we obtain the triplet of the hyperbolic, doubly hyperbolic, and antihyperbolic geometry (Figure 8.19). However, the cone yields new possibilities: Its vertex can lie on the plane. If the intersection is only one point, it yields the already known anti-Euclidean geometry. If the intersection is a pair of rays, the vertex is an absolute pole for all straight lines of the plane (Figure 9.6). On the other hand, the polars to all the points of the plane are rays through the pole. It is important that for the points on the two intersection rays the polars are themselves the intersection rays. That is, there again exist points that lie on their own polar. This case is dual to the Minkowski geometry, and is called the *anti-Minkowski geometry*. Here, we find curves of length zero. They are the rays through the pole. At last, the cone can touch the plane along one mantle line. This yields the Galilean geometry, as we will see now.

The remaining geometries are the ones with an absolute polar p. This absolute polar is the line at infinity or vanishing line, respectively. The geometries with such a universal polar satisfy the axiom of parallels. Two straight lines are defined to be parallel if they intersect on the polar. Parallel lines g have common perpendiculars that intersect at another point of the polar, the pole $P[g]$ of g. All poles lie on p. Consequently, the orthogonality defines an involution on the polar by assigning the intersection point gp to the pole $P[g]$. The polar with real or imaginary fixed points of this involution can be interpreted as a degenerate case of a conic section with infinite eccentricity.

In order to obtain these geometries in our framework, we now imagine that the quadric in the auxiliary space is flattened to a disk (Figure 9.7). The plane β of the disk intersects the drawing plane α in the line p. We now discover that the tangent cone of any point A of the drawing plane that does not lie on p touches the disk at its boundary, i.e., that the polar plane always coincides with β and that all polars $p[A]$ coincide with p, which is thus distinguished as an absolute polar. What can we do with the points A on p itself? Their tangent cone degenerates into a sector of the plane β that touches the disk at two points. The line connecting these two points meets the absolute polar p at a point $P[A]$. If A is the intersection pg of a line g with the absolute polar p, we obtain through $P[A] = P[g]$ an involutory map $P[A]$ of the points A of the line p. The polar structure associated with A is not a line but again a point $P[A]$, strictly the pencil of rays carried by this point. We obtain a geometry with the axiom of parallels and orthogonality defined by an involution on the absolute polar.

If the disk does not intersect the drawing plane, we obtain the *Euclidean geometry* (Figure 9.7). Orthogonality is defined by an involution of the polar without real fixed points. The orbits of the rotations (the circles) do not intersect the polar (at real points). The absolute conic is not real, and in addition it is degenerate. It contains two peculiar points, the imaginary points of the circle. Both lie on the polar in its complex extension. One cannot see them in the real plane but they are virtually present in the statement that the circle is already given by three real points (although a general conic section needs five points to be defined). We know that the characterization of a conic section as a circle requires to fix two periphery points in advance, i.e., the two fixed points of the involution on the absolute polar.

If the disk intersects the plane in a real segment, we obtain the *Minkowski geometry* or *pseudo-Euclidean geometry* (Figure 9.8): Again, all poles lie on the absolute polar. In contrast to the Euclidean geometry, the orthogonality is now determined by an involution with two *real* fixed points. These fixed points are the two *circular points at infinity* that lie on the periphery of any circle. The circular points at infinity (which are imaginary in the Euclidean geometry) are now present as the two asymptotic directions. In physical terms, the asymptotes of the circles are the lightlike lines in the space–time plane. They form two pencils of rays carried by the fixed points on the line at infinity, the polar. In the Minkowski geometry, the two circular points at infinity are real. Both geometries, the Minkowski geometry and the Euclidean geometry, exhibit the same universal polar (i.e., the horizon or the projective image of the line at infinity). Compared to the case of a general absolute conic, this is the sign

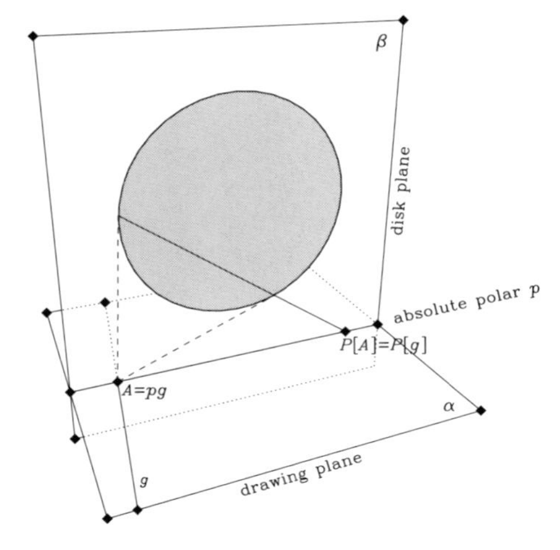

Figure 9.7: *Euclidean geometry.*

We consider the case of a sphere that degenerates into a disk. The plane of the disk is the absolute polar plane of the three-dimensional projective space, and its intersection with the drawing plane defines the horizon, i.e., the absolute polar p. This is the polar to any noncoincident point of the plane. At the same time, the poles of all straight lines of the plane lie on the absolute polar. The intersections gp define through these poles $P[g]$ an involution. Here, the disk does not intersect the drawing plane. Therefore, the involution has no real fixed point. Nevertheless, the involution can be obtained by a real construction using the disk. Here, the projective image of the Euclidean geometry is found.

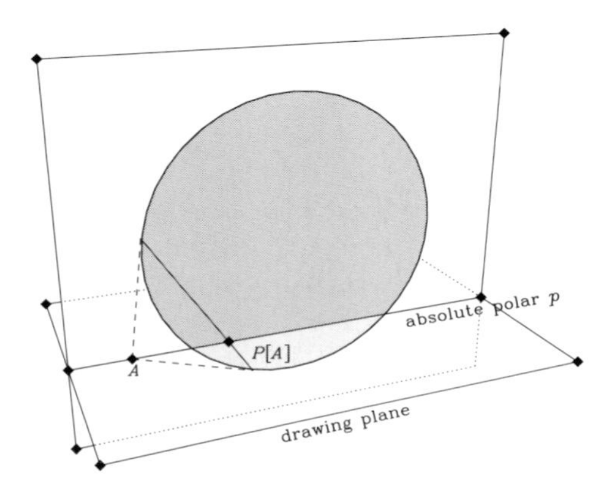

Figure 9.8: *Minkowski geometry.*

If the disk intersects the drawing plane, two fixed points of the orthogonality involution are found. This involution between the intersection points $A = pg$ and the poles $P[A] = P[g]$ shows the pattern of the Minkowski geometry.

of degeneracy. The straight lines have different poles, but these do not fill the plane. If we project the vanishing line to the infinity of our drawing plane, the pencil of perpendiculars to a straight line turns into a pencil of ordinarily parallel lines. The axiom of parallels and the degeneracy of the geometry are intimately related. Without the degeneracy of the projective-metric geometry, the axiom of parallels cannot hold.

If the disk touches the drawing plane in one point only, the orthogonal involution on the horizon has just *one* fixed point (to be counted twice because here the two fixed points of the former cases coincide). In this double point, the circles of the induced geometry touch the horizon. We find the projective representation of the *Galilean geometry* (see Figure 9.9). It is the third and the last geometry that conforms to the axiom of parallels. The first consequence of this axiom is the theorem of the equality of the step angles. Through this theorem, the sum of angles in a triangle is equal to a flat angle. All three geometries have no curvature. In

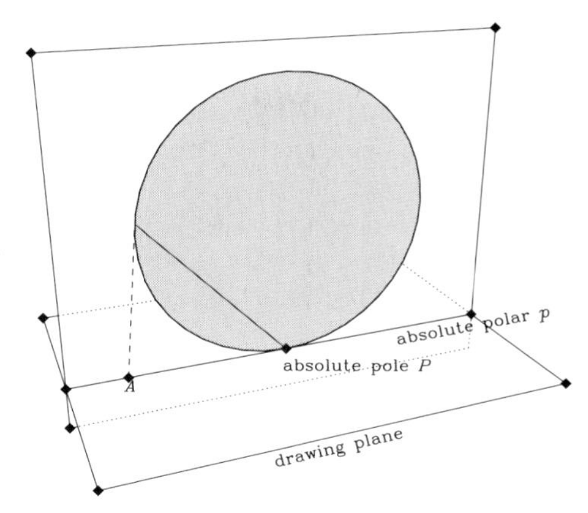

Figure 9.9: *Galilean geometry.*

Finally, we let the disk just touch the drawing plane. Now we not only find an absolute polar p for all points but also an absolute pole for all lines. This is the most degenerate geometry of the plane, the Galilean geometry.

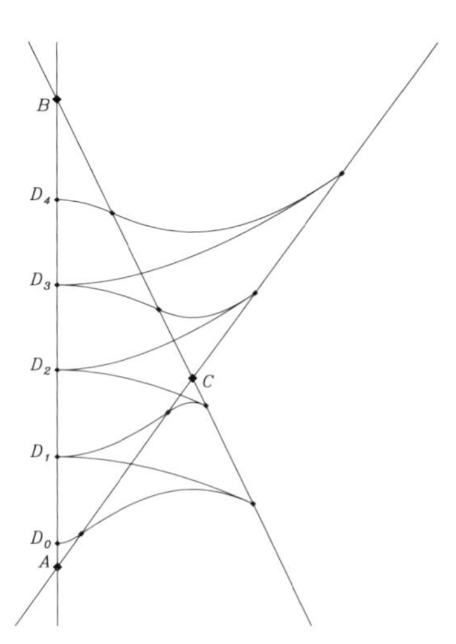

Figure 9.10: *The transport of lengths in pencils.*

This is the procedure that is dual to the transport of directions along the sides of a triangle (Figure 7.1). We draw a triangle of timelike lines in the Minkowski plane as in Figure 5.15. We begin with some point D_0. Its distance from A is transported to AC. Then the distance from C is transported to CB and, finally, the distance from B back to AB. We obtain D_1. The transport around the trilateral produces a translation from D_0 to D_1. It can be repeated, i.e., it is the same for all points on AB. The quantity of the translation is given by $d[D_0, D_1] = d[A, B] - d[A, C] - d[C, B]$.

addition, the Galilean geometry is the curvature-free exception between the anti-Euclidean and anti-Minkowski geometries. The involution in the pencil of polars (which is nondegenerate for them) here degenerates too.

The nine configurations built by the points and straight lines of the plane are summarized in Table 9.1. The relations between the straight lines g and the noncoincident points Q are used to characterize the geometry. In the geometries of the first column, we find those in which there are no parallels through a point Q to a line g when Q lies off g. The exterior angle is always smaller than the sum of the opposite interior angles of a triangle, and the sum of the internal angles is larger than the flat one. That is, the curvature is positive. In the geometries of the second column, the axiom of parallels holds. In each point Q, we find just one parallel h to a given line g that does not intersect g in the (metrically) finite. The external angle is equal to the sum of the opposite internal angles, and the curvature vanishes. In the geometries of the third column, we find many parallels to a line g through each point Q outside g that do not intersect g in the metrically finite. The external angle is larger than the sum of the opposite internal angles, and the curvature is negative.

In the geometries of the first row, the points of a line g all have a positive real distance to a point Q outside g. The ordinary triangle inequality holds; the sum of two sides is never smaller than the third. In the geometries of the second row, there is just *one* point on a line g that does *not* have a positive real distance to a point Q outside g, i.e., there is just one point on g simultaneous with Q. There exists a triangle equality: There is one side in any triangle that is equal to the sum of the other two. In the geometries of the third row, many points on a line g have no real positive distance to a given point Q outside g, i.e., there are *many* points (now relatively) simultaneous with Q. The pseudo-Euclidean triangle inequality holds, that is, in every triangle there is one side that is larger than the sum of the other two. The projective duality between points and lines determines a symmetry, which in Table 9.1 is indicated by the prefix *anti*.[3]

Table 9.1 shows how the classification through curvature and the classification through the triangle inequality are dual to each other. We have also seen that these two classifications can be characterized by the pairs of points and lines and the existence and uniqueness of parallels or point of zero distance. We supplement these considerations by giving a hint to the dual construction of the rotation of the tangent plane on a curved surface (known from Figure 7.1). The triangle with three connecting sides is considered now as a trilateral with three intersection vertices. Instead of transporting a *direction* along the connecting *lines*, we consider the transport of a *distance* around the *vertices* of intersection. The dual to the *rotation* of a direction is a *translation* of a point. Figure 9.10 shows the outcome in the Minkowski plane. The shift is given by the excess length of the longest side of the timelike triangle.

We now want to consider some details. First, we want to make sure of consistency by checking the theorem of the circumcenter and the orthocenter of a triangle. We will perform this check for a generic case, the de Sitter world. Figure 9.11 shows the necessary constructions of the theorem of the orthocenter, and Figure 9.12 for the circumcenter. Remarkably, each segment in our projective model has not only a proper midpoint, but also an improper one. We call it improper because it does not belong to the transitivity region of points chosen to constitute the particular geometry. Correspondingly, we find three improper circumcenters too and three supplementary circumscribed improper circles (all touching infinity). This is a fact dual to the well-known existence of three ex-circles besides the in-circle. In the geometries in which the axiom of parallels holds, the improper circles degenerate. The improper perpendicular bisectors all coincide with the horizon. We call the additional circles improper because they are mere orbits of rotations, and not loci of constant real distance to some center.

In the Euclidean and Minkowski geometries, the theorem of the angles at the circumference holds: The locus of all points C at which a given segment AB subtends a constant angle $\angle ACB$ is a circle. This theorem can hold only when the external angle is equal to the sum of the opposite angles. This is the case in geometries with the axiom of parallels, i.e., without curvature. We have already seen its modification in the Galilean geometry (Figure 6.10). In the generic case of nonvanishing curvature, the locus is a curve of fourth degree (Figure 9.13).

[3]The name anti-de Sitter world has another reason. It comes from cosmology and indicates the swapping of timelike and spacelike directions.

Table 9.1: The nine geometries of the plane

	Curvature positive No parallels[a] External angle less than the sum of the opposite internal angles Poles have no real tangents	Curvature zero One parallel External angle equal to the sum of the opposite internal angles Poles have one real tangent (the absolute polar)	Curvature negative Many parallels External angle larger than the sum of the opposite internal angles Poles have two real tangents
All distances[b] positive, Euclidean triangle inequality, polars do not intersect the absolute conic	Elliptical geometry (sphere)	Euclidean geometry (plane)	Hyperbolical (Lobachevski-) geometry (velocity space)
One distance zero, triangle equality, polars intersect the absolute conic in the absolute pole	Anti-Euclidean geometry	Galilean geometry (Newtonian mechanics)	Anti-Minkowski geometry
Many distances not positive, pseudo-Euclidean triangle inequality, polars intersect the absolute conic in two points	Anti-Lobachevski geometry (anti-de Sitter world)	Minkowski world (Einstein's mechanics)	Doubly hyperbolic geometry (de Sitter world)

[a]Through a point off a line.
[b]From the points of a line to a point off the line.

The proof of the property of the Feuerbach circle to pass through the nine points supposes the theorem of the angles at the circumference. Hence, it gets lost in this case.

We add a note on the conics that correspond to the Feuerbach circle. In any of the considered geometries, a triangle $\Delta A_1 A_2 A_3$ determines the intersection A_4 of the altitudes in such a way that each A_k is the intersection for the triangle of the other three. The four points form a complete quadrangle with six connecting lines (the sides and altitudes of any of the four possible choices) and three diagonal points (the foot points F_{12}, F_{23}, and F_{31} of the altitudes, which do not depend on the choice of the triangle). Let us now choose the triangle $\Delta A_1 A_2 A_3$. Its three sides $A_1 A_2$, $A_2 A_3$, and $A_3 A_1$ have two midpoints each (as indicated in Figure 9.12). We obtain the conic determined by the three feet F_{12}, F_{23}, and F_{31} and two midpoints of

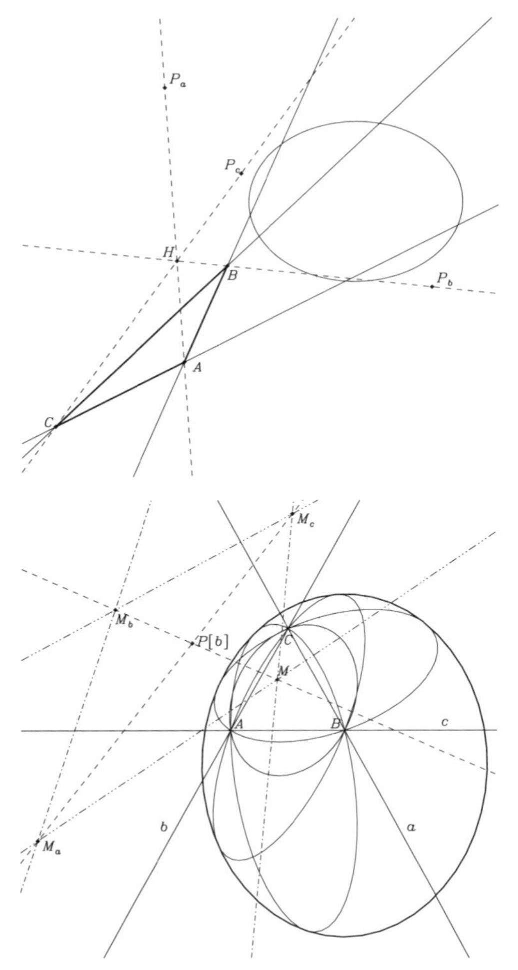

Figure 9.11: *The theorem of altitudes in general metrical geometry.*

To obtain only real construction elements, we draw a triangle ABC in the de Sitter geometry with its sides $c = AB$, $a = BC$, and $b = CA$, their poles P_c, P_a, P_b and the perpendiculars CP_c, AP_a, BP_b, which intersect at one point H. An algebraic proof is given in Appendix A, Figure A.7.

Figure 9.12: *The theorem of perpendicular bisectors in general metrical geometry.*

We see here a triangle ABC in the Lobachevski geometry. The broken lines are the perpendicular bisectors on the side $b = CA$, which intersect in the pole $P[b]$. The perpendicular bisectors (proper and improper) on c are chain lines, these on a are chain lines with triple dots. The poles of a and c lie outside the image. Apart from the expected circumcenter, there exist three other intersections of perpendicular bisectors; they belong to three other circumscribed (improper) circles that all touch infinity.

different sides pass through one of the midpoints of the third side. This can be proven, for instance, through the use of the calculus presented in Appendices C and D. That is, we obtain in general four such conics for each of the four triangles that can be chosen from the complete quadrilateral. No Feuerbach circle exists. Instead, the line connecting two bisection points on two different sides passes through one of the two bisection points on the third. The six midpoints are the vertices of a complete quadrilateral. In the geometries with absolute polar, all four triangles have one of the conics in common. This is the 11-point conic of Figure 8.12. In the geometries with absolute pole, one of the four points is always this pole. It remains only one triangle to choose and four conics of the kind that we considered. Figures 9.14 and 9.15 show the case for the anti-Euclidean geometry.

We now turn our attention to the pictures that we obtain by describing the orbits of rotations. A rotation about a center Q is found by consecutive reflection at two rays through Q. A point R keeps its distance to Q, i.e., it remains on the circle around Q on which it was

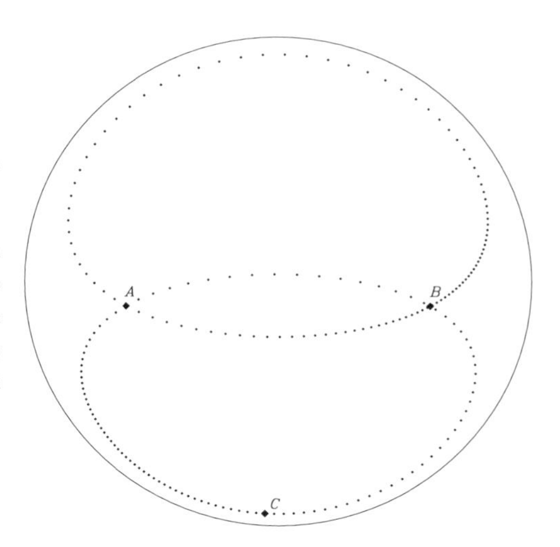

Figure 9.13: *The circle of the second kind in the general metrical geometry of the plane.*

We see here an example of a curve defined as the locus of the points at which the basis AB subtends a constant angle. The size of this angle can be defined, for instance, by a third point C. It is, in general, only a particular segment of the curve that admits the circumference property.

before the rotation. The angle of rotation is determined by the angle formed by the two mirror rays. In a rotation, the points move on circles around the rotation center Q (Figures 6.3 and 6.4). We obtain the image of these circles when we draw the orbits of rotation. We show here the projectively different cases. The completely nondegenerate real case is the de Sitter geometry (Figure 9.16). The center of rotation is a point Q outside the absolute conic section $\mathcal{K}$. The tangents from Q to this conic separate timelike from spacelike rays. The rotations cannot change the character of the rays and cannot move the tangents to the conic. Therefore, the circles cannot intersect these tangents. The absolute conic, the center Q, and its polar $p[Q]$ are preserved too. In addition, all circles pass through the points of contact $B[k[Q]]$. They touch there the conic and also the tangents. Summarizing, the circles around Q are conic sections with four given elements: They form a one-parameter congruence of curves, as expected. We could generate the conic sections pointwise using the construction of Figure C.6 ($AB = k_1[Q]$, $DE = k_2[Q]$, $QC =$ radius). No point will pass across the preserved curves $\mathcal{K}$, $p[Q]$, $k_1[Q]$, and $k_2[Q]$. There are no closed orbits. In fact, all rotations are translations along the orbits in the direction of one of the points of contact. The orbits inside $\mathcal{K}$ are circles with imaginary radius.

In the next case, we put the center of rotation in the interior of the absolute conic (Figure 9.17). All the orbits are now closed in the projective sense and can be traversed many times. We already know this property from the Euclidean geometry, but we find it in the elliptical and in the Lobachevski geometry too. The polar of the center is the only straight orbit. The Euclidean geometry (in a projection that represents the line at infinity as polar in the finite) shows the same configuration of orbits. The only difference is that the previously absolute conic is now an ordinary orbit.

We can also imagine a rotation about an infinitely distant point. Such a point Q lies on the absolute conic. The tangent to Q is its own polar $p[Q]$ (Figure 9.18). All orbits now touch the polar at the point Q.

In the case when the absolute conic has no real point, we obtain the orbits of elliptical geometry. The configuration is identical to that of hyperbolic geometry. The only difference

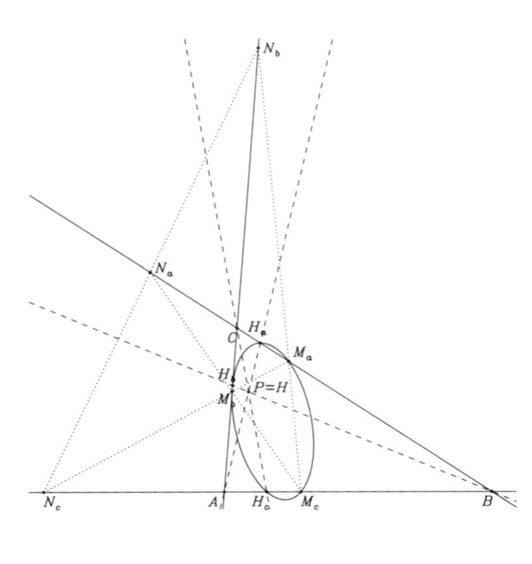

Figure 9.14: *A Feuerbach conic in the anti-Euclidean geometry.*

In the anti-Euclidean geometry, an absolute pole P exists. All perpendiculars pass this point. The altitudes of any triangle meet there ($P = H$). It is a simple task to find the altitudes. An interval (for instance, $c = AB$) is bisected by a pair of points (M_c and N_c) in harmonic position to the endpoints as well as to the intersection of the line with the absolute conic. The absolute conic is degenerate in our case (Figure 9.5), that is, the connections of M_c and N_c with $P = H$ have to be (by Euclidean measure) perpendicular. We obtain six midpoints. Three of them lie on the same intervals as the foot points of the altitudes. They lie with these foot points on a common conic. The midpoints of the altitudes are not defined (see Figure 8.12).

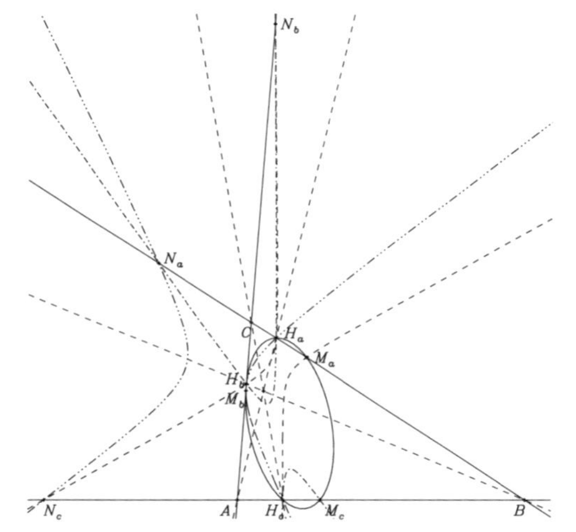

Figure 9.15: *The four Feuerbach conics of a triangle in the anti-Euclidean geometry.*

This is the same configuration as in the previous figure. The auxiliary lines are left out and the other three Feuerbach conics, all hyperbolas, are added. The expected conic is $[H_a, H_b, H_c, M_a, M_b, M_c]$, the other three are $[H_a, H_b, H_c, M_a, N_b, N_c]$ (broken line), $[H_a, H_b, H_c, N_a, M_b, N_c]$ (chain line with triple dots), and $[H_a, H_b, H_c, N_a, N_b, M_c]$ (chain line).

is again that the conic which represents there the absolute conic is now an ordinary orbit now. The configuration does not change even when the imaginary absolute conic degenerates into a line (on which it determines a polar involution without fixed points) and the Euclidean geometry results. The only difference is that the polar, which in elliptical geometry still varies with the center of rotation, is now the same absolute polar.

If, in addition, the center Q of rotation lies on the polar, the circular orbits degenerate into a pencil of lines through the point $P[Q]$ on p that is conjugate to Q (Figure 9.19). This conjugate point is determined by the absolute involution on the polar. It varies with Q as long as this involution is not degenerate (Euclidean and Minkowski geometries), and it is absolute in the Galilean geometry.

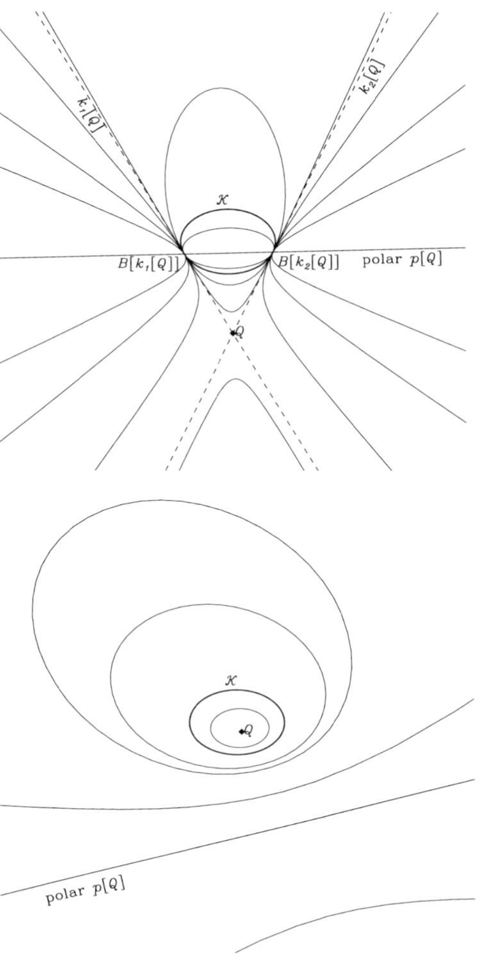

Figure 9.16: *Rotations in the de Sitter geometry.*

The rotations around the point Q are generated by the rays of the pencil carried by Q. They preserve the absolute conic section $\mathcal{K}$, the tangents $k[Q]$ to it, and the points of contact $B[k[Q]]$. The point Q and the points of contact are, respectively, a fixed point and singularities of the congruence of orbits. The rotation is a motion in the direction of one of these fixed points, i.e., a generalized translation. In the neighborhood of the center of rotation, the picture is that of a Lorentz transformation. That is, the farther away the absolute conic is, the less important is the requirement that it is nondegenerate.

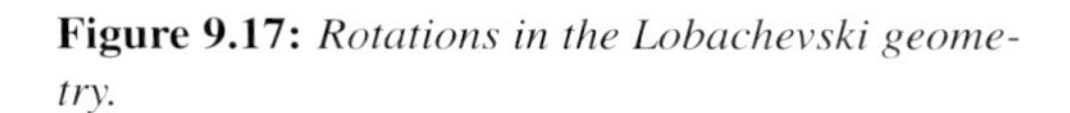

Figure 9.17: *Rotations in the Lobachevski geometry.*

In contrast to Figure 9.16, the center of rotation now lies inside the absolute conic. The tangents are no longer real and cannot restrict the orbits of rotation. The orbits are (projectively) closed, both in the inner region and in the outer region.

In the case of Minkowski geometry, the configuration of Figure 9.16 results. The only difference is that the absolute conic degenerates into the segment of the absolute polar between the absolute points of contact (i.e., the fixed points of the involution). If the two points coincide, we obtain the Galilean geometry and the orbits in the form of the pencil of rays through the pole that is now absolute (Figure 9.19). The same is true if the real absolute conic degenerates into a point (anti-Euclidean geometry) or into a pair of lines (anti-Minkowski geometry, see Table 9.2).

In this chapter, we have seen that we obtain a metric plane from a two-dimensional projective plane through a polarity that is defined by a quadric in a three-dimensional space. This fact suggested the idea that general relativity can be extended by a projective theory in a five-dimensional space [83–86]. In order to model a variable gravitational field, we must then conceive an *inhomogeneous* five-dimensional space in which different quadrics are attached to its points. The Einstein equations on this field of quadrics are equivalent to the Einstein–

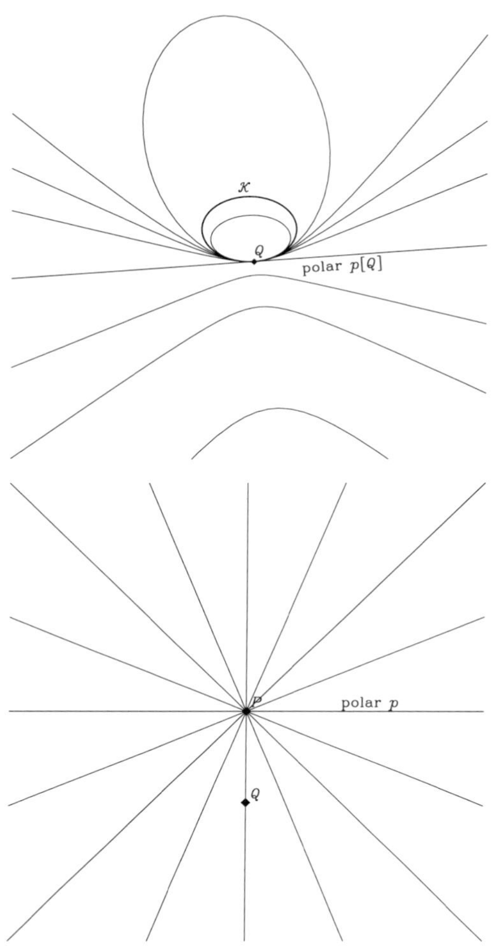

Figure 9.18: *Rotation about infinitely distant points.*

In the limiting case corresponding to the transition from Figures 9.16 to 9.17, we put the center of rotation on the absolute conic itself, that is at metrical infinity. In these rotations, the absolute conic, the center of rotation, and the polar are preserved. The orbits in the lower part do not seem to touch the polar. However, we must remember that these orbits are hyperbolas for which at most one branch touches a straight line. This second branch lies in the upper part and forms with the lower branch a projectively closed curve.

Figure 9.19: *Rotations about a point of the absolute polar.*

The orbits of rotations become a pencil through a pole if an absolute pole exists or if the center of rotation lies on an absolute polar.

Table 9.2: The orbits of rotations

	Elliptic geometry	Euclidean geometry	Hyperbolic geometry
Center in the finite:	Figure 9.17	Figure 9.17	Figure 9.17
Center at infinity:	Figure 9.18	Figure 9.19	Figure 9.18
	Anti-Euclidean geometry	Galilean geometry	Anti-Minkowski geometry
Center in the finite:	Figure 9.19	Figure 9.19	Figure 9.19
Center at infinity:	Figure 9.19	Figure 9.19	Figure 9.19
	Antihyperbolic geometry	Minkowski geometry	Doubly hyperbolic geometry
Center in the finite:	Figure 9.16	Figure 9.16	Figure 9.16
Center at infinity:	Figure 9.18	Figure 9.19	Figure 9.18

Maxwell equations for the coupled gravitational and electromagnetic fields. This confirms the relations that we have considered. A more detailed discussion would go beyond the aim of this volume.

10 General Remarks

10.1 The Theory of Relativity

In our excursion to the frontier between geometry and physics we were forced to cite many facts of physics without detailed observational motivation as well as many theorems of geometry without rigorous mathematical proof. The many details presented in the previous chapters shall be reviewed at last from a more general point of view.

The theory of relativity, which we cited many times as a physical counterpart to the Minkowski geometry, is a physical theory like many others, and it must be checked by an experiment that tests the applicability, not the consistency. The question of consistency is a matter of mathematics. We grant that the applicability can be shown at best in a given framework of assumptions, just as in mathematics consistency can be proven only in a given formal framework. The task of mathematics is not only correct calculation and logical deduction but also a program for establishing deeper and deeper foundations. Remarkably, the question of applicability remains unsolved as long as the limits of applicability are not known. An experiment with positive conclusion says something about the applicability in the given circumstances, in a given range of parameters such as velocities, energies, masses, charges, and temperatures. The extrapolation to other circumstances, which may occasionally be extreme, can be envisaged but has to be checked anew. Only an experiment with negative outcome can be final. This would have a positive aspect too. Namely, if we know the circumstances where the applicability begins to fail, we learn about the circumstances where it can be presumed without further doubt. For example, the negative result of the Michelson experiment informs us that the Newtonian mechanics cannot be applied for velocities comparable to the speed of light. It also tells us that we can trust the Newtonian mechanics for velocities much smaller than the speed of light and that the errors will be of the order $O[v^2/c^2]$.

Besides the quantitative consequences of a theory, which are tested by quantitative and consequently never ultimately exact experiments, we also know qualitative consequences that can be compared directly with fundamental experiences and which play a much more important role because they are not affected by small errors. For the theory of relativity, an example of such a qualitative statement is the existence of antiparticles. This existence was predicted by the theory. It is not a small effect but a structural necessity for consistently handling the fundamental equation for the energy of a free particle. This equation results from the fact that the rest mass is a characteristic of a particle and is proportional to the energy. The equivalence of mass and energy determines the time component of the momentum:

$$m^2c^2 - \boldsymbol{p}^2 = m_0^2c^2, \qquad E = mc^2 \quad \rightarrow \quad E^2 = c^2(m_0^2c^2 + \boldsymbol{p}^2).$$

The Geometry of Time. Dierck-E. Liebscher

ISBN: 3-527-40567-4

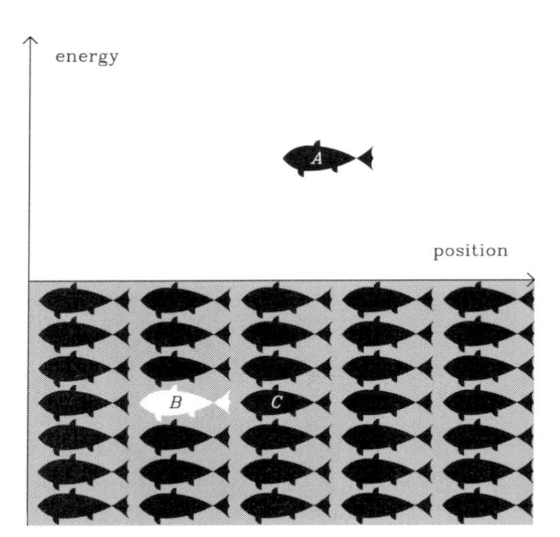

Figure 10.1: *Antiparticles as holes.*

In a dense swarm of fishes, the individual fish cannot move against the swarm. When we take out one (A), a hole (B) is created. The fish C can now swim into position B, and the hole will then move to position C, opposite to the direction of the attraction that caused the fish at C to swim into position B. The charges of fish and hole are opposite. The effective masses are the same: The acceleration of the hole is equal to that of the fish. If the fish A falls back into the swarm, he takes the position of the hole. The hole vanishes, and the fish looses its capability to move freely. So it can be found no longer as an individually moving object.

This is a quadratic equation and admits negative solutions for the energy. These solutions correspond to states with negative energy for free particles. When this equation is valid, the solutions of negative energy must be taken into account, too. The effects of the existence of such states are similar to what we considered in connection with tachyons. If states of arbitrarily large negative energy can be assumed by a particle, no stable equilibrium will ever be possible. However, we observe equilibria in many places. In fact, to perform slow and primitive measurements the existence of equilibria is indispensable. Consequently, there must be some reason why these states do not interfere freely with the observed states of positive energy. It was Dirac's conjecture that the states of negative energy are *occupied* and therefore passive. That seems to be an excuse, but it has testable consequences. States of negative energy that happen to be unoccupied (a kind of holes) then behave like particles of equal mass but opposite charge (Figure 10.1). We call these states antiparticles. If an ordinary particle with positive energy makes a transition to such an unoccupied state, the hole vanishes. The particle becomes passive and vanishes too. The energy is shifted to other degrees of freedom, for instance, to photons. Particle and antiparticle combine to "vanish," i.e., to be transformed into other particles that carry away their energy, momentum, and angular momentum with the speed of light. The factual observations that antiparticles exist, that their masses are equal to the masses of the corresponding particles [87], and that they have opposite charges with the consequence of the possible creation (see Figure 2.7) and annihilation of particle–antiparticle pairs constitute a qualitative confirmation of the theory of relativity and makes it an irrevocable theoretical insight of modern physics.

The limits of special relativity are known: They are determined by the gravitational field. Under the influence of gravitation, the flat Minkowski world becomes an approximation that holds only locally. We must reinterpret Figures 7.21 and 7.23, and so on: The Minkowski geometry holds only in the tangent planes to the hyperboloids that represent the curved universe. For the hyperboloids as well as for any curved space–time, the light-cone structure of the world remains, but the metric begins to vary with location and time by small amounts. The quantitative changes with respect to relativity without gravitation are of the order of the

A Reflections

There are two reasons why we should free all notions that we use in calculations from visualization in everyday life. First, visualization with all its many associations may mislead us into drawing an incorrect conclusion. Second, we would like to obtain logical structures that can be applied to genuinely different physical objects and situations, not only to the obvious. Paradoxically, the abstraction enlarges the realm of application.

Visualizations often lead us to accept as obvious relations that should not enter a logical conclusion. The history of relativity is a famous example which shows us that we have to abandon the visually suggestive prejudice of a mechanical aether as carrier of light to reach a clear understanding and correct predictions. If we intend to travel with our theories into realms beyond the dimensions of everyday life, many things should not be in the luggage. Usually, these things are found out only step by step. The axiomatic method, which substitutes implicit definitions for the visually explicit ones, is intended to eliminate hiding travellers from the start. The notion *implicit* means here to explicate a *set of properties* that characterize an object and to found the exploration solely on these properties. However, abstraction from visual impressions does not mean we cannot obtain its orientations from physical experiments or visual impression (footnote 2, Chapter 2). We have already used the sphere as well as the hyperboloid to find such an orientation beyond Euclidean geometry.

We investigate here the representation of operations that allow us to establish the equality of objects independently of the location and orientation. These operations are called motions. Motions provide the means to decide whether internal structures are equivalent. Because we intend to abstract from particular procedures, we characterize motion by only very few properties. First, we must assume that motions can be combined to give motions again. This is the counterpart of the logical transitivity of equivalence relations. The symmetry of these relations has a counterpart too: Because in an equivalence there is no abstract means to tell which is the moved object and which is the object to be moved, the reversal of a motion is also a motion. In a sequence of motions, any combination of successive motions must be possible. Taken all together, motions form a group. That is:

1. The composition of two motions is again a motion. If not, we could not speak of equivalence.

2. The fact that motions are conceived as the transition from one state to another means that in a series of successive motions factors can be combined at will if their order is not changed. Consecutive motions can be combined at will.

3. The motion back is included too; it serves as the inverse motion that after composition with the original one always yields the original state.

4. Consequently, "changing nothing" has to be considered as a motion too. This motion, which moves nothing at all, is the identity, or neutral element, of our group.

The Geometry of Time. Dierck-E. Liebscher

ISBN: 3-527-40567-4

Usually, we imagine motions as operations in some space in which embedded objects change their place and orientation. Abstracting from this visualization, we can interpret any group as a group of motions by considering the motion generated by the group on itself. The main example for such a procedure are the transformations. A transformation (of the group $\mathcal{G}$) with the element $a \in \mathcal{G}$ is the motion that maps each element $g \in \mathcal{G}$ to $\mathcal{T}_a[g] = a^{-1}ga$. This map preserves the full structure of the group: It is a motion.

The object that we actually move does not affect our definition or our calculation. From the mathematical point of view, the calculation reveals and unfolds the abstract structure of the group of motions. This group determines which objects are congruent and which are not, i.e., it determines the geometry. The application to physical objects requires an interpretation of this structure as well as its experimental verification, which are both subject to characteristic uncertainties.

Surprisingly, the notion of motion can be reduced to the notion of reflection (on straight lines in the plane, on planes in space) although reflections are not really motions in the ordinary visual sense. A reflection cannot be realized by the physical motion of a tangible object.[1] Nevertheless, this reduction is possible, and reflections generate the motions in the form of even products of reflections. The notion of reflection becomes fundamental for metric geometry. Reflections already permit the transport and comparison of segments and angles.

1. Let $S[A]$ be the image of the point A; then the mirror is the locus of all points that are equally distant from A as from $S[A]$.

2. Let Q be any point on the mirror; then the angles between the mirror and the lines QA and $QS[A]$ are equal.

This seems to be intuitively obvious. We shall construct our abstraction keeping this goal in mind. Curiously, the mirror of everyday life does not allow this comparison: We compare the reflected image with the reflected meter-stick, not with the meter-stick itself. In order to see that this does not affect the geometrical construction, we use *half-transparent* mirrors to perform a comparison of a virtual image with a real object. With this preparation, we can also see that any reflection produces the initial situation when repeated. We say that reflections are involutive maps. It is precisely this property that *defines* the reflections in the axiomatic approach [49]. The abstract definition concerns only the algebraic relations. We prepare the interpretation in ordinary geometry by prescribing which algebraic expressions are taken as points and straight lines.

The elementary reflection in the plane is the reflection on a straight line, and for every straight line a reflection should be defined. At least some part $\mathcal{S}$ of the reflections can be taken as the representative of straight lines. It turns out that this part $\mathcal{S}$ can be chosen as a generating system that generates all other reflections and the whole group of motions by successive multiplication. That is, the elements of the group of motions are just the products of the generating reflections. Points too can be defined as involutory elements because the product of two generating reflections can be involutory itself. In this case, a point is defined by this product. In the Euclidean plane, we can see immediately that the successive reflection

[1] If we extend the plane or space by an additional dimension, the reflection on a line or on a plane, respectively, changes into a rotation through an additional dimension, i.e., a real motion. The reflection on a line of a plane is thus embedded in a (involutory, of course) rotation of the space on a line about the flat angle.

on two perpendicular straight lines is a rotation about the intersection point through a flat angle (Figure A.1). There is an important theorem which states that all products of reflections can be represented by a product of at most three generators (a reflected rotation). Each product of an even number of points is the product of two generators (a rotation).

After these considerations, the subject of geometry is a group $\mathcal{G}$ of motions generated by a set $\mathcal{S}$ of involutory elements. This generating system $\mathcal{S}$ is to be invariant under the transformations defined above (because we intend to interpret the generators as *straight lines*, and straight lines remain straight under transformations of our geometry):

$$g \in \mathcal{S}, a \in \mathcal{G}: \quad \rightarrow \quad a^{-1}ga \in \mathcal{S}.$$

The elements of $\mathcal{S}$ can now be safely taken as the straight lines of the geometry. Hence, they are denoted by small letters.

Building on the notion of a straight line, we now define algebraically the *point*: The involutory maps in $\mathcal{G}$ that can be represented by the product of two generators in $\mathcal{S}$ are called points and denoted by capital letters. The visual apprehension tells us that the product of two reflections is a rotation. A rotation is involutory when the angle of rotation is flat. For the moment, the two straight lines define a point only when they are perpendicular. Therefore, we define perpendicularity of two lines by their product being involutory. Two straight lines g, h are perpendicular, $g \perp h$, precisely when their product is involutory, $ghgh = 1$. As we must expect, it follows from $h = h^{-1}$, $g = g^{-1}$ and $hghg = 1$ that $h = ghg = g^{-1}hg$: The straight line h coincides with the reflected image $S_g[h] = g^{-1}hg$ on g. This corresponds completely to our visual apprehension, which is shaped by Euclidean relations. If we connect the point A with its reflected image $g^{-1}Ag = gAg$, the connecting line $h = (A, gAg)$ is perpendicular to g because the reflected image ghg of h connects the same points and therefore coincides with h (if A and gAg are different). Precisely when A and $g^{-1}Ag$ are equal, A lies on g itself. This means nothing else but that $gA = Ag$, or that gA is again a reflection on a line h, which itself is perpendicular to g, i.e., $ghg = ggAg = Ag = gA = h$. Only if the point of rotation lies on the reflecting line we can expect that it does not matter whether the rotation about a flat angle is performed before reflection on the line or after (Figure A.2).

Let us now consider the treatment of straight lines that do not intersect in the finite. We abstain in the intersection theorems from the intersection point itself and define a pencil of lines instead: Lines lie in a pencil if they pass through a common point or have a common perpendicular.[2] If the product of three straight lines that lie in a pencil is again a straight line (i.e., a generator of the group), we obtain the theorem of perpendicular bisectors, which is so important for the consistency of length comparison. Therefore, we demand an axiom for the generating set $\mathcal{S}$: If three straight lines lie in a pencil, their product is again a straight line:

$$abc = d, \qquad ab = dc. \tag{A.1}$$

In this case, we say that b, d lies in a position symmetric under reflection with respect to a and c (Figure A.3). If $ab = da$, then a is the line on which b is reflected into d ($b = a^{-1}da = ada$). The geometry becomes trivial if all lines are perpendicular to each other. We need a

[2] In visual apprehension, we directly see a point, the carrier of the pencil. However, this point does not necessarily belong to the geometry, just as the points at infinity are only visible in the projective extension.

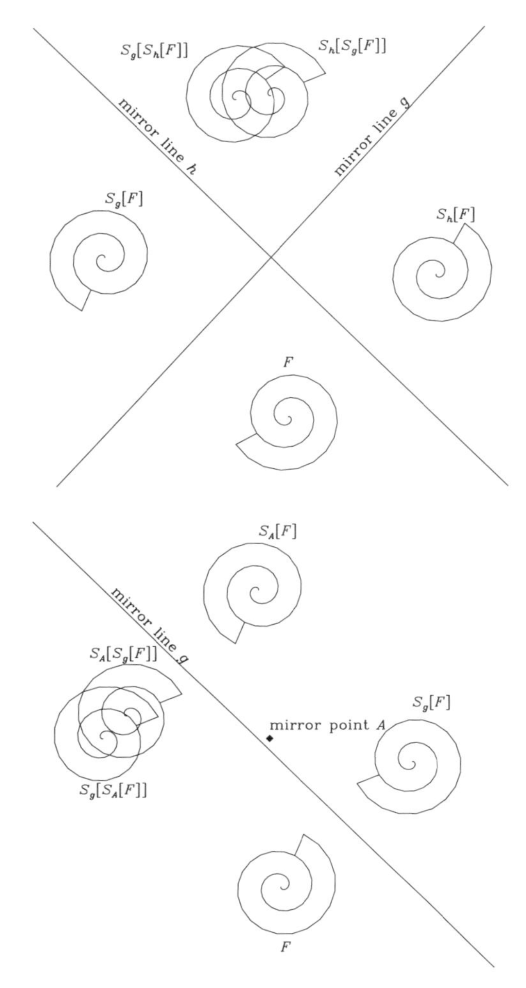

Figure A.1: *The point as a product of two straight lines.*

If in the Euclidean plane two straight lines are perpendicular, successive reflections rotate the image around the intersection point, while repetition of the operation reproduces the initial position of the objects. The intersection point is again a reflection but does not belong to the generating system. The product of two reflections on straight lines that are not perpendicular is not a reflection. In this figure, the objects $S_h[S_g[F]]$ and $S_g[S_h[F]]$ differ by the amount by which the lines h and g deviate from the perpendicular position (compare with Figure 3.1).

Figure A.2: *A point incident on a line.*

When the rotation point lies on the reflecting line, it does not matter whether the rotation about a flat angle is performed before reflection on the line or after. In this figure, the objects $S_A[S_g[F]]$ and $S_g[S_A[F]]$ fail to coincide by the amount by which A and g deviate from incidence. When A lies on g, $S_A[S_g[F]]$ and $S_g[S_A[F]]$ coincide.

supplementary axiom that tells us that there are lines *not* perpendicular to the lines of some perpendicular pair.

It is sufficient to choose five axioms of reflection to lay the foundations for the construction of the geometry [49].

1. Any two points A, B are supposed to determine a connecting line $g = (A, B)$ (i.e., $AgAg = 1$ and $BgBg = 1$, denoted by $A, B \mid g$).

2. If two points are connected by two lines, $A, B \mid g, h$, either the points or the lines coincide.

3. If three lines are concurrent, i.e., $a, b, c \mid A$, a line d exists with $abc = d$.

4. If three lines a, b, c have a common perpendicular, $a, b, c \mid g$, a line d exists with $abc = d$.

5. There exist three lines g, h, j for which g is perpendicular to h, but j is neither perpendicular to g nor to h.

These axioms ensure the existence of a unique connection between two points, the uniqueness of an intersection point, and the existence of a reflecting line for three lines in a pencil. Finally, nontriviality is also ensured.

The remaining part of this appendix is used to explain and show how to calculate formally with the defined reflections. We shall show this by calculating the intersection point of the perpendicular bisectors and the intersection point of the altitudes of a triangle. We recall that the intersection of the perpendicular bisectors ensures the consistency of the concept that reflections transport segments without change of length. In a hypothetical construction of involutive maps in which the perpendicular bisectors do not meet at one point, the interpretation that these maps transport segments without change of length cannot be consistent.

First, we show how to construct a line g that lies in a pencil with two others, m_a and m_c, and passes through a point B outside both lines (Figure A.4). To this end, we reflect B on m_a ($m_a B m_a = C$) and m_c ($m_c B m_c = A$). The connecting lines $a = (B, C)$ and $c = (B, A)$ are perpendicular to m_a and m_c, respectively, and define the points $M_a = a m_a$ and $M_c = c m_c$. We then reflect the point B on the connecting line $d = (M_a, M_c)$ and obtain the perpendicular $e = (B, dBd)$ from B to d. All three lines a, e, c pass through the point B. Axiomatically, their product $g = aec$ is a line passing through B too. We can show that this line g lies in a pencil with m_a and m_c [49]:

$$m_a g m_c = m_a a e c m_c = M_a e M_c = d d M_a e M_c = d M_a d e M_c = d M_a e d M_c.$$

This product is again a line,

$$d M_a e d M_c = h,$$

because dM_a, e, and dM_c are all perpendicular to the same line d, i.e., lie in another pencil too and produce a fourth line h. Consequently, the line g meets the aim of the construction.

Let us now assume that we have constructed the perpendicular bisectors m_a and m_c in the triangle ΔABC (Figure A.5, [49]). Then by definition

$$m_c A = B m_c, \qquad C m_a = m_a B.$$

We now draw the line g through B that lies in a pencil with both perpendicular bisectors (i.e., that passes through the common point of the two perpendicular bisectors in the simplest case). Then we obtain on one hand

$$gB = Bg,$$

while, on the other hand, the product of the three lines m_a, g, and m_c is a straight line by our axiom:

$$m_a \, g \, m_c = h. \tag{A.2}$$

This line h is the perpendicular bisector of the segment (A, C):

$$hA = m_a g m_c A = m_a g B m_c = m_a B g m_c = C m_a g m_c = Ch.$$

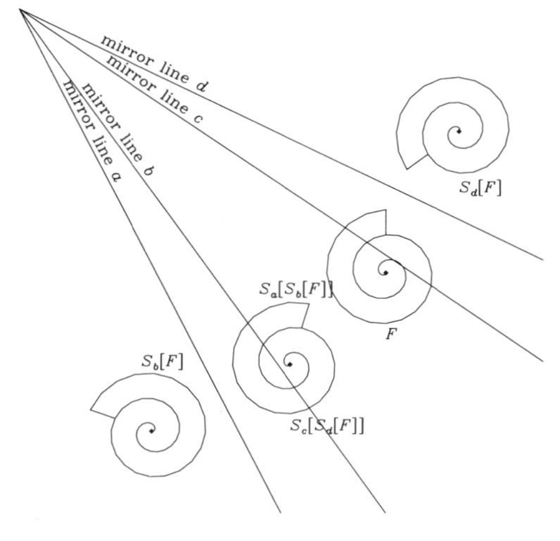

Figure A.3: *Straight lines in a position symmetric under reflection.*

The product of three reflections on lines lying in a pencil is again a reflection, and the corresponding straight line lies in the same pencil, $S_c S_a S_b = S_d$.

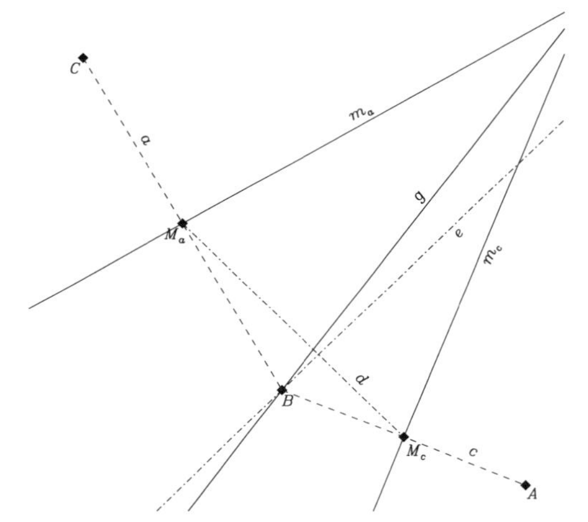

Figure A.4: *The connecting line to a virtual origin of a pencil.*

Two straight lines m_a and m_c are given. We seek a line g that lies in a pencil with both straight lines and passes through the point B. We determine first the reflection images $A = m_c B m_c$ and $C = m_a B m_a$ and continue by constructing the midpoints $M_a = a m_a$ (of the line a connecting B and C) and $M_c = c m_c$ (of the line c connecting B and A), which we connect by d. Now, we drop the perpendicular to d from B and find the line g in the position symmetric under reflection to e if we refer to the connecting lines $a = [B, C]$ and $c = [B, A]$, i.e., $ae = gc$, Eq. (A.1).

This completes the proof of the theorem of perpendicular bisectors. Conversely, we can show that the product of three lines in a pencil is again a line if we can assume this theorem. We simply repeat the construction in the opposite way.

Before arriving at the orthocenter theorem, we show that the product AgB of two points A and B and a line g yields a line h exactly if g is perpendicular to the connecting line of A and B (Figure A.6). Here, we consider only the nontrivial case when $A \neq B$. On the connecting line c, we draw the perpendiculars $q = Ac$ and $r = Bc$. If AgB is a line, $cAgBc = qgr$ is a line too. Both happen precisely when q, g, r lie in a pencil, i.e., if g is also perpendicular to h. Consequently, the product AgB is also perpendicular to c. An analogous statement can be proven for the product gAh. It yields a point B iff g and h have a common perpendicular and A lies on this perpendicular.

We can now prove the orthocenter theorem (Figure A.7, [49]). In a triangle ΔABC with $a = (B, C)$, $b = (C, A)$, $c = (A, B)$ we drop the altitudes h_a, h_b, h_c as connecting lines

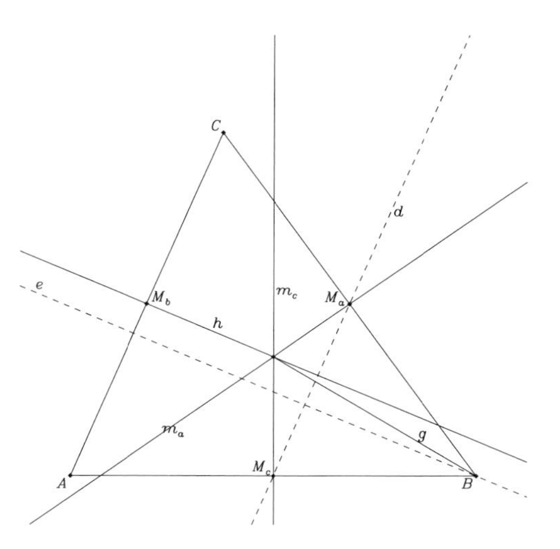

Figure A.5: *The theorem of perpendicular bisectors.*

For the two perpendicular bisectors m_c and m_a, we draw the line g through B that lies in a pencil with both. Axiomatically, the product of the three lines m_a, g, and m_c is again a straight line h. It can be shown that this line is the perpendicular bisector of $[A, C]$.

$h_a = (A, aAa)$, $h_b = (B, bBb)$, and $h_c = (C, cCc)$. The feet points are obtained as products, $F_a = ah_a$, $F_b = bh_b$, and $F_c = ch_c$. We now denote the products of the three lines that meet in each vertex by $k_a = bh_ac$, $k_b = ch_ba$, and $k_c = ah_cb$. The lines h_a and k_a lie reflection-symmetric with respect to b and c and so forth. Next, we determine five auxiliary lines by reflecting k_a on b and c, which yields the lines $r_b = bk_ab$ and $r_c = ck_ac$ and (if the sides of the triangle are not pairwise perpendicular, i.e., if $abc \neq 1$) by dropping from F_a the perpendiculars $s_b = (F_a, r_b)$ and $s_c = (F_a, r_c)$ to these two lines. The fifth line is given by the product $t_a = ak_ba$. We now begin the algebraic calculation. First, the perpendiculars s_b and s_c lie reflection-symmetric to a:

$$as_ba = aF_as_bF_aa = h_as_bh_a = h_a(F_a, r_b)h_a = (h_aF_ah_a, h_ar_bh_a) = (F_a, r_c) = s_c,$$

because

$$h_ar_bh_a = h_abk_abh_a = h_abbh_acbh_a = cbh_a = cbh_acc = ck_ac = r_c.$$

Secondly, we check by calculation that

$$F_ar_cF_c = ah_ar_cch_c = ah_ack_acch_c = abbh_ack_ah_c = abk_ak_ah_c = abh_c = ak_ca$$

is a straight line. Consequently, the point F_c lies on the perpendicular dropped from F_a to r_c, i.e., on s_c. Thirdly, we show that k_b is also perpendicular to s_c by calculating the product

$$r_bF_at_a = bk_abah_aak_ba = bk_abh_ak_ba = bk_ak_ack_ba = bck_ba = bh_b = F_b.$$

Consequently, the line $t_a = ak_ba$ is orthogonal to the perpendicular $(F_a, r_b) = s_b$, and the line $k_b = at_aa$ is perpendicular to $as_ba = s_c$. Now both F_a and F_c lie on s_c, and k_b is perpendicular to s_c. Hence, the product $F_ak_bF_c = u$ is a line. In addition, it is identical to the line

$$h_ch_bh_a = h_cch_baah_a = F_ck_bF_a = u.$$

Consequently, the three altitudes lie in a pencil. That has to be shown.

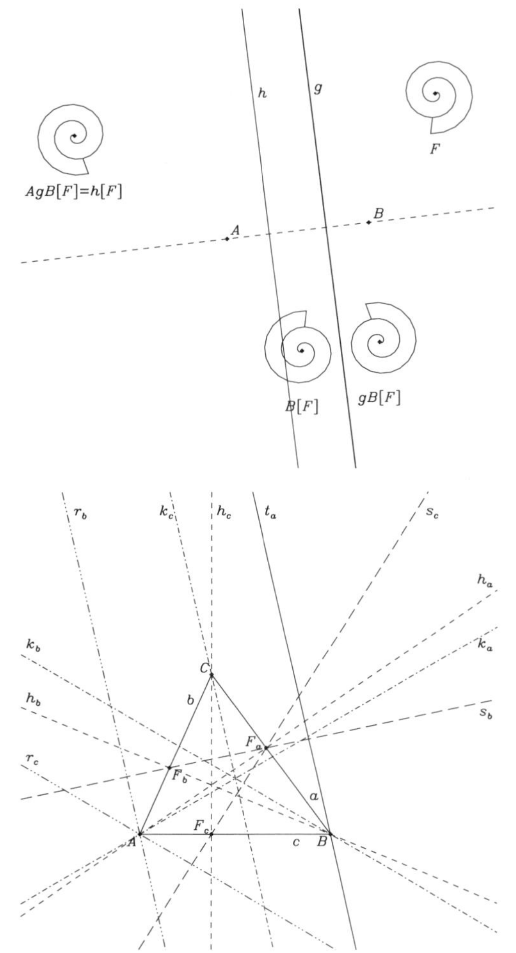

Figure A.6: *The product AgB.*

If the line g is perpendicular to the connecting line (A, B) the product AgB of the three corresponding reflections yields a line reflection h whereby h lies in a position symmetric to g under reflection with respect to AB.

Figure A.7: *The theorem of altitudes.*

In a triangle ΔABC we draw the altitudes h_a, h_b, h_c (broken lines) and their feet F_a, F_b, F_c. In each vertex, we draw the lines reflection-symmetric to the altitudes, $k_a = bh_ac$, $k_b = ch_ba$ and $k_c = ah_cb$ (chain lines). We continue by reflecting k_a on b and on c to obtain $r_b = bk_ab$ and $r_c = ck_ac$ (chain lines with triple dots) and drop the perpendicular on both r_b and r_c from F_a to obtain $s_b = (F_a, r_b)$ and $s_c = (F_a, r_c)$ (long dashed). Finally, we reflect k_b on a to find $t_a = ak_ba$. The product $F_ak_bF_c = u$ is a line identical to the product $h_ah_bh_c$.

Finally, we check that the point of intersection of the perpendicular bisectors is the orthocenter of the bisector triangle. This is evident only when the axiom of parallels holds: In this case, the line connecting the bisecting points M_a and M_c is parallel to the side b, and the perpendicular in M_b is parallel to this line (M_a, M_c) too, i.e., the altitude in the bisector triangle. If the axiom of parallels does not hold, it must be proven that the perpendicular bisector m_b is perpendicular to the segment (M_a, M_c). We see from the construction, Figure A.5, and from Eq. (A.2) that $m_am_bm_c = g$ is again a line and passes through the point B, i.e., $BgB = g$. The sides a and c pass through the point B too. Therefore, the product agc lies in a pencil and is also a line through B. This product can be transformed into

$$agc = am_am_bm_cc = M_am_bM_c.$$

Hence, the product $M_am_bM_c$ is a straight line. It follows that m_b is orthogonal to the connecting line (M_aM_c). It is the altitude in the midpoint triangle; our proof is complete. Dual to this derivation is the proof that the altitudes of a triangle are the bisectors of the angles in the foot point triangle.

For further presentations, we cite the book by Bachmann [49] and other sources [92–97]. After familiarization with the formalism, the algebraic method allows one to draw conclusions quickly and reliably even when the drawing becomes disagreeably complicated. In addition, it opens up new possibilities for calculating. Nevertheless, we recommend to have a glance at the books from the high time of synthetic geometry [88, 98].

B Transformations

B.1 Coordinates

Ordinarily, points in a space and events in a world are described by coordinates. These are numbers that are usually derived from spatial distances, orientations, and time intervals. Coordinates can be defined axiomatically or implicitly.

To begin with, coordinates are generated by some arbitrary but continuous and unambiguous assignment of combinations of numbers to the individual points of the space or events of the world, respectively (Figure B.1). We can freely substitute other coordinates, of course. It is a second matter how the description of the objects of our interest must be adapted to such substitutions. Except for general relativity, this arbitrariness is not used. Instead, we refer to the necessity to simply represent the elements of the existing group of motions. A motion shifts the points and figures to new positions and orientations. Let us suppose that we find a subgroup such that for any point A of the space we obtain exactly one element of the subgroup that shifts the origin O to A. Then we simply take the description of the elements of the subgroup as coordinates. Even when we adopt this procedure, the assignment of coordinates (called a coordinate system) is not unique. The transformations that do not disturb this representation are the automorphisms of the group. For such a transformation, the mere replacement of the old coordinates by the new ones (passive transformation) produces the same result as the physical motion of the objects into the positions given by the new coordinates, i.e., we obtain an interpretation of the new coordinates as new positions (active transformations). In a correct calculation, passive transformations are subject to no condition at all. From the point of view adopted here, the motions are the transformations that can be performed actively as well as passively and have the same result in both cases.[1]

In physics, we proceed with implicit methods. Newton's first law asserts the existence of coordinate systems in which force-free motions are straight in space and uniform in time. To begin with, we can construct a *linear system of coordinates* by referring to the world-lines of four independent free particles. After a linear transformation of such a system, we obtain coordinates of the same kind again. The third law (in Huygens' form) provides the *momentum conservation law* (Figure 3.10) and a measure in space and time. The momentum conservation law implies the existence of the (inertial) mass, which weights the velocities to yield a conservation of total momentum. The inertial mass turns out to be a characteristic property of

[1] In general relativity, the group of motions is trivial in most cases, i.e., it contains only the unit element, which does not move at all. Only the so-called algebraically special solutions allow nontrivial groups of motion, and only in particular cases does the group of motions contain a transitive subgroup that can be used to define coordinates uniquely. The de Sitter universes (Chapter 7) admit a maximal group of motions.

The Geometry of Time. Dierck-E. Liebscher

ISBN: 3-527-40567-4

all bodies. The linear reference systems in which the masses are isotropic, i.e., independent of the direction of motion, are then called inertial frames of reference. The requirement of isotropic masses introduces a comparison of the length of segments of different direction. The collision figure defines a circle, i.e., the locus of points equidistant from some center. When we actually define distances like this, masses do not depend on the orientation of the motion of the body. In general, linear transformations destroy the inertial character of a system of reference. However, a subgroup of linear transformations (the geometrical motions in the world) does not. The structure of this subgroup of transformations that are still allowed depends on the way the masses vary with velocity.

Linear coordinates can even exist when we have no force-free particles. Let us consider a homogeneous gravitational field. All particles are subject to the same acceleration. The world-lines appear as parabolas. However, this set of parabolas has the intersection properties of a set of straight lines. So we can choose a reference frame that has this acceleration too, and then the world-lines appear to be ordinary straight lines, i.e., governed by *linear* relations between the coordinates of the points on the line (Figure B.2). This is the concept of the freely falling observer. In the general theory of relativity, it allows the introduction of (local) inertial reference systems in a gravitational field and the application of special relativity so long as the system can be considered linear. In the mathematical abstraction, the world-lines are straight lines in any coordinate system because the defining structures do not depend on the choice of coordinates.

B.2 Inertial Reference Systems

If the masses do not depend on velocity (as we experience in everyday life), it is the *Galilean transformations* that mediate between the inertial systems. In the two-dimensional world, we write

$$\begin{pmatrix} t^* \\ x^* \end{pmatrix} = \mathcal{G}[v] \begin{pmatrix} t \\ x \end{pmatrix} + \begin{pmatrix} t_0 \\ x_0 \end{pmatrix},$$

that is

$$\begin{aligned} t^* &= t + t_0, \\ x^* &= x - vt + x_0. \end{aligned}$$

Here, v denotes the relative velocity of the two systems, t_0 is a translation of the time origin, and x_0 is a translation of the origin in space. The matrix

$$\mathcal{G}[v] = \begin{pmatrix} 1, & 0 \\ -v, & 1 \end{pmatrix}$$

is the formal expression of a Galilean rotation in the space–time that we constructed explicitly in the preceding chapters. The product of two such matrices is again of the same form: The matrices represent the elements of a group of motions. On multiplying two elements, we obtain a composition of velocities that in this case is additive ($\mathcal{G}[v_1]\mathcal{G}[v_2] = \mathcal{G}[v_1 + v_2]$). In a four-dimensional space–time, we must allow the ordinary rotations in space as well, and the

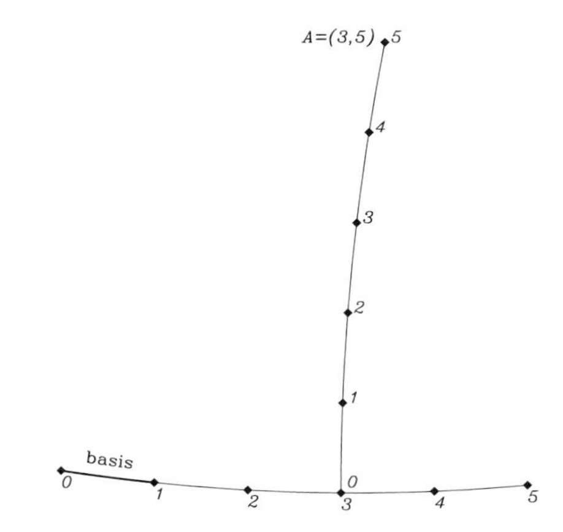

Figure B.1: *Simple construction of coordinates.*

Starting from a given basis, we move in two defined directions until the point to be coordinated is reached. The coordinates are the numbers of steps. However, such a procedure presupposes a metrical space: It must be possible to translate a segment and to find perpendiculars, including the orientation.

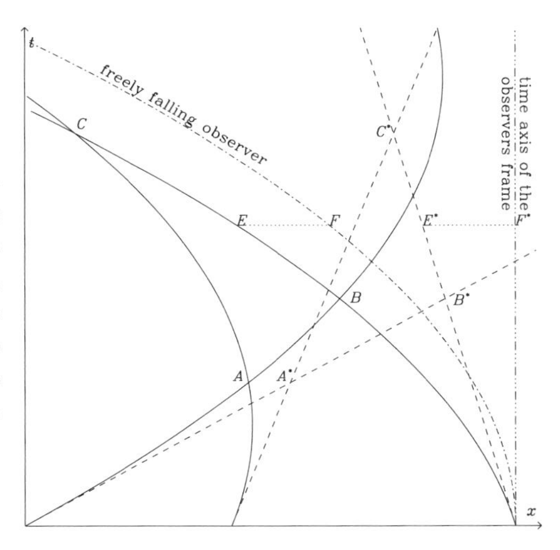

Figure B.2: *Freely falling reference systems.*

We draw three world-lines with equal acceleration. They are parabolas. If we let an observer fall with the same acceleration (dashed–dotted line) and refer the space coordinate to his or her world-line, we find straight lines in the reference system of the observer. For instance, when F is transferred to F^*, E is transferred to E^*. Correspondingly, the curvilinear triangle ΔABC becomes the rectilinear triangle $\Delta A^*B^*C^*$.

Galilean transformations are given by

$$\begin{pmatrix} t^* \\ \boldsymbol{x}^* \end{pmatrix} = \mathcal{G}[\mathcal{A}, \boldsymbol{v}] \begin{pmatrix} t \\ \boldsymbol{x} \end{pmatrix} + \begin{pmatrix} t_0 \\ \boldsymbol{x}_0 \end{pmatrix},$$

that is

$$\begin{aligned} t^* &= t + t_0, \\ \boldsymbol{x}^* &= \mathcal{A}\boldsymbol{x} - \boldsymbol{v}t + \boldsymbol{x}_0. \end{aligned}$$

These transformations constitute the group of motions of Newtonian mechanics, that is, the Galilean group. Transformations without rotation are called the special Galilean transformations. They form a commutative subgroup,

$$\mathcal{G}[\mathcal{E}, \boldsymbol{v}_1]\mathcal{G}[\mathcal{E}, \boldsymbol{v}_2] = \mathcal{G}[\mathcal{E}, \boldsymbol{v}_2]\mathcal{G}[\mathcal{E}, \boldsymbol{v}_1] = \mathcal{G}[\mathcal{E}, \boldsymbol{v}_1 + \boldsymbol{v}_2].$$

Velocities are composed additively.

The wave equation for some excitation Φ,

$$\left(\frac{1}{c^2}\frac{\partial^2}{\partial t^2} - \frac{\partial^2}{\partial x^2} - \frac{\partial^2}{\partial y^2} - \frac{\partial^2}{\partial z^2}\right)\Phi = \text{source}, \tag{B.1}$$

changes its form under the Galilean transformations. Mixed derivatives appear, and the isotropy described by Eq. (B.1) disappears (Chapter 4). The requirement of a universal isotropy of the propagation of light translates into the requirement that the wave equation must be invariant if its propagation velocity c is the speed of light. Transformations between inertial reference systems must leave the wave equation unaffected. One can again show that the transformations that are now allowed are linear. In the case of plane, the (general homogeneous) linear transformation

$$ct^* = \gamma\,(\alpha\, ct - \beta\, x), \qquad x^* = \gamma\,(x - vt)$$

is a general motion of the new origin $x^* = 0$ with velocity v in the old reference frame. The coefficients α, β, and γ are to be determined. The invariance of the two-dimensional wave equation yields

$$\gamma = \frac{1}{\sqrt{1-\frac{v^2}{c^2}}}, \qquad \alpha = 1, \qquad \beta = \frac{v}{c}.$$

We can also obtain these values by a detour through the light coordinates introduced in Figure 5.1. The product $\xi[P]\eta[P] = \xi[S[P]]\eta[S[P]]$ has been shown to be preserved. This is equivalent to our formulas of the (special) *Lorentz transformation*[2]

$$\begin{pmatrix} ct^* \\ x^* \end{pmatrix} = \mathcal{L}[v]\begin{pmatrix} ct \\ x \end{pmatrix},$$

that is

$$\begin{aligned} ct^* &= \frac{ct - \frac{v}{c}x}{\sqrt{1-\frac{v^2}{c^2}}}, \\ x^* &= \frac{x - \frac{v}{c}ct}{\sqrt{1-\frac{v^2}{c^2}}}. \end{aligned}$$

The matrix $\mathcal{L}[v]$ is again the formal expression of a rotation in the world, the transformations again form a group, and we again find a composition of velocities by multiplication: $\mathcal{L}[v_1]\mathcal{L}[v_2] = \mathcal{L}[V]$. However, we find that the composition law is no longer linear. We obtain Einstein's addition of velocities,

$$V = \frac{v_1 + v_2}{1 + \frac{v_1 v_2}{c^2}}. \tag{B.2}$$

[2]We omit the translations of the origin, $x^* = x + x_0$, $t^* = t + t_0$. They form a subgroup, which can be included. The Lorentz group extended by these shifts is called the inhomogeneous Lorentz group, or the *Poincaré group*.

It follows that the relative velocity is no longer an angle (which is defined as the additive parameter of the rotations) but the hyperbolic tangent of an angle. Equation (B.2) is the addition theorem for this function of an angle:

$$\tanh[\theta_1 + \theta_2] = \frac{\tanh[\theta_1] + \tanh[\theta_2]}{1 + \tanh[\theta_1]\tanh[\theta_2]}.$$

In a four-dimensional space–time, the special Lorentz transformations,

$$\begin{pmatrix} t^* \\ \boldsymbol{x}^* \end{pmatrix} = \mathcal{L}[\mathcal{E}, \boldsymbol{v}] \begin{pmatrix} t \\ \boldsymbol{x} \end{pmatrix}, \tag{B.3}$$

are given by

$$ct^* = \frac{ct - \frac{1}{c}\boldsymbol{v}\boldsymbol{x}}{\sqrt{1 - \frac{\boldsymbol{v}^2}{c^2}}}, \qquad \boldsymbol{x}^* = \boldsymbol{x} - \frac{\boldsymbol{v}}{\boldsymbol{v}^2}\boldsymbol{v}\boldsymbol{x} + \boldsymbol{v}\frac{\frac{\boldsymbol{v}\boldsymbol{x}}{\boldsymbol{v}^2} - t}{\sqrt{1 - \frac{v^2}{c^2}}}.$$

The special Lorentz transformations in a four-dimensional space–time do *not* constitute a subgroup and do *not* commute except for parallel velocities. The product of two special Lorentz transformations always contains a rotation in space. This fact expresses the curvature of the velocity space and is the origin of the Thomas precession. The velocity space acquires a hyperbolic geometry. It is locally Euclidean but exhibits negative curvature in the large. In the plane, the velocity coordinates realize Klein's model of non-Euclidean geometry (Section D.3). Using the hyperbolic angle, in that model we introduce polar coordinates analogous to the polar coordinates on the sphere. A difference is that the circles of fixed polar distance θ on the sphere start to decrease again beyond the equator and that the complete sphere has been covered when twice the polar distance of the equator has been reached. On the time shell, or the mass shell, or the velocity space, the polar distance increases without limit. It must be noted that the azimuthal projection of the unbounded hyperbolic space covers only a finite region of the projective plane, while for the finite sphere the infinite plane is not large enough to obtain a uniquely invertible azimuthal projection.

The invariance of the wave equation implies the invariance of the line element

$$\mathrm{d}s^2 = c^2\,\mathrm{d}t^2 - \mathrm{d}x^2 - \mathrm{d}y^2 - \mathrm{d}z^2 = \sum_{ik} \eta_{ik}\,\mathrm{d}x^i\,\mathrm{d}x^k$$

with

$$\eta_{ik} = \begin{pmatrix} 1, & 0, & 0, & 0 \\ 0, & -1, & 0, & 0 \\ 0, & 0, & -1, & 0 \\ 0, & 0, & 0, & -1 \end{pmatrix}. \tag{B.4}$$

This is the expression of Pythagoras's theorem in the Minkowski geometry (Figure 5.2). A line element is used to determine the arc length of a curve. In the Euclidean geometry, we take a curve $Q[\lambda] = [x[\lambda], y[\lambda]]$ and divide it into infinitesimally small segments. These segments are evaluated by Pythagoras's theorem (Figure 3.5),

$$\mathrm{d}[Q[\lambda], Q[\lambda + \mathrm{d}\lambda]] = \left((x[\lambda + \mathrm{d}\lambda] - x[\lambda])^2 + (y[\lambda + \mathrm{d}\lambda] - y[\lambda])^2\right)^{\frac{1}{2}},$$

and integrated subsequently. The expression

$$\mathrm{d}s^2 = (x[\lambda + \mathrm{d}\lambda] - x[\lambda])^2 + (y[\lambda + \mathrm{d}\lambda] - y[\lambda])^2 = \mathrm{d}x^2 + \mathrm{d}y^2$$

is called the Euclidean line element. If we substitute general coordinates ($x = x[\xi^1, \xi^2]$ and $y = y[\xi^1, \xi^2]$), it yields a general quadratic form

$$\mathrm{d}s^2 = \sum_{ik} g_{ik}\, \mathrm{d}\xi^i\, \mathrm{d}\xi^k.$$

This form can be generalized to higher dimensions and variable coefficients.

On the world-line of a body at rest, x, y, and z do not vary. Consequently, the line element describes the flow of proper time, the clocks that measure it moving together with the body on the world-line in question.

$$\mathrm{d}\tau = \frac{1}{c}\,\mathrm{d}s.$$

We obtain the four-velocity as normalized tangent to a world-line $x^i = x^i[\lambda]$ ($i = 0, \ldots, 3$). It is the increment of all four coordinates with the proper time,

$$u^i = \frac{\mathrm{d}x^i}{\mathrm{d}\tau} = \frac{1}{\sqrt{1 - \frac{v^2}{c^2}}}[c, v_x, v_y, v_z].$$

Its first component describes the time dilation. The intervals of proper time are always smaller by the factor γ than the corresponding intervals of system time.

In order to obtain the four-momentum, we must multiply the four-velocity by the rest mass,

$$p^i = m_0 u^i = \frac{m_0}{\sqrt{1 - \frac{v^2}{c^2}}}[c, v_x, v_y, v_z]. \tag{B.5}$$

Newton's third law in Huygens' form is to be formulated as

$$\sum_{A=1}^{M} p_A^i = \sum_{B=1}^{N} p_B^i. \tag{B.6}$$

In the considered process, M particles with the four-momenta p_A^i collide to form N particles with the four-momenta p_B^i. Newton's second law relates the detailed changes of momenta to forces,

$$F^i = \frac{\mathrm{d}p^i}{\mathrm{d}\tau}.$$

The collision condition, Eq. (B.6), and the conservation of identity, $\eta_{ik}p^ip^k = m_0^2c^2 =$ constant, are to be read as conditions on the forces F^i.

The inertial mass, measured in three-dimensional collisions, is given by

$$m[v] = \frac{m_0}{\sqrt{1-\frac{v^2}{c^2}}}.$$

This is the variation of mass with velocity, Eq. (5.2). It must be observed if we are right in adapting the invariance of mechanics to the invariance of the wave equation. If the inertial mass does not vary with velocity, mechanics can be invariant only with respect to the Galilean transformations.

B.3 Riemannian Spaces, Einstein Worlds

If we accept that the light ray is a paradigm of a straight line and that the gravitational field modifies the propagation of light, all hope for linear coordinates must be abandoned. In this case, the homogeneous or even the Euclidean geometry of space or the Minkowski geometry of the world, can only be a local approximation, and can only be applied as far as the inhomogeneity of the gravitation field is not felt. When we consider experiments in which this inhomogeneity is important, for instance, experiments involving motion through interplanetary space, neither space nor space–time can be assumed to be homogeneous. We must accept that the wave equation now has variable (hence general) coefficients,

$$\sum_{ik} g^{ik}[P]\frac{\partial^2}{\partial x^i \partial x^k}\Phi = \text{first-order derivatives, sources.} \tag{B.7}$$

Such an invariant wave equation implies the existence of an invariant line element,

$$\mathrm{d}s^2 = \sum_{ik} g_{ik}[P]\,\mathrm{d}x^i\,\mathrm{d}x^k \tag{B.8}$$

with

$$\sum_l g_{il}g^{lk} = \delta_i^k = \begin{matrix} 1 & \text{for} & i = k \\ 0 & \text{for} & i \neq k \end{matrix}.$$

We must accept that coordinates can no longer be restricted by a group of motions, as in homogeneous spaces. Therefore, general substitutions of coordinates are to be admitted, and the description of our objects must be adapted to this situation. This adaptation is called *general covariance*. When one starts the construction of a geometry with arbitrary coordinates, the line element, Eq. (B.8), is the central notion. It is the prescription for determining the separation of neighboring points ($P = [x^1, \ldots, x^n]$, $Q = P + \mathrm{d}P$, $\mathrm{d}P = [\mathrm{d}x^1, \ldots, \mathrm{d}x^n]$) when their coordinates are given. The line element remains a modified form of Pythagoras's theorem; strictly speaking, it is a quadratic form of the coordinate differences $\mathrm{d}x^i$. The array of coefficients g_{ik} varies with location: Because the distance must not depend on the arbitrarily chosen coordinates, substitutions of new coordinates must not change the value of $\mathrm{d}s^2$. Consequently, any substitution of new coordinates requires a definite transformation law for the array g_{ik} , which is then called the *metric tensor*. The case of reference, Eq. (B.4), is called the Minkowski tensor.

The two tensors g_{ik} and g^{ik} are inverse to each other but no longer numerically equal (as in the Minkowski geometry). They behave differently under coordinate transformations. Substitution of new coordinates in Eq. (B.7) yields the transformation law

$$g^{*ik} = \sum_{lm} g^{lm} \frac{\partial x^{*i}}{\partial x^l} \frac{\partial x^{*k}}{\partial x^m}.$$

The substitution also yields first-order derivations. They pose an additional question not considered here. From Eq. (B.8), we obtain

$$g^*{}_{ik} = \sum_{lm} g_{lm} \frac{\partial x^l}{\partial x^{*i}} \frac{\partial x^m}{\partial x^{*k}}.$$

The overwhelming importance of the transition to general coordinate substitutions led to the definition of vectors and tensors solely on the basis of the implied transformation laws. Vectors and tensors are transformed by substitution combined with multiplication by a set of Jacobi matrices, $\mathcal{J} = [\partial x^*/\partial x]$, or $\mathcal{J}^{-1} = [\partial x/\partial x^*]$. Each index indicates a matrix factor in substitutions, a lower index the matrix $\frac{\partial x}{\partial x^*}$, an upper index $\frac{\partial x^*}{\partial x}$. Corresponding to lower and upper indices, one also speaks of covariant and contravariant vectors[3] and tensors. The general transformation law is homogeneous:

$$\begin{aligned} A^{*i\ldots}_{\ \ \ldots} &= \sum_{l\ldots} A^{l\ldots}_{\ \ \ldots} \frac{\partial x^{*i}}{\partial x^l} \ldots, B^{*\ldots}_{\ \ k\ldots} = \sum_{m\ldots} B^{\ldots}_{m\ldots} \frac{\partial x^m}{\partial x^{*k}} \ldots, \\ \sum_k A^{*k} B^*{}_k &= \sum_m A^m B_m. \end{aligned}$$

Objects of the same transformation type form a vector algebra: They can be compared, added, and multiplied by a (scalar) factor. A scalar is a quantity that is subject to mere substitution without being subject to multiplication by some transformation matrix. We must clearly distinguish between upper and lower indices, i.e., between the objects of different transformation types. In any covariant equation, all terms must belong to the same type.

The canceling of transformation matrices that happens in summation over pairs of lower and upper indices is very important. In addition, it is the reason for Einstein's summation convention: We omit the symbol for sum in these cases and declare automatic summation if, in a term, the same letter appears as lower and upper indices. Instead of $\sum_m A^m B_m$ we write simply $A^m B_m$, for instance. From now on, the summation convention will be used.

The primary contravariant vector is the position increment $\mathrm{d}x^i$. Its transformation law is the chain rule of differentials. The characteristic derivative is the four-velocity, $u^i = \mathrm{d}x^i/\mathrm{d}\tau$. The primary covariant vector is the gradient $\partial\Phi/\partial x^k$ of a potential Φ. Its transformation law is to be interpreted as the chain rule for partial derivatives. The product

$$\mathrm{d}\Phi = \frac{\partial \Phi}{\partial x^k} \mathrm{d}x^k \qquad \left(= \sum_k \frac{\partial \Phi}{\partial x^k} \mathrm{d}x^k, \quad \text{to remember!} \right)$$

is the scalar differential of the potential Φ.

[3] In our picture, vectors carry just one index.

In Newtonian mechanics, a conservative force (gravitational attraction, electrostatic force) is given by the gradient of a potential. In canonical mechanics too, the increment of the momenta is the gradient of the Hamiltonian. Hence, the force and the momentum are primarily covariant vectors. The relation between momentum and velocity includes the metric

$$\frac{\partial \Phi}{\partial x^i} = F_i = \frac{\mathrm{d}p_i}{\mathrm{d}\tau} = \frac{\mathrm{d}}{\mathrm{d}\tau}(m_{ik}u^k) = \frac{\mathrm{d}}{\mathrm{d}\tau}(g_{ik}m_0 u^k).$$

In general coordinates, the inertial mass, which is given by $p_i = m_{ik}u^k$, can be formally anisotropic. It is a contingent fact that this anisotropy can be absorbed into the metric of space–time. This we already did by defining the circle through symmetric collisions in Chapter 3. The anisotropy becomes unobservable if the field equations for the forces contain only this metric [99]. For instance, the potential could be subject to a wave equation,[4] Eq. (B.7), with the inverse of g_{ik} taken as coefficients g^{ik}. Any anisotropy between inertial mass and wave equations can be tested [100]. In spite of the precison of 10^{-24}, nothing has been found.

The characteristic signature and the curvature are contained within the structure of the metric tensor and its variation with location. The general treatment of such a construction leads to the *Riemannian geometry* of inhomogeneously curved spaces and the *Einsteinian geometry* of inhomogeneously curved space–times. The choice of coordinates and the vector algebra must be formally separated. The former is kept totally free (we construct in general covariance), and the latter remains absolutely local, that is, the vector spaces remain tangent spaces and retain the Euclidean or Minkowski geometry. That is, the operations of vector algebra like multiplication of vectors with scalars, Eq. (B.5), and addition of vectors, Eq. (B.6), can be performed at each event of the world separately. However, it is now a new task to compare vectors at different events, for instance, to find the force from the change of momentum. The subtraction of a vector $p^i[\tau]$ at $P = [x^i[\tau]]$ from another one ($p^i[\tau + \mathrm{d}\tau]$) at a different event ($Q = P + \mathrm{d}P = [x^i[\tau] + \mathrm{d}x^i]$) is an operation composed of the (parallel) transport of the vector $p^i[\tau]$ from P to Q followed by ordinary subtraction at that point. The formal expression of this fact is that the simple increment $\mathrm{d}p^i$ cannot be used: Its transformation law is not homogeneous, so it does not belong to the vector algebra. The parallel transport of vectors through space or space–time requires the knowledge of a specific prescription, which in Einstein's theory of general relativity is derived from the orientation of geodesics, as we saw in Chapter 7. All this can be found in introductions to the theory of general relativity; it is not the subject of this book.

[4] The equations for both the electromagnetic field and the gravitational field are more complicated. Nevertheless, they contain, apart from the field, only the metric g_{ik} and quantities constructed from it.

C Projective Geometry

C.1 Algebra

We do not intend to consider at large the axiomatic foundation of projective geometry but try to find as simply as possible a formalization that allows an arithmetic reexamination of geometrical relations. Considering the characteristic transfer of the cross-ratio from one line to another (Figure 8.5), we use three-dimensional pencils of rays to find an appropriate coordinate representation of the projective plane. We embed the plane in a three-dimensional space, choose an origin outside the plane, and substitute rays and planes through this origin for the points and lines of the planes (Figure C.1). We keep in mind that we use the summation convention.

We have already seen in the transfer of the cross-ratio from line to line that the points of a line are appropriately represented by the rays through a pencil vertex. For the plane, we characterize the points by rays in a three-dimensional space (without loss of generality in Cartesian coordinates) whose direction coefficients are determined up to a common factor. A point A is then given by a triple $[A^1, A^2, A^3]$, while a factor does not matter: $[A^1, A^2, A^3]$ and $[\lambda A^1, \lambda A^2, \lambda A^3]$ represent the same point in the drawing plane. Without loss of generality, this plane can be given by $A^3 = 1$. On the plane, the points have the Cartesian coordinates

$$\xi = \frac{A^1}{A^3}, \qquad \eta = \frac{A^2}{A^3} \tag{C.1}$$

which really are independent of any factor λ that multiplies the triple. Correspondingly, the triple $[A^1, A^2, A^3]$ gives homogeneous coordinates of the point.

The lines of the projective plane are intersections with planes through the three-dimensional origin. Rays in such a plane correspond to points on the line in the projective plane. The equation for such rays is

$$g_1 A^1 + g_2 A^2 + g_3 A^3 = g_k A^k = 0, \tag{C.2}$$

where the plane through the origin is given by its direction coefficients $g = [g_1, g_2, g_3]$ and is just as independent of a common factor as the homogeneous point coordinates. We index the line coordinates with lower indices and the point coordinates with upper indices. After agreeing upon this procedure, we can use Einstein's summation convention without any danger of confusion and omit all summation signs.

Because both homogeneous point coordinates and homogeneous line coordinates are triples, one can imagine an interchange. If one reads in the arithmetic representation of a

The Geometry of Time. Dierck-E. Liebscher

ISBN: 3-527-40567-4

construction all homogeneous point coordinates as line coordinates and vice versa, an equally valid construction is generated: This is called duality in the projective geometry. This duality has already been mentioned earlier.

The relations in a two-dimensional projective plane are described by a three-dimensional linear vector algebra. The vectors of this algebra are the triples of homogeneous coordinates. Points are denoted by capital letters. If Eq. (C.1) is applied, the line at infinity contains all points with $A^3 = 0$. The projective coordinates are obtained from the usual Cartesian coordinates $[\xi, \eta]$ of the plane by interpreting it as the plane $\zeta = 1$ in a space and by assigning to each point the ray $[\lambda\xi, \lambda\eta, \lambda]$, where λ remains undetermined. The value of λ does not change the point of intersection of the ray with the projective plane but determines only a position on the ray, which is irrelevant for our purpose. Straight lines are denoted by small letters. The undetermined factor can be chosen so that Hesse's normal form of the equation is generated, but it is not necessary to fix the factor. A point $A = [A^1, A^2, A^3]$ lies on the line $g = [g_1, g_2, g_3]$ if $g_k A^k = 0$ holds.

The vector algebra also admits addition. The addition is defined component by component, for instance,

$$[g_1, g_2, g_3] + [h_1, h_2, h_3] = [g_1 + h_1, g_2 + h_2, g_3 + h_3]. \tag{C.3}$$

This addition is invariant with respect to homogeneous linear transformations, i.e.,

$$\mathcal{T}(g + h) = \mathcal{T}g + \mathcal{T}h.$$

However, we should not expect that Eq. (C.3) can be interpreted as addition of two lines. Lines are represented by triples g, but they do not define them uniquely. A common factor of all components is free. Consequently, we calculate with the addition under formally homogeneous linear transformations but we keep in mind that projectively meaningful expressions are always *linearly homogeneous* in all variables. This is a nice test of correct calculation too. After we have found an expression

$$P = P[A, \ldots, g, \ldots, \mathcal{B}],$$

we should check that there exist some exponents with the property

$$P[\lambda_A A, \ldots, \lambda_g g, \ldots, \lambda_B \mathcal{B}, \ldots] = \lambda_A^{n_A} \ldots \lambda_g^{n_g} \ldots \lambda_B^{n_B} \ldots P[A, \ldots, g, \ldots, \mathcal{B}].$$

If no such set of exponents exists, we must look for a bug in the calculation. If we have calculated correctly but found no such homogeneity, the result P is an object of the linear space but not of the projective plane.

We can construct a net of projective coordinates (Figure C.2) in such a way that the coordinates, on every line represent the cross-ratio with some three points. If four points are given on a line and if the cross-ratio of the four first coordinates, the four second coordinates and the four third coordinates can be calculated, all will found to be equal to the cross-ratio of the four points derived recursively by construction from the harmonic range. The means to transport segments for constructing coordinates is this harmonic range. Successively applied, it yields a uniform grading of the line that can be refined by harmonic division. For the definition of projective coordinates, it is not necessary to use the Cartesian coordinates of the plane.

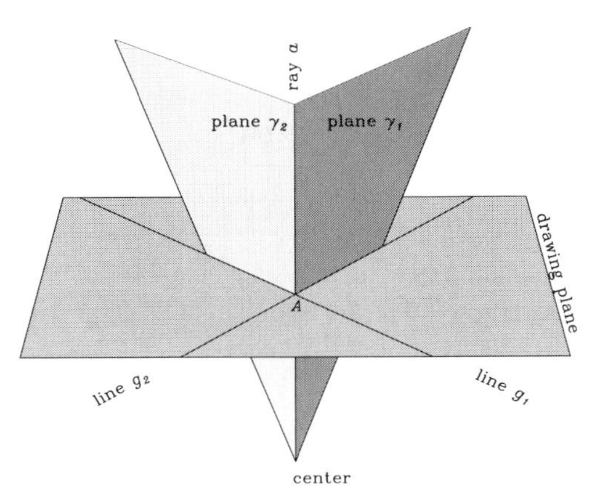

Figure C.1: *Homogeneous point and line coordinates.*

Two planes γ_1 and γ_2 through the center intersect along a ray a that also passes through the center. The planes and the ray intersect the drawing plane in two lines g_1 and g_2 and the point A. We characterize the lines in the drawing plane by planes in space, which are determined by their normal vectors alone. We characterize the point in the drawing plane by the ray in space, which is again characterized only by its direction. The norm of the direction vectors is irrelevant.

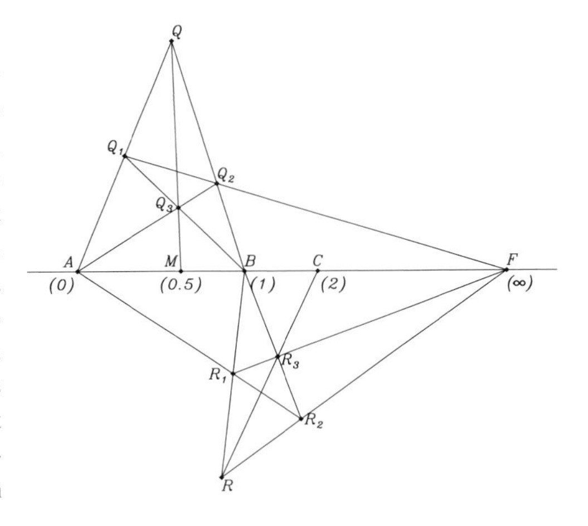

Figure C.2: *Projective coordinates.*

Projective coordinates are constructed with the harmonic range, whose form and invariance follow from the axioms alone. On a line, we define the coordinates of three points, for instance, 0 for A, 1 for B, and ∞ for F. When we now intend to multiply a basis $[A, B]$, we find the point C that with A divides the segment $[B, F]$ harmonically. Point C acquires the coordinate 2. Continuing, we transfer the basis over the whole line. The obtained segments can also be bisected. For instance, the point M with the coordinate 0.5 is found as the point that together with F divides the segment $[A, B]$ harmonically. The coordinates are now chosen so that they yield the cross-ratio. We obtain $\frac{2-1}{2-\infty}\frac{0-\infty}{0-1} = -1$ as well as $\frac{0.5-0}{0.5-1}\frac{\infty-1}{\infty-0} = -1$.

We now define the products of the vector algebra necessary for the following calculations. We have already used the *scalar product*. This is the expression for the form

$$\langle g, A\rangle \stackrel{\text{def}}{=} g_k A^k. \tag{C.4}$$

A point and line can always form a scalar product. (Things are not so simple in the case of two points or two lines. A scalar product can be invariant only if the factors are transformed with matrices that are inverse to each other. The two factors cannot belong to the same linear space. In the Euclidean geometry, the transformation matrices are orthogonal, and the necessity of the distinction becomes less obvious. i.e., the scalar product can be identified with the metric.) The scalar product tests the incidence of a point A with a line g. If A lies on g, the scalar product vanishes. One often omits the brackets and the comma and, in a sequence, understands gA or Ag as a scalar product. Of course, brackets are virtually present and cannot be moved or set otherwise. We shall use brackets in order to avoid misinterpretations in connection with group operations.

By the *cross product*, a line (the connecting line) is assigned to two points of a plane, and a point (the intersection point) to two lines. We correspondingly define

$$a \times b \stackrel{\text{def}}{=} [a_2b_3 - a_3b_2,\ a_3b_1 - a_1b_3,\ a_1b_2 - a_2b_1], \tag{C.5}$$

and we write for this

$$(a \times b)^k = \epsilon^{klm} a_l b_m, \qquad (A \times B)_k = \epsilon_{klm} A^l B^m.$$

The threefold indexed symbols ϵ_{klm} and ϵ^{klm} are the signs of the corresponding permutations. The permutation symbol ϵ is zero if two indices are equal, it is equal $+1$ if the permutation is even, and it is equal -1 if the permutation is odd. So $\langle A, A \times B\rangle = \langle B, A \times B\rangle = 0$. Both points A and B lie on the line $A \times B$, and the dual is equally true. Both lines g and h pass through the point $g \times h$. The cross product is antisymmetric. We find the important formula

$$(A \times (g \times h)) = g \cdot \langle A, h\rangle - h \cdot \langle A, g\rangle. \tag{C.6}$$

One often omits the brackets and the cross and, in a sequence, understands AB or gh as a cross product. Again, brackets are virtually present and cannot be moved or set otherwise. The cross product AB yields the coefficients of the connecting line, while the cross product gh yields the coefficients of the intersection point.

Cross product and scalar product can be combined to yield the *triple product*. A point P is collinear with two other points Q and R if it lies on the connecting line. The triple product of the three points,

$$[P, Q, R] \stackrel{\text{def}}{=} \epsilon_{klm} P^k Q^l R^m = \langle P, Q \times R\rangle, \tag{C.7}$$

tests this collinearity, while the triple product of three lines,

$$[f, g, h] \stackrel{\text{def}}{=} \epsilon^{klm} f_k g_l h_m = \langle f, g \times h\rangle,$$

tests for intersection at a point (the pencil property). The triple product of three points vanishes if they are collinear, and the triple product of three lines vanishes if they lie in a pencil. We obtain the rule

$$\langle P, (Q \times R)\rangle \stackrel{\text{def}}{=} [P, Q, R] = [R, P, Q] \stackrel{\text{def}}{=} \langle R, (P \times Q)\rangle \tag{C.8}$$

and so on. The rule (C.6) can be extended to yield

$$(P \times Q) \times (R \times S) = R \cdot [P, Q, S] - S \cdot [P, Q, R]. \tag{C.9}$$

If two factors are equal (that is if they are equal for the projective plane), both the cross product and the triple product vanish. The triple product can be interpreted as the volume of the parallelepiped spanned by the vectors given with the three coordinate triples in the ordinary three-dimensional space. When we calculate constructions with several intersection points and connecting lines, we obtain many nested cross products, and these can be simplified with the formulas (C.6), (C.9) and

$$\langle g \times h, A \times B \rangle = \langle g, B \rangle \langle h, A \rangle - \langle g, A \rangle \langle h, B \rangle.$$

We now introduce the *direct product* of two triples. It is a matrix that is indexed above and below corresponding to the character of the factors:

$$(P \circ Q)^{ij} \stackrel{\text{def}}{=} P^i Q^j, \qquad (P \circ g)^i_k = P^i g_k, \qquad (g \circ h)_{kl} = g_k h_l. \tag{C.10}$$

It follows that

$$(P \circ Q) g = P \cdot \langle Q g \rangle.$$

Finally, we note the important separation formula

$$g \circ (a \times b) + a \circ (b \times g) + b \circ (g \times a) = \mathcal{E}\ [a, b, g], \tag{C.11}$$

which is valid for all g, a, b.

C.2 Projective Maps

The introduction of homogeneous coordinates allows us to represent projective relations by linear maps in a three-dimensional space. The simplification of the calculation is paid for with the increase in dimension. The simplest example of a linear map is parallel projection. Lines remain lines, infinity remains infinity, and all ratios on the lines are preserved. In general, the projective maps of the plane become the linear homogeneous maps of the space when the points of the plane become central rays of the space and when the lines of the plane become central planes of the space.

In our case, we find four kinds of linear maps:

1. Maps of the points to points. The corresponding matrices carry one index above, and one index below: $Q^* = \mathcal{T}[Q] = \mathcal{T}Q,\ \ Q^{*k} = T^k_l Q^l$.

2. Maps of the lines to lines. The corresponding matrices are of the same kind: $g^* = \mathcal{U}[g] = \mathcal{U}g,\ \ g^*_k = g_l {U^l}_k$.

3. Maps of the points to lines. The corresponding matrices carry both indices below: $g = \mathcal{A}[Q] = \mathcal{A}Q,\ \ g_k = A_{kl} Q^l$.

4. Maps of the lines to points. The corresponding matrices carry both indices above: $Q = B[g] = \mathcal{B}g,\ \ Q^k = B^{kl} g_l$.

If the determinants of the matrices are different from zero (i.e., if they are nondegenerate), a map of the point space defines by its inverse a map of the line space too. We then get[1]

$$T^m_{\ k} U^l_{\ m} = \delta^l_k \,, \quad T^k_{\ m} U^m_{\ l} = \delta^k_l \quad \text{and} \quad A_{km} B^{ml} = \delta^l_k \,, \quad B^{lm} A_{mk} = \delta^l_k \,.$$

With the matrix U as the inverse of T, the map of the point space to itself induces a map of the line space to itself. The defining relation requires the image Q^* of the point Q to lie on the image g^* of the line g if and only if Q lies on g. Thereby, it yields the scalar products

$$g^*_k Q^{*k} = g_l U^l_{\ k} T^k_{\ m} Q^m = g_l \delta^l_{\ m} Q^m = g_l Q^l .$$

A projective map is precisely such a pair of linear maps of the point space onto itself and the line space onto itself for which the scalar product (i.e., the incidence) is preserved. The map of the lines is obtained with the inverse matrix of the map of the points: $g^* = g\mathcal{T}^{-1}$,

$$(\mathcal{T}Q)^k = T^k_l Q^l, \qquad (g\mathcal{T}^{-1})_k = g_l (\mathcal{T}^{-1})^l_k. \tag{C.12}$$

As discussed in Appendix B, we say that the point coordinates are contravariant to the line coordinates because the transformations are inverse. That was the reason why we decided to write the indices above. The line coordinates are called covariant, and their indices remain below. We read off what to do in projective transformations. The transformation matrices carry one index above and one below to make sure that the multiplication with a point yields a point and the multiplication with a line yields a line. This is analogous to the considerations in Section B.3.

A point transformation $Q^* = \mathcal{T}Q$ is accompanied by a line transformation $g^* = g\mathcal{T}^{-1} = g\mathcal{U}$. The scalar product is invariant, $\langle g, Q \rangle = \langle g^*, Q^* \rangle$, and for the cross product we find the formulas

$$(g_1\mathcal{U} \times g_2\mathcal{U})\mathcal{U} = (\det \mathcal{U}) g_1 \times g_2, \qquad \mathcal{T}(\mathcal{T}P_1 \times \mathcal{T}P_2) = (\det \mathcal{T}) P_1 \times P_2 \tag{C.13}$$

and

$$\mathcal{T}(g_1 \times g_2) = (\det \mathcal{T})(g_1\mathcal{T}^{-1}) \times (g_2\mathcal{T}^{-1}). \tag{C.14}$$

In addition, we obtain corollaries for linear maps $\mathcal{A}$ of points onto lines and maps $\mathcal{B} = \mathcal{A}^{-1}$ of lines onto points:

$$\mathcal{A}(g_1 \times g_2) = (\det \mathcal{A})(\mathcal{B}g_1 \times \mathcal{B}g_2) \,, \quad \mathcal{A}(g \times \mathcal{A}Q) = (\det \mathcal{A})(\mathcal{B}g \times Q) \,,$$
$$\langle (Q_1 \times Q_2), \mathcal{B}(Q_3 \times Q_4) \rangle = (\det \mathcal{B})(\langle Q_1, \mathcal{A}Q_3 \rangle \langle Q_2, \mathcal{A}Q_4 \rangle - \langle Q_1, \mathcal{A}Q_4 \rangle \langle Q_2 \mathcal{A}Q_3 \rangle) \,,$$
$$Q_1 \times \mathcal{B}(Q_2 \times Q_3) = (\det \mathcal{B}) \mathcal{A}(Q_2 \langle Q_1, \mathcal{A}Q_3 \rangle - Q_3 \langle Q_1, \mathcal{A}Q_2 \rangle) \,.$$

Projective maps of the plane leave the cross-ratio of four points of a line or four lines of a pencil unchanged. Taking an arbitrary point S not collinear with $[A, B, C, D]$, we can describe the general cross-ratio as a cross-ratio of volumes in the three-dimensional space:

$$\mathcal{D}[A, B; C, D] = \frac{[A, C, S]}{[A, D, S]} \frac{[B, D, S]}{[B, C, S]}. \tag{C.15}$$

[1]The symbols δ^k_l denote the unit matrix: $\delta^k_l = 1$ if $k = l$ and $\delta^k_l = 0$ if $k \neq l$.

The cross-ratio does not depend on the position of the auxiliary point S. To see this, we write

$$\mathcal{D}[A,B;C,D] = \frac{\langle S,(A\times C)\rangle\langle (B\times D),S\rangle}{\langle S,(A\times D)\rangle\langle (B\times C),S\rangle}.$$

The four lines $A\times C$, $A\times D$, $B\times C$, $B\times D$ all coincide projectively because the four points lie on one carrier line. Only their norm is different. The comparison of vector lengths is accomplished by multiplication with some arbitrary matrix, which is chosen in the form $S\circ S$ in our case. S can be interpreted as the coordinate triple of a point.

Three points Q_1, Q_2, Q_3 can be taken as the basis of homogeneous coordinates. The points Q_1, Q_2, Q_3 must not lie on a line because they must be linearly independent. The reciprocal basis in the line space is now given by $g_1 = [Q_1,Q_2,Q_3]^{-1}Q_2\times Q_3$, $g_2 = [Q_1,Q_2,Q_3]^{-1}Q_3\times Q_1$, $g_3 = [Q_1,Q_2,Q_3]^{-1}Q_1\times Q_2$. Equation (C.11) yields

$$Q_1\circ(Q_2\times Q_3) + Q_2\circ(Q_3\times Q_1) + Q_3\circ(Q_1\times Q_2) = [Q_1,Q_2,Q_3]\,\mathcal{E}. \tag{C.16}$$

We obtain the corollary

$$\begin{aligned}(Q_1\times Q_2)\circ(Q_3\times Q_4)+(Q_3\times Q_4)\circ(Q_1\times Q_2) &\quad +\\ (Q_1\times Q_3)\circ(Q_4\times Q_2)+(Q_4\times Q_2)\circ(Q_1\times Q_3) &\quad +\\ (Q_1\times Q_4)\circ(Q_2\times Q_3)+(Q_2\times Q_3)\circ(Q_1\times Q_4) &\quad = \quad 0\,.\end{aligned} \tag{C.17}$$

A projective transformation of the plane is determined if one knows the images of four points and no three of them lie on a common line. That is, the four points must form a nondegenerate complete quadrangle, that is, the basic figure of all our projective constructions. Correspondingly, five points of the projective plane already define projective invariants, for instance $f[A,B,C,D;S] = \mathcal{D}_S[A,B;C,D]$ (Eq. (C.15)). If we regard this invariant as a function of the point E, it distinguishes between the conic sections that can be drawn through the four points A, B, C, and D. These conic sections form a one-parameter pencil of conics that can be labeled by the invariant. For the degenerate conics (pairs of lines), it takes the values 0, 1, and ∞. If the four points A, B, C, D lie on a straight line, the function f is simply the cross-ratio of the four points and does not vary with the position of S (as long as S does not lie on the same line). The projective transformation ($\mathcal{T}: [Q_1,Q_2,Q_3,Q_4] \rightarrow [Q_1^*,Q_2^*,Q_3^*,Q_4^*]$) can be written in the form

$$\mathcal{T} = \lambda Q_1^*\circ(Q_2\times Q_3) + \mu Q_2^*\circ(Q_3\times Q_1) + \nu Q_3^*\circ(Q_1\times Q_2).$$

with as yet undefined coefficients λ, μ, and ν. These coefficients must be found through the equation of the fourth point ($Q_4^* = \mathcal{T}Q_4$). We obtain

$$\mathcal{T} = \frac{[Q_2^*Q_3^*Q_4^*]}{[Q_2Q_3Q_4]}Q_1^*\circ(Q_2\times Q_3) + \frac{[Q_3^*Q_1^*Q_4^*]}{[Q_3Q_1Q_4]}Q_2^*\circ(Q_3\times Q_1) + \frac{[Q_1^*Q_2^*Q_4^*]}{[Q_1Q_2Q_4]}Q_3^*\circ(Q_1\times Q_2).$$

The simplest theorem to be checked by the above rules is the theorem of Pappos. (Figure C.3). The points of intersection of the opposite sides of a hexagon are given by $Q_1 = (A_1\times A_2)\times(A_4\times A_5)$, $Q_2 = (A_2\times A_3)\times(A_5\times A_6)$, and $Q_3 = (A_3\times A_4)\times(A_6\times A_1)$. It must be shown that $[Q_1,Q_2,Q_3] = 0$ if $[A_1,A_3,A_5] = 0$ and $[A_2,A_4,A_6] = 0$. This can be done by direct calculation using the rules explained above.

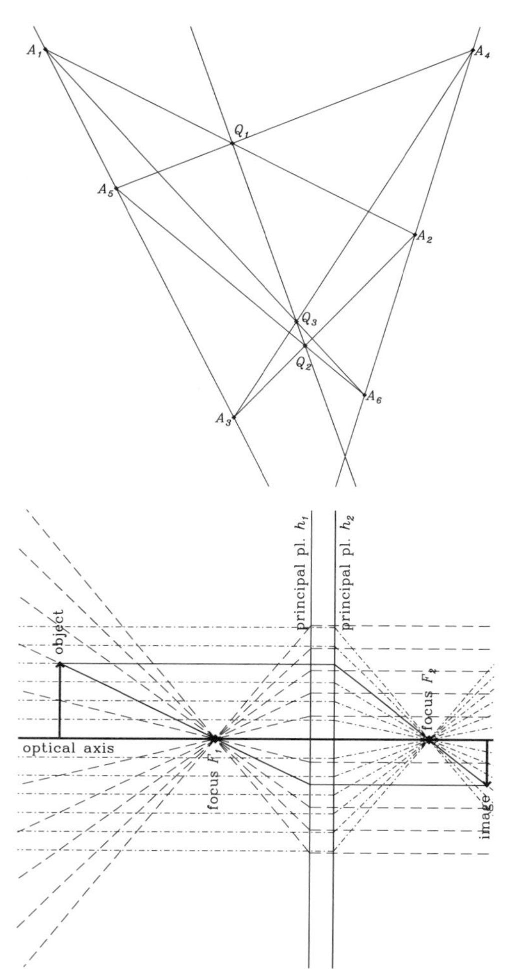

Figure C.3: *The theorem of Pappos.*

If the vertices of a hexagon $A_1, A_2, A_3, A_4, A_5, A_6$ lie alternately on two lines g and h, the points of intersection Q_1, Q_2, Q_3 of the opposite sides are also collinear, i.e., they lie on one third line.

Figure C.4: *The thick lens.*

A thick lens is defined by its two principal planes $h_1 = [-1, 0, x_1]$ and $h_2 = [-1, 0, x_2]$ and the focal length f. The rays parallel to the axes correspond to the rays that pass through the adjoint focal points.

A peculiar example of projective maps is given by the thick lens (Figure C.4). In the approximation of rays near to the optical axis, the map can be constructed through two principal planes and the focal points. When the principal planes coincide, we speak of a thin lens. We here choose the optical axis $y = 0$, put the principal planes in $x = x_1$ and $x = x_2$, and denote the focal length by f. The focus for the object has the coordinates $F_1 = [x_1 - f, 0, 1]$; the focus on the side of the image $F_2 = [x_2 + f, 0, 1]$. The homogeneous coordinates of the principal planes are $h_1 = [-1, 0, x_1]$ and $h_2 = [-1, 0, x_2]$; the infinite point of the optical axis is given by $O = [1, 0, 0]$. The pencil carried by F_1 is mapped on the pencil carried by O, and this one is mapped on the pencil carried by F_2 as indicated in the figure. We obtain the matrix of the projective map $\mathcal{T} A \propto (O \times (h_1 \times (F_1 \times A))) \times (F_2 \times (h_2 \times (O \times A)))$ in the form

$$\mathcal{T} = \begin{pmatrix} f + x_2 & 0 & (f - x_1)(f + x_2) - f^2 \\ 0 & f & 0 \\ 1 & 0 & f - x_1 \end{pmatrix}. \tag{C.18}$$

The matrices of this kind form a subgroup of the projective maps. For rays near enough to the optical axis, each lens system with common optical axis is equivalent to a thick lens. It is particularly curious that the refracting power f^{-1} of an ideal telescope[2] vanishes and that the principal planes pass through infinity.

C.3 Conic Sections

Conic sections can be defined in various ways. The simplest is to say that they are curves of second degree, that is, they are solutions of a quadratic equation. We regard a conic section as the set of points that solve an equation $\langle Q, \mathcal{A}Q\rangle = A_{kl}Q^kQ^l = 0$ or as the tangent bundle that solve an equation $\langle t, \mathcal{B}t\rangle = B^{kl}t_kt_l = 0$. The matrices $\mathcal{A}$ and $\mathcal{B}$ are chosen to be symmetric because any antisymmetric component would not enter the equations. Both A_{kl} and B^{kl} can be rescaled, so they have five essential parameters. Consequently, the conic is given by five points.

Four points define a pencil of conic sections. Its equation is constructed with the cross product. We use the fact that the triple product vanishes if two arguments coincide. Each conic section

$$\mathcal{K} = \lambda(g_{12} \circ g_{34} + g_{34} \circ g_{12}) + \mu(g_{13} \circ g_{24} + g_{24} \circ g_{13}) + \nu(g_{23} \circ g_{14} + g_{14} \circ g_{23}) \tag{C.19}$$

with $g_{ik} = Q_i \times Q_k$ passes through the four points Q_j: $\langle Q_j, \mathcal{K}Q_j\rangle = 0$ for all Q_j. Each of the three terms is constructed deliberately to yield this result. The matrix $\mathcal{K}$ is a linear combination of three terms. Only two of them are linearly independent (Eq. (C.17)). In addition, one of the parameters is irrelevant because of the homogeneity of the equations. A one-parameter pencil of conics is the result. If we require the conic section (C.19) to pass through the fifth point Q_5, we obtain an equation for the free parameter. If we put $\lambda = 1$ and $\nu = 0$, we obtain with $\langle Q_5, \mathcal{K}Q_5\rangle = 0$ an equation for μ. This coefficient corresponds to the invariant, Eq. (C.15). The matrix $\mathcal{K}$ turns out to be

$$\begin{aligned} \mathcal{K} &= \frac{Q_1 \times Q_2 \circ Q_3 \times Q_4 + Q_3 \times Q_4 \circ Q_1 \times Q_2}{[Q_1, Q_2, Q_5][Q_3, Q_4, Q_5]} \\ &- \frac{Q_1 \times Q_3 \circ Q_2 \times Q_4 + Q_2 \times Q_4 \circ Q_1 \times Q_3}{[Q_1, Q_3, Q_5][Q_2, Q_4, Q_5]}. \end{aligned} \tag{C.20}$$

The quadratic equation $\langle Q, \mathcal{K}Q\rangle = 0$ with the solutions Q_i ($i = 1, \ldots, 5$) can be given in the form

$$[(Q_1 \times Q_2) \times (Q_4 \times Q_5), (Q_2 \times Q_3) \times (Q_5 \times Q), (Q_3 \times Q_4) \times (Q \times Q_1)] = 0. \tag{C.21}$$

---

[2]In an ideal telescope, the second focus of the objective lens coincides with the first of the ocular system.

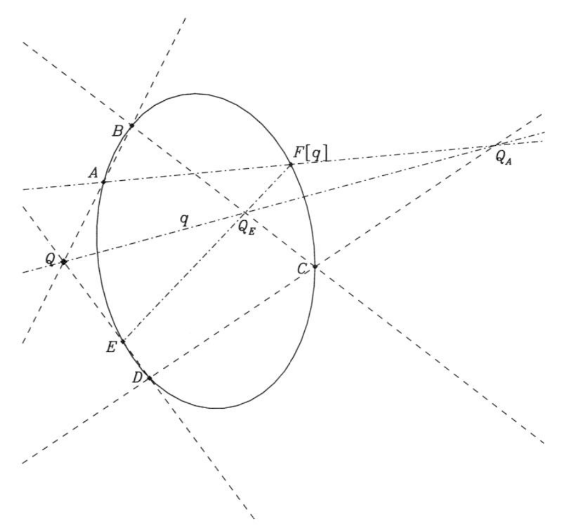

Figure C.5: *The theorem of Pascal.*

We inscribe the hexangle $ABCDEF$ in a conic. The pairs of opposite sides (AB and DE, BC and EF, CD and FA) intersect at Q_1, Q_2, and Q_3, respectively. These three points lie on a common line, Pascal's line. This fact is expressed by Eq. (C.21).

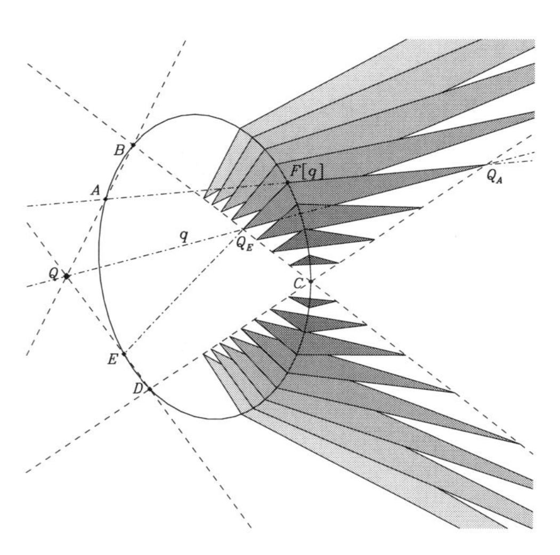

Figure C.6: *The projective definition of a conic section.*

We choose five points ($ABCDE$) and seek a sixth (P) in such a way that the points of intersection of the opposite sides of the hexagon $ABCDEP$ lie on a straight line, i.e., $ABCDEP$ is a Pascal hexagon. To begin with, the lines AB and DE intersect in Q. This point lies on Pascal's line. Next, we draw the lines $BC = e$ and $CD = a$ and expect that their intersections with EP and PA, respectively, lie on a common line with Q. Now we draw such a line q through Q arbitrarily. Its intersection with a is to lie on the side AP through the as yet unknown sixth point P, its intersection with e on the side EP. Therefore, we can now construct both sides. Their intersection is the desired point P on the conic section. To each straight line q through Q there belongs such a point.

By substituting the five points Q_i for Q one can check that they are solutions. The interpretation of this equation is the theorem of Pascal: The opposite sides of a hexagon inscribed in a conic section intersect at points that lie on a line, Pascal's line (Figure C.5). The theorem of Pappos is only a special case of this theorem—the case in which the conic degenerates into a pair of straight lines. In Eq. (C.21), $(Q_1 \times Q_2) \times (Q_4 \times Q_5)$ is the intersection of the two sides $(Q_1 \times Q_2)$ and $(Q_4 \times Q_5)$. Three such intersection points lie on a line, i.e., their triple product vanishes. The Pascal configuration can be used to construct the conic pointwise (Figure C.6). A conic can be defined as a curve for which Pascal's theorem holds. Equation (C.21) then shows that a conic is a curve of second degree.

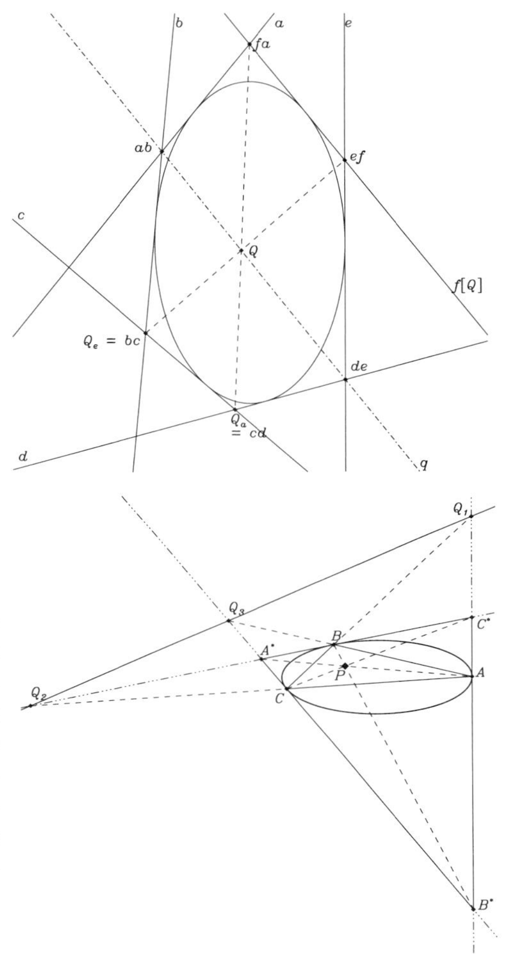

Figure C.7: *Brianchon's theorem.*

We circumscribe a hexilateral $ABCDEF$ on a conic. The pairs of opposite vertices define connecting lines that are concurrent, i.e., they pass through one point P. This theorem is dual to Pascal's theorem: Points are replaced by lines, intersection by connection, and collinearity by the property to be concurrent.

Figure C.8: *Triangle and conic.*

When in Figure C.5 three pairs of adjacent points or in Figure C.7 three pairs of adjacent tangents coincide, we obtain an inscribed triangle and a circumscribed trilateral. The intersections of the sides of the circumscribed trilateral with the opposite sides of the inscribed triangle are collinear. Dually, the lines that connect the vertices of the inscribed triangle with the opposite vertices of the circumscribed trilateral are concurrent two projective pencils of rays.

Finally, Figure C.6 can be read as the construction of two projective pencils of rays. A conic can also be defined as the product of two projective pencils: It contains the points at which the corresponding rays intersect. We can see this as follows. The pencil carried by A is perspectively mapped by the line a on the pencil carried by Q, and this is again perspectively mapped by the line e on the pencil carried by E. This establishes a projective relation of the pencils A and E. Conics can be defined as the product of two projectively related pencils of rays. One then easily shows that Pascal's theorem is implied and that one obtains equivalence to the other definitions. The dual of Pascal's theorem is the theorem of Brianchon (Figure C.7), and both combine in the particular case of a triangle with vertices on a conic and its tangent trilateral (Figure C.8). A straight line intersects a conic section at two points, which are not necessarily real. Here, this is a trivial consequence of the fact that $\langle Q, \mathcal{K}Q\rangle = 0$ is a quadratic equation. For instance, if the line is given by $Q = P_1 + \lambda P_2$, the values of λ are found by solving $\langle Q, \mathcal{K}Q\rangle = \langle P_1, \mathcal{K}P_1\rangle + 2\lambda\langle P_1, \mathcal{K}P_2\rangle + \lambda^2\langle P_2, \mathcal{K}P_2\rangle = 0$. Solving this equation, we get two real or two complex solutions or one double solution. In the last case, the line touches the conic.

If the two intersection points K_1, K_2 of a line $g = Q_1 \times Q_2$ are real and if we know one of them, we obtain a linear equation for the second. Its solution is

$$K_2 = K_1 - 2\frac{\langle K_1, \mathcal{K}Q_1\rangle}{\langle Q_1, \mathcal{K}Q_1\rangle}Q_1. \tag{C.22}$$

D The Transition from the Projective to the Metrical Plane

D.1 Polarity

Projective maps keep points and lines apart: There is no a priori linear map of points to lines and vice versa (the cross product is bilinear; it maps *pairs* of lines to points and *pairs* of points to lines). The introduction of a linear map between points and lines turns projective space into the metrical world of physics, where orthogonality is known and lengths and angles can be compared and multiplied, i.e., can be measured. In the following, the matrices $\mathcal{A}$, $\mathcal{B}$, etc. are assumed to be symmetric.[1] We use the summation convention.

We recall the axiom that all perpendiculars to a line lie in a pencil.[2] To any straight line there is a particular point that carries this pencil. As regards the generalization of perpendicularity, we now consider the relations between lines g and the corresponding points $P[g]$. When these relations are linear, we call it a *polarity*. We write

$$P[g] \stackrel{\text{def}}{=} \mathcal{B}g, \qquad P^k[g] = B^{kl}g_l. \tag{D.1}$$

We now intend to show that such a map can be constructed through three pairs of lines and poles when the orthocenter theorem is taken as an axiom. This map will turn out to be unique (Figure D.1), and the orthocenter theorem makes the map a projective one. We proceed in three steps. First, we obtain a general formula for the orthocenter theorem. Second, we show that the matrix $\mathcal{B}$ in Eq. (D.1) is symmetric, and finally the matrix $\mathcal{B}$ is formally constructed.

Three pairs of lines and poles fulfill the orthocenter theorem if

$$[(P_1 \times (g_2 \times g_3)), (P_2 \times (g_3 \times g_1)), (P_3 \times (g_1 \times g_2))] = 0.$$

When we expand the triple product using the known rules, we obtain

$$\langle g_1, P_2\rangle\langle g_2, P_3\rangle\langle g_3, P_1\rangle = \langle g_1, P_3\rangle\langle g_2, P_1\rangle\langle g_3, P_2\rangle. \tag{D.2}$$

Let us assume three such pairs of lines (not concurrent) and poles that fulfill the orthocenter theorem. Then the pole of any other line is determined by the orthocenter theorem (Figure D.1). The three altitudes $h_{4_{CDF}} = ((D \times P_2) \times (F \times P_1)) \times C$, $h_{4_{AEF}} = ((E \times P_2) \times (F \times P_3)) \times A$, and $h_{4_{BDE}} = ((E \times P_1) \times (D \times P_3)) \times B$ meet at one point $P[g_4]$ (i.e., $[h_{4_{CDF}}, h_{4_{AEF}}, h_{4_{BDE}}] = 0$). The map $g_4 \to P[g_4] = h_{4_{CDF}} \times h_{4_{AEF}}$ is linear in

[1] We could generalize the following constructions to nonsymmetric matrices, of course. This generalization would lead to spaces with torsion.

[2] This can be interpreted as a peculiar case of the orthocenter theorem.

The Geometry of Time. Dierck-E. Liebscher

ISBN: 3-527-40567-4

g_4 (not of fourth order, as one might expect). When we expand the cross products and remove a common factor (of third order in g_4) we obtain

$$\begin{aligned} P[g_4] &= h_{4_{CDF}} \times h_{4_{AEF}} \quad \propto \quad \mathcal{B}\, g_4 \\ \mathcal{B} &= \lambda_1\,(g_2 \times g_3) \circ P_1 + \lambda_2\,(g_3 \times g_1) \circ P_2 + \lambda_3\,(g_1 \times g_2) \circ P_3 \\ \lambda_1 &= \langle g_1, P_2\rangle\langle g_2, P_3\rangle,\ \ \lambda_2 = \langle g_2, P_1\rangle\langle g_2, P_3\rangle,\ \ \lambda_3 = \langle g_3, P_2\rangle\langle g_2, P_1\rangle. \end{aligned} \tag{D.3}$$

Again through the orthocenter theorem, this matrix is symmetric. This can be shown by simply expanding the equations $\langle g_i, \mathcal{B}g_k\rangle = \langle g_k, \mathcal{B}g_i\rangle$ for the three lines g_i. We have to use Eq. (D.2), i.e., the orthocenter theorem shows that a polarity is a projective map with symmetric matrix (Figure D.2).

Any symmetric matrix, i.e., any polarity, defines a conic, and vice versa. The conic can be defined as the set of lines that contain their own pole, $B^{kl}t_k t_l = 0$ (or as the set of points that carry their own polar). These lines constitute the bundle of tangents $[t_k]$ to the conic. The (abstract) polarity is identical with the ordinary polarity attributed to the conic. The poles of the tangents are the points of contact, $P^k[t] = B^{kl}t_l$. The tangents are orthogonal to themselves: The defining relation can be read this way. We know such a property from the discussion of the lightlike lines in Minkowski space–time. We can formulate it as follows: The polarity is a configuration with respect to the absolute conic. If $\mathcal{B}$ can be inverted, $\mathcal{AB} = \mathcal{E}$, we obtain for each point a polar, $p[Q] = \mathcal{A}Q,\ \ p_k[Q] = A_{kl}Q^l$. In this case, the polarity uniquely maps points to lines and lines to points. Special cases occur when this invertibility does not hold. An example of this are the geometries without curvature (i.e., with the axiom of parallels), for which all poles lie on one line (the line at metrical infinity). The two-parameter set of lines is then mapped on a one-parameter point set. In this case, $\mathcal{B}$ is of rank smaller than 3. In addition, it ceases to determine all of the geometry. We observe a distinct copolarity, defined by $\mathcal{A}$, that maps points P on lines $g[P] = \mathcal{A}P$. The matrix $\mathcal{A}$ too is of rank smaller than 3. We here obtain $A_{ik}B^{kl} = 0$.

Many relations of linear algebra acquire a simple geometrical interpretation. We only recall Eq. (C.13). It means the intersection of two straight lines is the pole of the line connecting the poles of the two lines (when the polarity is nondegenerate). Dually, the line connecting two points is the polar that is the intersection of the polars of the two points. In particular, the intersection of two tangents to the absolute conic is the pole of the connection of the points of contact. This construction is shown in Figure 8.16.

The pencil of perpendiculars to a line is determined by such a projective assignment $\mathcal{B}$ of its carrier. Two lines g and h are said to be perpendicular if the product $\langle h, \mathcal{B}g\rangle = h_k B^{kl} g_l = 0$ vanishes. In other words, if two lines are perpendicular, one passes through the pole of the other.

The polarity $P^k = B^{kl}g_l$, which assigns to each line g its pole $P[g]$, defines a metrical geometry. We need only an appropriate interpretation. First, we say that two lines are perpendicular if $g_k B^{kl} h_l = 0$. This notion of orthogonality refers to the polarity alone. Having defined bisection and multiplication by the harmonic range, we can start to verify the axioms of metric geometry.

For any polarity of full rank, polar triangles can exist in which each vertex is the pole of the opposite side. A first example was drawn in Figure 7.1. We begin with a line g, determine its pole $P[g]$, and draw a line h through it. The pole $P[h]$ again lies on g. The foot point

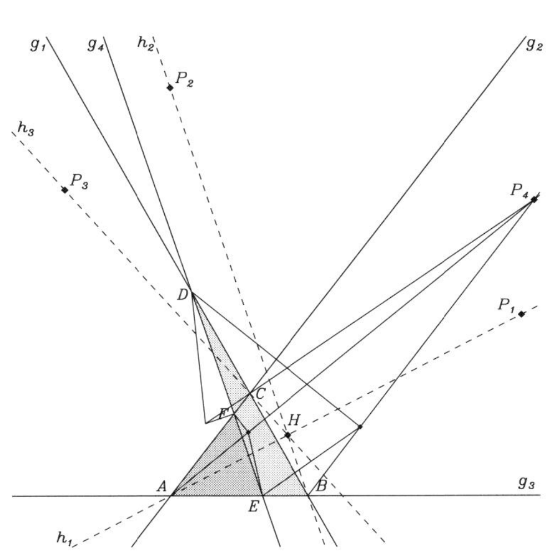

Figure D.1: *Three pairs determine the polarity.*

In general, three polar pairs $[g_i, P_i]$, $i = 1, \ldots, 3$, determine the polarity. Even these three pairs obey a restriction: The altitudes of the trilateral $[g_1, g_2, g_3]$ must intersect at one point (Figure A.7). The orthocenter is defined by two poles already. The third pole must lie on the third altitude. The pole of a fourth line g_4 is found as an intersection of the altitudes on g_4 in the trilaterals $[g_2, g_3, g_4]$, $[g_3, g_1, g_4]$, and $[g_1, g_2, g_4]$. These three altitudes meet at one point P_4. When the three poles are collinear, they define an absolute polar. When they coincide, they coincide at the absolute pole.

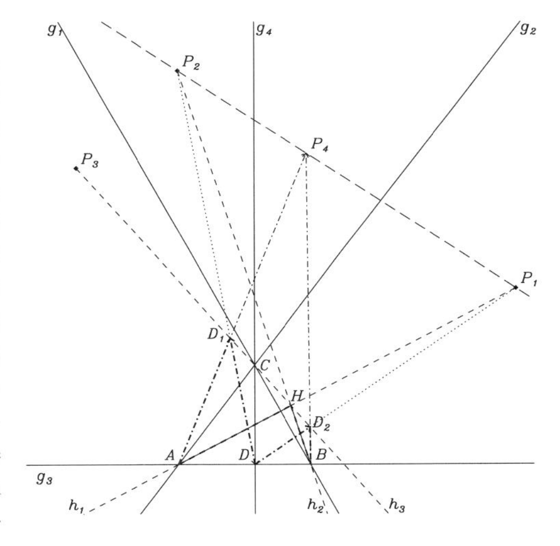

Figure D.2: *The polarity is a projective map.*

When the polarity is given by three polar pairs, the orthocenter theorem states that the poles of all lines that are concurrent with g_1 and g_2 lie on the line that connects P_1 with P_2. The map is projective, i.e., the pencil C is mapped on g_3, and g_4 on D in particular, subsequently g_3 from P_2 onto h_3, in particular D on D_1, and finally h_3 from A onto the connection P_1P_2, in particular D_1 on P_4. The line AD_1 is the altitude on g_4 in the trilateral $[g_2, g_3, g_4]$. Analogously, one constructs BD_2, that is, the corresponding altitude in the trilateral $[g_1, g_3, g_4]$. The intersection of both altitudes (P_4) is collinear with P_1 and P_2, because the three points are intersections of opposite sides in the hexagon $[A, D_1, D, D_2, B, H]$ (Pappos theorem, Figure C.3).

$F_g[h] = g \times h$ of h on g turns out to be the pole of the line $P[g] \times P[h]$ connecting $P[g]$ and $P[h]$. By assumption, $h_k g_l B^{kl} = 0$ holds: Just as $P[g]$ carries the line h, the point $P[h]$ lies on g. The pole of $P[g] \times P[h]$ is now determined by the rule (C.13). This requires $\mathcal{B}[B[g] \times B[h]] = \det[\mathcal{B}]\ g \times h$, i.e., the desired pole is the intersection of the lines g and h. In our triangle, each vertex is the pole of the opposite side (Figure D.3). Corresponding to our definition of orthogonality, we obtained a triangle with three right angles, which we know from the sphere (Figure 7.1). In fact, on a sphere each great circle can be replaced by its pair of poles. The two poles coincide in the central projection onto the plane. The situation is modified when we consider the case of a real absolute conic. Here, the sets of points and lines are divided into different, separately invariant subsets. The polar triangles do not then lie in *one* transitivity region. The projective plane allows us to draw a polar triangle in this case too (Figure D.3), but now not all the three vertices and not all the three sides are elements of the group of motions of the induced geometry.

If the rank of $\mathcal{B}$ is maximal (i.e., equal to 3), the conic section $A_{kl}Q^kQ^l = 0$ is identical to the envelope of the tangent bundle $B^{kl}t_kt_l = 0$. Each straight line g intersects the conic at two not necessarily real points $K_1[g]$ and $K_2[g]$. The tangents at these points intersect at the pole $P[g]$, which is always real, of the line. The poles of a pencil of rays through Q lie on the polar $p[Q] = \mathcal{A}Q$ of its origin:

$$\langle Q, g\rangle = \langle \mathcal{A}Q, \mathcal{B}g\rangle.$$

If Q lies on g, i.e., $\langle Q, g\rangle = 0$, we obtain $\langle \mathcal{A}Q, \mathcal{B}g\rangle = 0$ too, i.e., the image $\mathcal{A}Q$ of the point lies on the image of the line, $\mathcal{B}g$.

If $\mathcal{B}$ is of rank 3, the matrix $\mathcal{A}$ is determined as the inverse of $\mathcal{B}$. If $\mathcal{B}$ is only of rank 2, one must take for $\mathcal{A}$ the matrix of the subdeterminants $A_{mn} = \epsilon_{ijm}\epsilon_{kln}B^{ik}B^{jl}$. This matrix satisfies the equation $A_{ik}B^{kl} = 0$ nontrivially and is of rank 1, of course. Through the equation $\mathcal{B}p = 0$, the matrix $\mathcal{B}$ determines an absolute polar (the line at infinity or its image, respectively), and hence one can write $\mathcal{A}$ in the form of a direct product $\mathcal{A} = p \circ p$. If a straight line g intersects this polar at the point $F[g] = g \times p$, the pairs $[F[g], \mathcal{B}g]$ define an involution on the absolute polar. The fixed points of this involution are real if the matrix $\mathcal{B}$ is indefinite (Minkowski geometry). Next, if $\mathcal{B}$ is only of rank 1, the matrix $\mathcal{A}$ can be of rank 1 or of rank 2. In the latter case, the relations between $\mathcal{A}$ and $\mathcal{B}$ are inverted. The equation $\mathcal{A}P = 0$ defines an absolute pole P, and $\mathcal{B}$ can be written in the form $\mathcal{B} = P \circ P$. The points Q define rays $f[Q] = Q \times P$ through P that together with $\mathcal{A}Q$ define an involution in the pencil of polars. The fixed rays of this involution are real if $\mathcal{A}$ is indefinite (anti-Minkowski geometry). Finally, the matrix $\mathcal{A}$ can be of rank 1 (Galilean geometry). In this case, $\mathcal{A}P = 0$ defines a linear point row (carried by the absolute polar), and $\mathcal{B}p = 0$ defines a pencil of rays (carried by the absolute pole).

D.2 Reflection

We now construct the formulas for the reflections. We begin with the formulas for the reflection on a line g. First, the line and each of its points are reflected into itself. The line connecting an arbitrary point Q to its reflected image $S[Q]$ must be perpendicular to the mirror g. Hence, the connecting line passes through the pole $P[g]$ of the mirror line. The pole is mapped to itself as is the foot point F in which the mirror line and the perpendicular meet. Consequently, the cross-ratio of $P[g]$ and F with Q and $S[Q]$ does not change if $S[Q]$ and Q are interchanged: It is equal to minus one. We obtain $S[Q]$ by a harmonic separation on the perpendicular dropped from Q onto the mirror line g. The expression

$$\mathcal{S}_g[Q] = \mathcal{S}Q = (\mathcal{E}\langle g, \mathcal{B}g\rangle - 2\mathcal{B}g \circ g)Q, \qquad S^k_l = \delta^k_l g_r B^{rs} g_s - 2B^{kr} g_r g_l \tag{D.4}$$

is a solution of these conditions. The problem that is dual to that of finding the reflection on a line would be to construct the reflection on a point Q. The objects of this reflection are now lines. By the duality of the construction, both Q and its polar are preserved in the reflection, and they provide the fixed pair for the harmonic figuration of a line g with its image $\mathcal{S}_A[h]$. We obtain the dual formula

$$\mathcal{S}_Q[h] = h\mathcal{S} = h(\mathcal{E}\langle Q, \mathcal{A}Q\rangle - 2Q \circ Q\,\mathcal{A}), \qquad S_l^k = \delta_l^k Q^r A_{rs} Q^s - 2Q^k Q^r A_{rl}. \quad \text{(D.5)}$$

It is obvious that in both cases $\mathcal{S}^2 = 1$. If the polarity is nondegenerate, i.e., $\mathcal{AB} = \mathcal{E}$, then the two reflections coincide, $\mathcal{S}_{Bg} = \mathcal{S}_g$. In this case, the reflection on a line and on its pole are identical operations.[3] If $\mathcal{B}$ is not of full rank, degenerate cases are possible. In any case, a reflection in the projective-metric plane is determined by a polar pair of a point Q and a line g.

$$\mathcal{S} = \mathcal{E}\langle Q, g\rangle - 2Q \circ g. \qquad \text{(D.6)}$$

If g and Q are a polar pair, the polarity is conserved, and the map is a reflection member of our group of motions. Otherwise, Eq. (D.6) describes only a general involution.

The involutions that we call reflections preserve $\mathcal{A}$ as well as $\mathcal{B}$. If a point Q lies on the absolute cone $\mathcal{A}$, i.e., $\langle Q, \mathcal{A}Q\rangle = 0$, then its reflected image lies on this cone too. The image of any tangent to the absolute conic $\mathcal{B}$, i.e., $\langle g, \mathcal{B}g\rangle = 0$ is again a tangent. We check this by calculating the transforms of $\mathcal{A}$ and $\mathcal{B}$, respectively:

$$S_m^k A_{kl} S_n^l = A_{mn}(g_r B^{rs} g_s)^2, \qquad S_m^k B^{mn} S_n^l = B_{mn}(g_r B^{rs} g_s)^2.$$

Points and tangents of the conic are mapped on points and tangents of the conic again (Figure 8.15).

The next problem is the reflection of a given point Q_1 at another given point Q_2. We solve it by constructing the perpendicular bisector (Figure D.4). First, we determine the pole $P[Q_1 \times Q_2]$ of the connecting line. The pole $P[m]$ of the required perpendicular bisector must divide the chord (Q_1, Q_2) harmonically with the foot point $F = m \times (Q_1 \times Q_2)$. The foot F and pole $P[m]$ are the solutions of the equation $F = Q_1 + \lambda Q_2$, $P = Q_1 + \mu Q_2$ with

$$\frac{\mu}{\mu - \infty}\frac{\lambda - \infty}{\lambda} = -1, \qquad \mathcal{A}P = F \times \mathcal{B}(Q_1 \times Q_2).$$

On the one hand, $\mu = -\lambda$, and

$$\mathcal{A}(Q_1 + \mu Q_2) = (Q_1 - \mu Q_2) \times (\mathcal{B}(Q_1 \times Q_2)) \quad \to \quad \langle (Q_1 - \mu Q_2), \mathcal{A}(Q_1 + \mu Q_2)\rangle = 0.$$

on the other. Hence, we obtain for the coefficient $\mu^2\langle Q_2, \mathcal{A}Q_2\rangle = \langle Q_1, \mathcal{A}Q_1\rangle$ and for the perpendicular bisector

$$m[Q_1, Q_2] = B[Q_1 \times Q_2] \times \left(Q_1 \pm Q_2 \sqrt{\frac{\langle Q_1, AQ_1\rangle}{\langle Q_2, AQ_2\rangle}} \right). \qquad \text{(D.7)}$$

Naturally, we obtain two solutions for a segment $[Q_1, Q_2]$ because the cross-ratio is a relation between four points. One of them must be chosen as bisector and foot point F. Then the other turns out to be the pole $P[m]$ of the perpendicular bisector. The reflection is now given by

$$\mathcal{S} = \mathcal{E}\langle P[m], m\rangle - 2P[m] \circ m.$$

[3] We consider here reflections in the projective plane. The reflection on a pole is usually not to be understood as the reflection on a point in the sense of Appendix A if the pole does not belong to the group of motions. The projective plane used to represent the group of motions sometimes contains more points and lines than the group does.

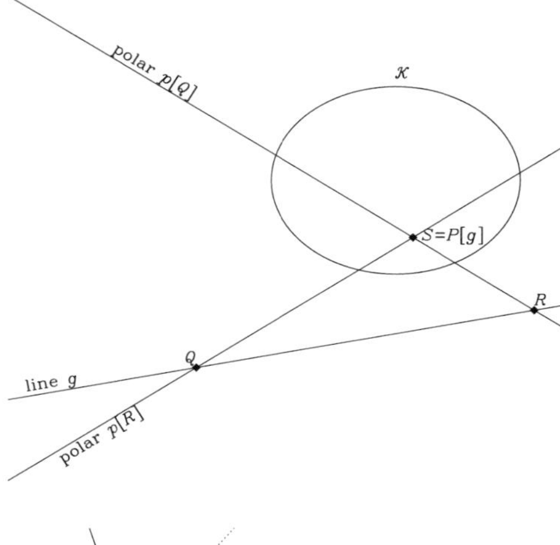

Figure D.3: *Polar triangles.*

We put a point Q and through it a line g into the projective plane. The polar $p[Q]$ intersects g at the point R. Its polar $p[R]$ intersects g at Q again and $p[Q]$ at a third point S. The latter turns out to be the pole $P[g]$ of g. In this way, we obtain a triangle in which the sides are the polars of the vertices. The figure shows the case of a real absolute conic section.

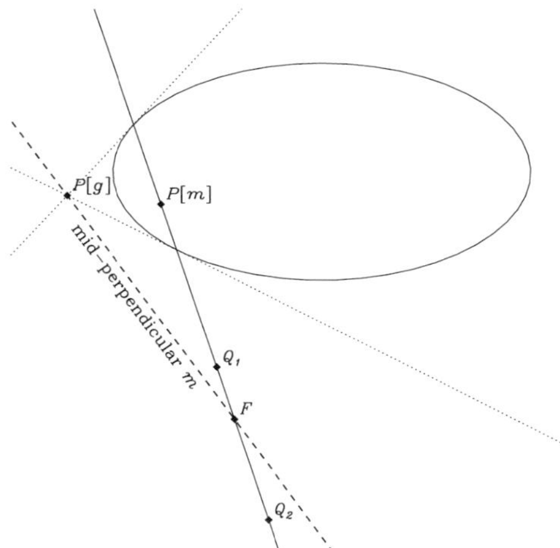

Figure D.4: *The perpendicular bisector.*

In order to obtain the perpendicular bisector between Q_1 and Q_2 we draw the connecting line g and its pole $P[g]$. The perpendicular bisector m passes through this pole and intersects g at some point F. The pole $P[m]$ and F must divide both the segment $[Q_1, Q_2]$ as well as the chord of the absolute conic harmonically. $P[m]$ and F are projectively equivalent bisectors of Q_1Q_2.

When we try to find the reflection of two lines on each other, we obtain the angle bisector. The construction is dual to the one above. The angle bisectors are

$$w = h_1 \pm h_2 \sqrt{\frac{\langle h_1, \mathcal{B}h_1 \rangle}{\langle h_2, \mathcal{B}h_2 \rangle}}.$$

The two reflections are

$$\mathcal{S}_\pm = \mathcal{E}\langle \mathcal{B}w_\pm, w_\pm \rangle - 2\mathcal{B}w_\pm \circ w_\pm.$$

In the nondegenerate case, $\mathcal{A}[M_\pm] \propto M_\mp \times B[Q_1 \times Q_2]$ and $B[w_\pm] \propto w_\mp \times \mathcal{A}[h_1 \times h_2]$.

The product $h = agb$ of three rays of one pencil is found in the form

$$g = a + \lambda b \quad \rightarrow \quad agb = h, \qquad h = a + \frac{1}{\lambda} \frac{\langle a, \mathcal{B}a \rangle}{\langle b, \mathcal{B}b \rangle} b. \tag{D.8}$$

If the product $QgR = h$ is a line, it is found analogously. Because the line g is perpendicular to the connecting line $Q \times R$ (Figure A.6), the pole of g lies on $Q \times R$:

$$\mathcal{B}g = Q + \lambda R \quad \rightarrow \quad QgR = h, \qquad h = \mathcal{A}\left(Q + \frac{1}{\lambda}\frac{\langle Q, \mathcal{A}Q\rangle}{\langle R, \mathcal{A}R\rangle} R\right). \tag{D.9}$$

The product $aQb = R$ is a point if Q lies on the common perpendicular of a and b, i.e., the polar of Q belongs to the pencil defined by a and b:

$$\mathcal{A}Q = a + \lambda b \quad \rightarrow \quad R = \mathcal{B}\left(a + \frac{1}{\lambda}\frac{\langle a, \mathcal{B}a\rangle}{\langle b, \mathcal{B}b\rangle} b\right). \tag{D.10}$$

The product $QER = F$ is a point if the polars lie in a pencil:

$$\mathcal{A}F = \mathcal{A}Q + \lambda \mathcal{A}R \quad \rightarrow \quad F = Q + \frac{1}{\lambda}\frac{\langle Q, \mathcal{A}Q\rangle}{\langle R, \mathcal{A}R\rangle} R. \tag{D.11}$$

We add two useful formulas. Corresponding to Eq. (C.11), we obtain in the case $[g, a, b] = 0$ the formulas

$$g \ \propto \ a\langle b \times g, \mathcal{A}(a \times b)\rangle + b\langle g \times a, \mathcal{A}(a \times b)\rangle,$$

and we obtain for the reflection $h = agb$ the relation

$$h \ = \ b\langle a, \mathcal{B}a\rangle\langle b \times g, \mathcal{A}(a \times b)\rangle + a\langle b, \mathcal{B}b\rangle\langle g \times a, \mathcal{A}(a \times b)\rangle.$$

All this is valid for the nondegenerate case.

We obtain formulas for a *rotation* as a composition of two reflections whose mirror lines intersect at the rotation center. Such a product has the form

$$\mathcal{S}_{g_2}\mathcal{S}_{g_1} = 4\langle g_1, \mathcal{B}g_2\rangle \mathcal{B}g_2 \circ g_1 - 2\langle g_1, \mathcal{B}g_1\rangle \mathcal{B}g_2 \circ g_2 - 2\langle g_2, \mathcal{B}g_2\rangle \mathcal{B}g_1 \circ g_1 + \langle g_1, \mathcal{B}g_1\rangle\langle g_2, \mathcal{B}g_2\rangle \mathcal{E}.$$

If we want to rotate a given line g_1 into another given line g_3, we combine a reflection on g_1 with a reflection on the angle bisector $g_2 = g_1 + g_3\sqrt{\langle g_1 \mathcal{B}g_1\rangle / \langle g_3, \mathcal{B}g_3\rangle}$.

D.3 Velocity Space

The space of relative velocities is the simplest and most famous example of hyperbolic (Lobachevski) geometry. The figures in the velocity space are called hodographs. Each point symbolizes a velocity that the considered object can have. The norm of the velocities is bounded by the speed of light (strictly speaking, by the absolute velocity); all figures lie within limit circles or limit spheres, respectively. Translations in a hodograph are generated by composition with global velocities. The limit circle is preserved, i.e., the light velocity changes at most its direction (see aberration). This composition is a projective transformation with preserved (i.e., absolute) conic. Consequently, the velocity space is endowed with the Lobachevski geometry.

Our example is the billiard collision (one ball at rest is struck by one in motion). The figure of the positions reached after a certain time in a billiard collision (Figure 5.6) must be generated from the symmetric collision (Figure 3.7) by a procedure based on relativity. In accordance with Einstein's addition theorem of velocities, we obtain it by inserting the relative velocity between the billiard collision and the symmetric collision into the figure of the latter. The circle of the end positions is turned into a figure that is an ellipse by Euclidean standards but a circle in the Lobachevski geometry. The projection of this circle's center is found experimentally by the intersection of the line connecting the pairs of the end points of individual collisions. For calculation, we first determine the projective map that generates this translation. We calculate it in Klein's model with the boundary

$$x^2 + y^2 - z^2 = 0.$$

The matrix of translation in the x direction is given by

$$\mathcal{T} = \begin{bmatrix} 1 & 0 & r \\ 0 & \sqrt{1-r^2} & 0 \\ r & 0 & 1 \end{bmatrix}.$$

This transformation maps the boundary onto itself. The circle $r^2z^2 - x^2 - y^2 = 0$ is shifted into an ellipse: The points $Q = [r\cos\varphi, r\sin\varphi, 1]$ are mapped to

$$\mathcal{T}Q = \left[1 + \cos\varphi, \sqrt{1-r^2}\sin\varphi, \frac{1}{r} + r\cos\varphi\right].$$

As we must expect, the point $[-r, 0, 1]$ is shifted to $[0, 0, 1]$. The diameter $\beta = 2/(r + 1/r)$ is equal to the velocity of the projectile particle (normalized by the speed of light). Hence, the employed parameter r is given by the formula

$$r = \frac{1}{\beta}\left(1 - \sqrt{1-\beta^2}\right).$$

The projection $[r, 0, 1]$ of the center $[0, 0, 1]$ divides the diameter in the ratio

$$\gamma = \frac{r}{\frac{2}{r+\frac{1}{r}} - r} = \frac{1}{\sqrt{1-\beta^2}}.$$

This is the known ratio of the *moving mass* (inertial mass) to the *rest mass*, Eq. (5.2).

A translation in the velocity space is a composition with some given velocity. The points on the boundary line remain there (the speed of light is not changed) but are moved on the line (Figure D.5). This motion is aberration (Figure 4.3). Completed to the sphere, it yields a conformal map (Figure 5.24). The group of conformal maps on a sphere is isomorphic to the (homogeneous) Lorentz group.

Figure D.5 reminds us that two special Lorentz transformations with velocities in different directions cannot yield another special Lorentz transformation (Figure D.6). This is due to the curvature of the velocity space (Figures 7.11 and 7.12). We see this formally as follows. The

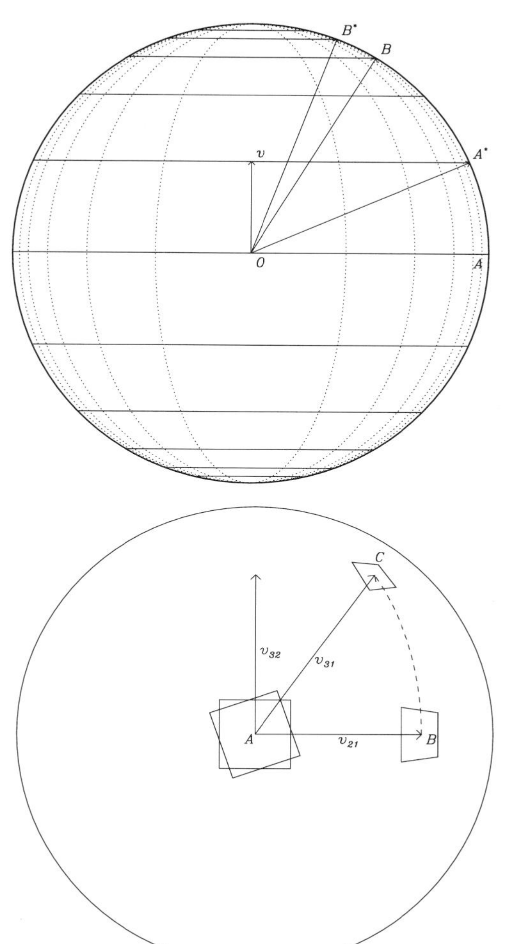

Figure D.5: *Velocity translation and aberration.*

The circle represents a two-dimensional velocity space in the form of Klein's model. The vector $\boldsymbol{v}$ indicates a translation, i.e., a composition of all velocities with $\boldsymbol{v}$. Through this translation, the set of straight lines is mapped onto itself, with A into A^* and B into B^*. The dotted lines are orbits of the translation (see also Figure 9.16).

Figure D.6: *Velocity translations do not form a group.*

In general, the combination of two velocity translations does not result in another velocity translation. With the orbits of a translation indicated in Figure D.5, we see that the composition of two translations $\boldsymbol{v}_1$ and $\boldsymbol{v}_2$ contains a rotation with respect to the translation with the combined velocity $\boldsymbol{v}$. This rotation yields the Thomas precession.

general translation in velocity space is given by

$$\mathcal{T}[u,v] = \begin{bmatrix} \gamma + v^2 \frac{1-\gamma}{u^2+v^2} & -uv\frac{1-\gamma}{u^2+v^2} & \gamma u \\ -uv\frac{1-\gamma}{u^2+v^2} & \gamma + u^2 \frac{1-\gamma}{u^2+v^2} & \gamma v \\ \gamma u & \gamma v & \gamma \end{bmatrix}. \tag{D.12}$$

The components of the translation (normalized to the speed of light) are u and v, and the coefficient γ is as usual $\gamma = 1/\sqrt{1-u^2-v^2}$. We easily check that $\mathcal{T}[u,v]\mathcal{T}[-u,-v] = \mathcal{E}$. We check as well that $\mathcal{T}[0,v]\mathcal{T}[u,0]$ is a transformation that shifts the center, $O = [0,0,1]$, to the point $P = [u\sqrt{1-v^2}, v, 1]$. The combination $\mathcal{T}[-u\sqrt{1-v^2}, -v]\mathcal{T}[0,v]\mathcal{T}[u,0]$ then has the fixed point O, and is a rotation about the angle φ with

$$\sin\varphi = \frac{uv}{1+\sqrt{1-u^2}\sqrt{1-v^2}}, \qquad \cos\varphi = \frac{\sqrt{1-u^2}+\sqrt{1-v^2}}{1+\sqrt{1-u^2}\sqrt{1-v^2}}.$$

Pure velocity translations, Eq. (D.12), do *not* form a subgroup of the motions in the velocity space. This is a corollary to the fact that, in a four-dimensional world, the special Lorentz transformations, Eq. (B.3), do not form a subgroup of the Lorentz group.

We use the Lobachevski translations in velocity space to show its relation to the variation of mass with velocity [102]. In Figure D.7 we show the velocities in a particular collision in the plane. Two particles reflect each other in one direction while one passes the other in the perpendicular one. We compare the reference frames in which the velocity of one of the two particles merely changes its sign. The orientation can be chosen in such a way that this occurs just in the v_x component. We obtain the paths marked in the left part of the figure. On the right, the corresponding velocities are drawn in the velocity space. The onflight velocities are EA and EB in the lower left. They are shifted to EC and ED when we go over to the reference frame of the upper left. In both cases, the total momentum must be conserved. We now suppose that the masses vary with velocity $\boldsymbol{v}$ by a factor $\gamma[v]$ that does not depend on the direction, i.e., $m[\boldsymbol{v}] = m\cdot\gamma[v]$. We can choose $\gamma[0]=1$. The momentum conservation for the first component is now given twice:

$$\begin{aligned} m_1\,\gamma[A]\,v_x[A] + m_2\,\gamma[B]\,v_x[B] &= 0, \\ m_1\,\gamma[C]\,v_x[C] + m_2\,\gamma[D]\,v_x[D] &= 0. \end{aligned}$$

We use the equation for the ellipses, i.e.,

$$\frac{v_x^2[C]}{v_x^2[A]} + \frac{v_y^2[C]}{c^2} = 1, \qquad \frac{v_x^2[B]}{v_x^2[D]} + \frac{v_y^2[B]}{c^2} = 1.$$

The momentum conservation law now gets the form

$$\begin{aligned} m_1\,\gamma[A]\,v_x[A] + m_2\,\gamma[B]\,v_x[D]\sqrt{1-\frac{v_y^2[B]}{c^2}} &= 0, \\ m_1\,\gamma[C]\,v_x[A]\sqrt{1-\frac{v_y^2[C]}{c^2}} + m_2\,\gamma[D]\,v_x[D] &= 0. \end{aligned}$$

We recall that $v_y[B] = -v_y[C]$ and write shortly $v_y[B] = u$. The momentum conservation is now a system of linear homogeneous equations for $m_1 v_x[A]$ and $m_2 v_x[D]$. It can be solved only in the case of

$$\gamma[A]\gamma[D] - \gamma[B]\gamma[C]\left(1-\frac{u^2}{c^2}\right) = 0.$$

In the limit of very small components v_x we find $\gamma[A]=\gamma[D]=1$, $\gamma[B]=\gamma[C]=\gamma[u]$, and, finally, $\gamma[u] = 1/\sqrt{1-\frac{u^2}{c^2}}$. This is the well-known Lorentz factor. We easily check that this function $\gamma[u]$ also solves the system for finite components v_x. The result is that Einstein's composition of velocities implies the variation of mass with velocity (Table 5.1).

D.4 Circles and Peripheries

As we have often remarked, it is helpful to imagine a sphere in order to guess formulas, which, of course, must be proven later. In such a way, we determine the formula of a circle on the unit sphere. In accordance with its three-dimensional embedding, the circle is the locus of the unit vectors (from the center Z of the sphere to the periphery points Q) that make a constant angle with the unit vector to the center M of the circle:

$$\cos(\angle MZP) = \frac{\langle \vec{ZM}, \mathcal{A}\,\vec{ZQ}\rangle}{\sqrt{\langle \vec{ZM}, \mathcal{A}\,\vec{ZQ}\rangle}\sqrt{\langle \vec{ZM}, \mathcal{A}\,\vec{ZQ}\rangle}} = \text{const} \qquad \text{with } \mathcal{A} = \begin{pmatrix} 1 & 0 & 0 \\ 0 & 1 & 0 \\ 0 & 0 & 1 \end{pmatrix}.$$

We now use the fact that the vector coordinates are already the projective point coordinates (the sphere is projected from its center). In addition, a scalar product of two points contains the map $\mathcal{A}$ of points to lines. The matrix of this map is the unit matrix in the case of a sphere and so is not explicitly visible. To obtain a formula that is valid generally and not only for spherical geometry, we have to take into account the map $\mathcal{A}$. For the points Q of a circle around M, we obtain

$$\frac{\langle Q, \mathcal{A}M\rangle^2}{\langle Q, \mathcal{A}Q\rangle\langle M, \mathcal{A}M\rangle} = \text{const}.$$

The constant can be determined by a point Q_1 if, for instance, we know that it lies on the circumference. Consequently, it yields

$$\langle Q, \mathcal{A}M\rangle^2\langle Q_1, \mathcal{A}Q_1\rangle = \langle Q_1, \mathcal{A}M\rangle^2\langle Q, \mathcal{A}Q\rangle.$$

In other words, the point Q_1 determines the radius and the circle. Finally, we obtain the quadratic form

$$\langle Q, \mathcal{K}Q\rangle = 0, \qquad \mathcal{K} = \mathcal{A}\langle Q_1, \mathcal{A}M\rangle^2 - \langle Q_1, \mathcal{A}Q_1\rangle\; \mathcal{A}M \circ \mathcal{A}M. \tag{D.13}$$

Given $\mathcal{A}$ and M, one essential parameter is free for definition of a circle. Figures 9.16–9.19 show such one-parametric pencils of circles.

In the Euclidean geometry, a circle is determined by three points Q_1, Q_2, Q_3 on its circumference. Here, we have to represent it as a particular conic. Both the elements that are lacking for a conic are supplied by the relation to the absolute conic. In the general case, we construct from three given points the circumcenter as an intersection of the perpendicular bisectors and then use Eq. (D.13). In another method, we could first construct some new points by reflection on the perpendicular bisectors and then use Eq. (C.20).

Let us assume that a quadratic form $\mathcal{K}$ is given. What is the condition for it to determine a circle? We must find an origin M carrying rays that cut out segments bisected by M itself. The pencil of rays through M is the pencil of diameters. The poles of the rays g through M lie on its polar $p[M] = \mathcal{A}M$. Because of the division property of the diameters, this is also the polar with respect to the circle, $p[M] = \mathcal{K}M$. All rays g of the diameter pencil satisfy $\mathcal{K}^{-1}g \propto \mathcal{B}g$ (Figure D.8). The equation $\langle Q, \mathcal{K}Q\rangle = 0$ defines a circle if there exists a pencil

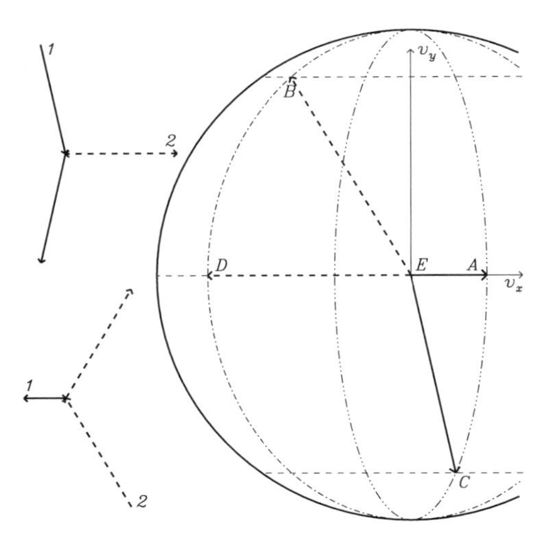

Figure D.7: *Composition of velocities and variation of mass with velocity.*

On the left, we see two space diagrams of a glancing transverse collision in which a particle coming from the left collides with another coming from the right. In the lower diagram, the reference frame is chosen in such a way that the velocity of the left particle merely changes its sign. The same holds for particle coming from the right in the upper part. In order to transit from one reference frame to the other, we must superpose an appropriate velocity. Here, this velocity has only a component in the v_y direction. On the right, we show the corresponding Lobachevski translations of the velocity vectors. The points A and B represent the velocities of both particles before the collision in the frame of the lower left. They are shifted by the superposition into C and D.

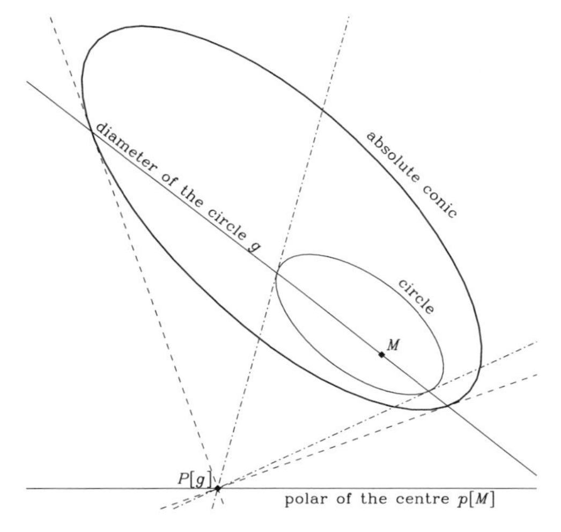

Figure D.8: *The circle.*

The metric relations are given by an absolute conic. We choose the center M of the circle and draw its polar $p[M]$. Now we consider some diameter g. Its pole $P[g]$ lies on the polar of M. The (broken line) tangents to the absolute cone are perpendicular to g. The (chain line) tangents to a circle around M in the intersections with the diameter g are perpendicular to this diameter. So they intersect at $P[g]$, too. That is, if g is a diameter, its pole $\mathcal{B}g$ referring to the absolute conic coincides with the pole $\mathcal{K}^{-1}g$ referring to the circle itself. Circles of this general form are the orbits of rotations in Chapter 9.

M whose rays g form an invariant subspace of $\mathcal{KB}$:

$$\mathcal{KB}g = \lambda g \qquad \text{for fixed } \lambda \text{ for all } g \text{ with } \langle g, M\rangle = 0.$$

In the dual construction, there exists a point row q (which is equal to $p[M]$) whose points Q fulfill the condition

$$\mathcal{K}Q = \lambda\mathcal{A}Q \qquad \text{for fixed } \lambda \text{ for all } Q \text{ with } \langle q, Q\rangle = 0.$$

Circles are conic sections that touch the absolute conic twice. The points of contact lie on the polar of $p[M]$ of the circle's center M (and can be imaginary, if this polar does not intersect the circle in real points). The center M and its polar $p[M]$ are polar not only with respect to the absolute conic but also with respect to the circle.

Finally, we determine a curve for which the theorem of the circumference angles is valid. Such a curve is the locus of the points at which a given segment subtends a constant angle. In both the Euclidean and pseudo-Euclidean geometries, this is a circle. We shall show why in general we find a curve of fourth order. We derive the formula by looking at a sphere again. Two points on the sphere (which form the segment) are given. Any third point defines two planes with the coefficients $g_1 = Q_3 \times Q_1$ and $g_2 = Q_3 \times Q_2$ for the normals. The angle between the planes is the angle formed at the third point in the triangle with the given segment (Q_1, Q_2). This gives

$$\cos\gamma = \frac{\langle g_1, \mathcal{B}g_2\rangle}{\sqrt{\langle g_1, \mathcal{B}g_1\rangle\langle g_2, \mathcal{B}g_2\rangle}}.$$

Consequently, the equation for the desired circumference is given by

$$\frac{\langle g_1, \mathcal{B}g_2\rangle}{\sqrt{\langle g_1, \mathcal{B}g_1\rangle\langle g_2, \mathcal{B}g_2\rangle}} = \frac{\langle g_{31}, \mathcal{B}g_{32}\rangle}{\sqrt{\langle g_{31}, \mathcal{B}g_{31}\rangle\langle g_{32}, \mathcal{B}g_{32}\rangle}} = \cos\gamma \text{ (constant)}. \tag{D.14}$$

This is an equation of fourth order. The circumference is *not* a conic section in the generic case (Figure 9.13), in contrast to the degenerate cases of the Euclidean or pseudo-Euclidean geometry. Of course, Eq. (D.14) can be written with the matrix $\mathcal{A}$ too. By substituting $B^{ij} = \epsilon^{ikl}A_{km}\epsilon^{jmn}A_{ln}$, we obtain the nice formula

$$\begin{aligned} & \left\langle Q, \left(\mathcal{A} - \tfrac{\mathcal{A}Q_1 \circ Q_2\mathcal{A}}{Q_1\mathcal{A}Q_2}\right) Q\right\rangle \left\langle Q, \left(\mathcal{A} - \tfrac{\mathcal{A}Q_1 \circ Q_2\mathcal{A}}{Q_1\mathcal{A}Q_2}\right) Q\right\rangle \\ = \; & \lambda \left\langle Q, \left(\mathcal{A} - \tfrac{\mathcal{A}Q_1 \circ Q_1\mathcal{A}}{Q_1\mathcal{A}Q_1}\right) Q\right\rangle \left\langle Q, \left(\mathcal{A} - \tfrac{\mathcal{A}Q_2 \circ Q_2\mathcal{A}}{Q_2\mathcal{A}Q_2}\right) Q\right\rangle . \end{aligned}$$

The value of λ must be determined by the fact that the curve must pass through the third point Q_3. If we decompose Q into the linear combination $Q = \mu_1 Q_1 + \mu_2 Q_2 + \mu_3 Q_3$, then each value of μ_1/μ_3 yields a quadratic equation for μ_2/μ_3. Each ray through Q_1 contains two points of the curve (Figure 9.13).

D.5 Two Examples

The relationship of the geometries of the plane will be illustrated by two more elaborate examples that have a peculiar geometric appeal. We omit the proofs. They require thinking really hard. We only demonstrate the formulation through projective concepts.

Our first example is an amazing intersection point. Let us consider three conics. Any two of them shall have one common focus. We then have a triangle of three foci, and each side carries a conic. Any two conics then intersect at just two points. The three connecting lines are concurrent; they intersect at one point (Figure D.9).

First, we must know how to determine a focus in a general projective-metric geometry. We already know that a circle is a conic $\mathcal{K}$ with some point M (its center) that has the following property. Any line g through the center M (i.e., $\langle g, M\rangle = 0$) has a pole $P_{\mathcal{K}}[g]$ with respect to $\mathcal{K}$ and a pole $P_{\mathcal{B}}[g]$ with respect to $\mathcal{B}$. Both poles coincide with a circle. A general conic

$\mathcal{H}$ has two corresponding points, i.e., the foci. When the absolute conic $\mathcal{B}$ is given, we can find two points, F_1 and F_2, with the property that for any line g through F, the three points F, $P_{\mathcal{H}}[g]$, and $P_{\mathcal{B}}[g]$ are collinear. If these two points coincide, the conic $\mathcal{H}$ is a circle with respect to the absolute conic $\mathcal{B}$. Because the collinearity must exist for both foci, we obtain for the conic $\langle Q, \mathcal{H}Q\rangle = 0$ the conditions

$$\mathcal{H}_{12}^{-1} g = \mathcal{B}g + \alpha F_1 \qquad \text{for} \quad \langle g, F_1\rangle = 0$$

and

$$\mathcal{H}_{12}^{-1} g = \mathcal{B}g + \beta F_2 \qquad \text{for} \quad \langle g, F_2\rangle = 0.$$

The result is

$$\mathcal{H}^{-1}[\kappa, F_1, F_2] = \mathcal{B} + \frac{\kappa}{\langle F_1, \mathcal{B}^{-1}F_2\rangle}(F_1 \circ F_2 + F_2 \circ F_1). \tag{D.15}$$

The inverse is given by

$$\mathcal{H}[\kappa, F_1, F_2] = \mathcal{A} - \lambda\mu\frac{\mathcal{A}F_1 \circ \mathcal{A}F_2 + \mathcal{A}F_2 \circ \mathcal{A}F_1}{\langle F_1, \mathcal{A}F_2\rangle} + \lambda^2\nu\left(\frac{\mathcal{A}F_1 \circ \mathcal{A}F_1}{\langle F_1, \mathcal{A}F_1\rangle} + \frac{\mathcal{A}F_2 \circ \mathcal{A}F_2}{\langle F_2, \mathcal{A}F_2\rangle}\right)$$

with

$$\begin{aligned} \mu &= \frac{\langle F_1, \mathcal{A}F_2\rangle^2}{\langle F_1, \mathcal{A}F_2\rangle^2 - \lambda^2\langle F_1, \mathcal{A}F_1\rangle\langle F_2, \mathcal{A}F_2\rangle}, \\ \nu &= \frac{\langle F_1, \mathcal{A}F_1\rangle\langle F_2, \mathcal{A}F_2\rangle}{\langle F_1, \mathcal{A}F_2\rangle^2 - \lambda^2\langle F_1, \mathcal{A}F_1\rangle\langle F_2, \mathcal{A}F_2\rangle}, \end{aligned}$$

where $\lambda = \kappa/(1+\kappa)$. When F_1 and F_2 coincide, we obtain the equation of the circle (D.13). $\kappa < 1$ corresponds to an ellipse of the Euclidean geometry, $\kappa > 1$ to a hyperbola.

Let us now consider three foci F_1, F_2, F_3 and three conics $\mathcal{H}_1 = \mathcal{H}[\kappa_1, F_2, F_3]$, $\mathcal{H}_2 = \mathcal{H}[\kappa_2, F_3, F_1]$, and $\mathcal{H}_3 = \mathcal{H}[\kappa_3, F_1, F_2]$ that share each focus with another conic of the set. The conics have two directices each, which are given as the polars of the foci ($d_{12} = \mathcal{H}_1 F_2$ and so on). Any two conics have two intersection points that define now a connecting line (for instance, we find g_1 for the intersection of $\mathcal{H}_2$ and $\mathcal{H}_3$). Analytically, we can write Q_{11} and Q_{12} for the common real solutions of $\langle Q, \mathcal{H}_2 Q\rangle = 0$ and $\langle Q, \mathcal{H}_3 Q\rangle = 0$, and define g_1 in the form $g_1 = Q_{11} \times Q_{12}$. The three lines intersect at one common point S, i.e., $[g_1, g_2, g_3] = 0$.

The second example is a famous problem of ancient Greek mathematics. The *partition of an angle into three equal parts* cannot be solved by mere application of ruler and compass. Instead, one finds solutions by constructing the intersection of circles and hyperbolas. Here, we show a construction in the Minkowski plane that divides a pseudo-Euclidean angle into three equal parts (Figure D.10). Surprisingly, the construction is perfectly dual to a well-known construction in the Euclidean geometry. The duality is the motive to formulate the result in projective terms. From our point of view, the circle $\mathcal{K}$ is the most elementary construction. It obeys the simply structured Eq. (D.13) when the absolute conic $\mathcal{B}$ is known and of full rank. When $\mathcal{B}$ is only of rank 2, Eq. (D.13) degenerates to a void condition. We have to supplement $\mathcal{B}$ with a small term that restores full rank, invert $\mathcal{B}_{\text{suppl}}$ to obtain some $\mathcal{A}_{\text{suppl}}$, and perform in Eq. (D.13) the limiting process that makes the supplementary components infinitesimally small. The leading term becomes trivial but the next yields a valid equation. When $\mathcal{B}$ is only of rank 2, it defines an absolute polar p. By means of this polar, we can understand $\mathcal{B}$ as derived from some matrix $\mathcal{B}^*$ of full rank:

$$\mathcal{B} = \mathcal{B}^* - \frac{\mathcal{B}^* p \circ \mathcal{B}^* p}{\langle p, \mathcal{B}^* p \rangle}.$$

This matrix $\mathcal{B}^*$ can be inverted as usual: $\mathcal{A}^* \mathcal{B}^* = \mathcal{E}$. We now define through

$$\mathcal{A}_1 = \mathcal{A}^* - \frac{p \circ p}{\langle p, \mathcal{B}^* p \rangle}$$

again a matrix $\mathcal{A}_1$ of rank 2. This matrix enables us to write the equation of the circle in the form $\langle Q, \mathcal{K} Q \rangle = 0$. We have to choose

$$\mathcal{K} = \mathcal{A}_1 + 2\frac{\langle Q_1, \mathcal{A}_1 M \rangle}{\langle p, Q_1 \rangle \langle p, M \rangle} p \circ p - \frac{1}{\langle p, M \rangle}(p \circ M\mathcal{A}_1 + \mathcal{A}_1 M \circ p) - \frac{\langle Q_1, \mathcal{A}_1 Q_1 \rangle}{\langle p, Q_1 \rangle^2} p \circ p.$$

In Figure D.10, we calculate with the Minkowski geometry. We know that $\mathcal{B}$ is of rank 2 and indefinite. The conic $\mathcal{K}$, with the intersection O of the two straight lines g_0 and g_1 as the center, is a hyperbola by its Euclidean appearance but a circle by the measure implicit in the absolute conic $\mathcal{B}$. It intersects the line g_1 at some point P. We now choose the midpoint Q of the segment OP as the center of another conic $\mathcal{H}$ that passes through P. In the Euclidean case this conic $\mathcal{H}$ is a hyperbola but here a circle (by Euclidean appearance). It is *not* a circle of the geometry defined by $\mathcal{B}$, but we may construct a second $\mathcal{B}_1$ so that $\mathcal{H}$ obtains the form of a circle in the geometry of $\mathcal{B}_1$. Formally, we construct $\mathcal{B}_1$ with the help of $\mathcal{B}$ and the one line g_0. This goes as follows. We try the combination

$$\mathcal{B}_1 = \mathcal{B} - \frac{\lambda}{\langle g_0, \mathcal{B} g_0 \rangle} \mathcal{B} g_0 \circ g_0 \mathcal{B}.$$

It is chosen to be homogeneous in $\mathcal{B}$ as well as in g_0. The parameter λ remains to be determined appropriately. The condition to be met is that the two isotropic for $\mathcal{B}_1$ directions (defined by $\langle e, \mathcal{B}_1 e \rangle = 0$, not real in our case of Figure D.10) are perpendicular for $\mathcal{B}$ (i.e., $\langle e_1, \mathcal{B} e_2 \rangle = 0$). We find $\lambda = 2$. When we intersect the conic $\mathcal{K}$ (i.e., the circle to $\mathcal{B}$ through P about O) with the conic $\mathcal{H}$ (i.e., the circle to $\mathcal{B}_1$ through P about Q), we obtain new real intersection points D (one in the pseudo-Euclidean case, three in the Euclidean case). One can show that the lines $g_2 = O \times D$ are moved to coincide with g_1 when they are reflected first at g_0 and next at the image $g_3 = \mathcal{S}_{g_0}[g_2]$. This is the intricate point. If this result is reached, we simply get $3\angle[g_2, g_0] = \angle[g_0, g_1]$. In the Euclidean case, this is to be taken modulo 2π.

Figure D.10 illustrates the procedure. We draw the pseudo-Euclidean circle $\mathcal{K}$ (determined by $\mathcal{B}$) in the form of a hyperbola about the vertex O and obtain the point P on the second leg g_2. About the center Q of the segment OP, we draw the second conic $\mathcal{H}$ in the form of a circle and find the intersection point D. In order to show that the angle between $g_2 = O \times D$ and g_0 is the third part of the angle between g_0 and g_1, we draw the connecting line of P and D. It intersects the isotropic lines through O at the two points E and F. We know the property of the hyperbola that the bisection point H of PD also bisects EF. The line $g_3 = O \times H$ bisects the angle $\angle POD$ in the pseudo-Euclidean measure. It remains to show

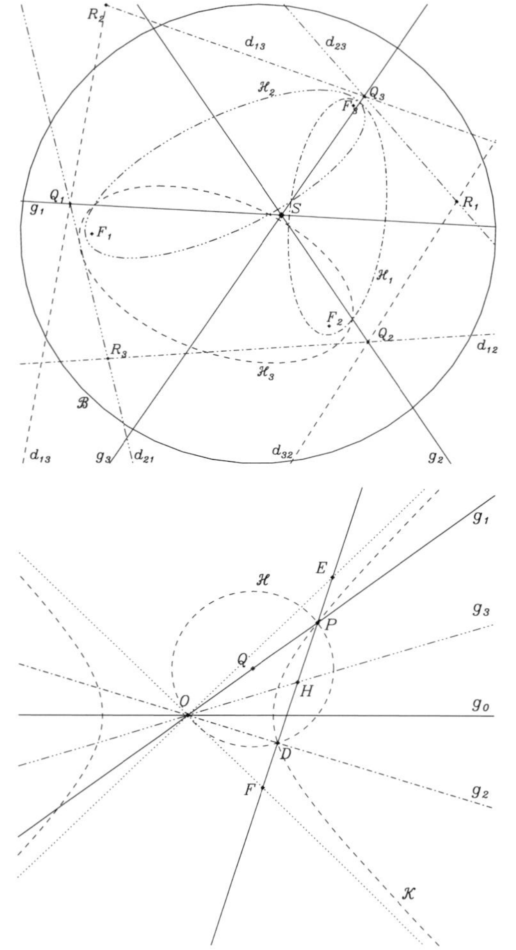

Figure D.9: *The intersection of the segments of conics about three foci.*

The figure shows three points F_1, F_2, F_3. Any two of them are foci of a conic. The directrices of the conics are drawn as lines in the same style. The segments g_1, g_2, g_3 through the intersection points of any pair of conics are concurrent, i.e., they intersect at one point S.

Figure D.10: *The partition of an angle into three equal parts.*

We show the partition in the Minkowski geometry. The isotropic directions are chosen as usual. The angle is put in an orientation in which g_0 is horizontal. In this case, the conic $\mathcal{H}$ defined in the text turns out to be a circle by the Euclidean appearance.

that g_0 bisects the angle $\angle[g_2, g_1]$. Now, g_0 is the horizontal, the (Euclidean) symmetry axis of the isotropic lines. It bisects exactly if it also bisects the angle in the Euclidean measure. Hence, we can argue with the Euclidean geometry. On the one hand, the angle $\angle ODP$ is a right angle (by Thales's theorem). The angle $\angle FOE$ is right by construction (more precisely, by the choice of $\mathcal{B}$ and $\mathcal{B}_1$). Therefore, we obtain $\angle FOD = \angle FEO$. On the other hand, the segments OH and HE are equally long (still by the Euclidean measure) so that $\angle FEO = \angle HOE$. We conclude that $\angle FOD = \angle HOE$, i.e., g_0 bisects $\angle FOE$ as well as $\angle DOH$. We obtain $\angle[g_2, g_0] = \angle[g_0, g_3]$ by the Euclidean as by the pseudo-Euclidean measure. By the pseudo-Euclidean measure, we had $\angle[g_2, g_3] = \angle[g_3, g_1]$. Therefore, by the pseudo-Euclidean measure, $3\angle[g_2, g_0] = \angle[g_0, g_1]$.

The division of an angle into three equal parts is a task beyond the reach of ruler and compass. The deeper reason for this is the lack of an absolute pole in the Euclidean as in the pseudo-Euclidean geometry. In the Galilean geometry, an absolute pole exists, and the

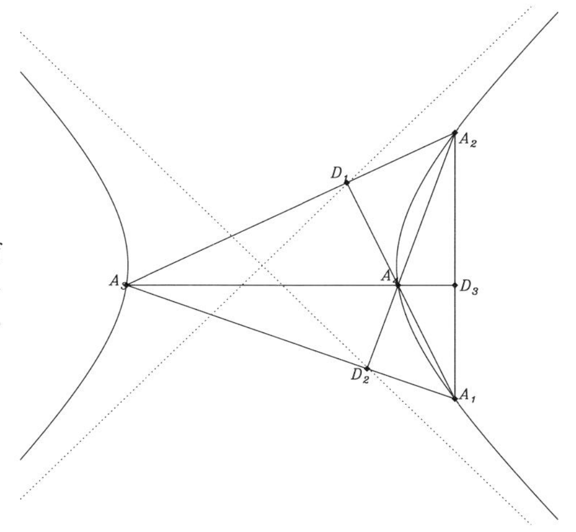

Figure D.11: *Hyperbola and Euclidean altitudes.*

When we construct the intersection of altitudes of a triangle by Euclidean rules, it lies on the circumscribed pseudo-Euclidean circle, if this is an equilateral hyperbola.

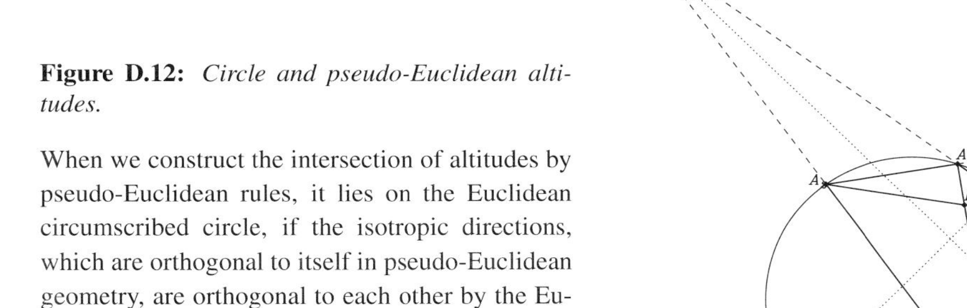

Figure D.12: *Circle and pseudo-Euclidean altitudes.*

When we construct the intersection of altitudes by pseudo-Euclidean rules, it lies on the Euclidean circumscribed circle, if the isotropic directions, which are orthogonal to itself in pseudo-Euclidean geometry, are orthogonal to each other by the Euclidean rule.

division of an angle is equivalent to the division of a segment. It can be merely performed by an application of ruler and compass. In perfect duality, we can state that the division of a segment into three equal parts (easily done by ruler and compass in the three geometries: Galilean, Euclidean, and pseudo-Euclidean) cannot be performed any longer merely by an application of ruler and compass when no absolute polar exists, for instance, on the sphere (i.e., in the elliptic geometry).

There are simpler examples of the curious connection between circles and equilateral hyperbola than the trisection of the angle. Figures D.11 and D.12 show that a hyperbola is found in a simple Euclidean construction and a circle is found in the pseudo-Euclidean dual. When we construct the intersection of altitudes of a triangle by Euclidean rules, it lies on any equilateral hyperbola through the tree vertices. Dually, when we construct this intersection by pseudo-Euclidean rules, it lies on the Euclidean circumscribed circle. We leave the proof to the reader.

E The Metrical Plane

E.1 Classification

In this appendix, we intend to develop the formal foundation for the exposition in Chapter 9. We first expand the consideration to the three-dimensional projective space. It is described by the pencils of rays, planes, and three-dimensional hyperplanes through the origin of a four-dimensional linear space. A point in the three-dimensional projective space is now given by four homogeneous coordinates $Q = (Q^1, Q^2, Q^3, Q^4)$, and the ordinary coordinates are obtained by $(\xi, \eta, \zeta) = (Q^1/Q^4, Q^2/Q^4, Q^3/Q^4)$, for example. The element dual to a point is no longer a line but a plane. A line is a connection of two points as well as an intersection of two planes. Correspondingly, a plane is given by four coordinates, $e = (e_1, e_2, e_3, e_4)$. A point Q lies in the plane e if the scalar product of the coordinate quadruples vanishes, $\langle e, Q\rangle = e_k Q^k = 0$. We use the summation convention.

The figure corresponding to a conic section in the plane is called a quadric in the three-dimensional projective space. Such a quadric can be an ellipsoid or a hyperboloid by Euclidean standards. Projectively, the quadrics differ by their signature, that is, the difference between the numbers of positive and negative diagonal elements in a diagonal representation of the matrix K_{lm}. This difference is also of no interest here. Without loss of generality, we choose here the special form

$$\frac{(x-\xi t)^2}{a^2} + \frac{(y-\eta t)^2}{b^2} + \frac{(z-\zeta t)^2}{c^2} = \frac{t^2}{d^2}.$$

The midpoint has the coordinates $Z = [\xi, \eta, \zeta, 1]$, and the squares of the principal half-axes are a^2/d^2, b^2/d^2, and c^2/d^2. If these values are all positive, the quadric is an ellipsoid in the chosen representation. The three-dimensional polarity is given by the matrices

$$\mathcal{A}^* = \begin{pmatrix} \frac{1}{a^2} & 0 & 0 & -\frac{\xi}{a^2} \\ 0 & \frac{1}{b^2} & 0 & -\frac{\eta}{b^2} \\ 0 & 0 & \frac{1}{c^2} & -\frac{\zeta}{c^2} \\ -\frac{\xi}{a^2} & -\frac{\eta}{b^2} & -\frac{\zeta}{c^2} & -\frac{1}{d^2} + \frac{\xi^2}{a^2} + \frac{\eta^2}{b^2} + \frac{\zeta^2}{c^2} \end{pmatrix}, \qquad \det \mathcal{A} = -\frac{1}{a^2 b^2 c^2 d^2}, \tag{E.1}$$

The Geometry of Time. Dierck-E. Liebscher

ISBN: 3-527-40567-4

and

$$\mathcal{B}^* = \begin{pmatrix} a^2 - d^2\xi^2 & -d^2\xi\eta & -d^2\xi\zeta & -d^2\xi \\ -d^2\eta\xi & b^2 - d^2\eta^2 & -d^2\eta\zeta & -d^2\eta \\ -d^2\zeta\xi & -d^2\zeta\eta & c^2 - d^2\zeta^2 & -d^2\zeta \\ -d^2\xi & -d^2\eta & -d^2\zeta & -d^2 \end{pmatrix}. \tag{E.2}$$

A point Q has the polar plane $\pi[Q] = \mathcal{A}^*Q$; a plane γ has the pole $P[\gamma] = \mathcal{B}^*\gamma$.

We now put $d = 1$. The intersection with a plane $z = 0$ is a conic section with the equation

$$\frac{(x - \xi t)^2}{a^2} + \frac{(y - \eta t)^2}{b^2} = t^2 \left(1 - \frac{\zeta^2}{c^2}\right).$$

If $\zeta < c$, the quadric does not intersect the plane in a real curve. We can write

$$\mathcal{A} = \begin{pmatrix} \frac{1}{a^2} & 0 & -\frac{\xi}{a^2} \\ 0 & \frac{1}{b^2} & -\frac{\eta}{b^2} \\ -\frac{\xi}{a^2} & -\frac{\eta}{b^2} & -1 + \frac{\xi^2}{a^2} + \frac{\eta^2}{b^2} + \frac{\zeta^2}{c^2} \end{pmatrix}, \qquad \det\, \mathcal{A} = -\frac{c^2 - \zeta^2}{a^2 b^2 c^2}.$$

For the matrix $\mathcal{B}$, this turns out to be a bit more complicated, but in the nondegenerate case we only have to find the inverse of $\mathcal{A}$, i.e.,

$$\mathcal{B} = -\frac{c^2}{c^2 - \zeta^2} \begin{pmatrix} a^2\left(-1 + \frac{\xi^2}{a^2} + \frac{\zeta^2}{c^2}\right) & \xi\eta & \xi \\ \xi\eta & b^2\left(-1 + \frac{\eta^2}{b^2} + \frac{\zeta^2}{c^2}\right) & \eta \\ \xi & \eta & 1 \end{pmatrix}.$$

We convince ourselves that the polar plane $\pi[Q] = \mathcal{A}^*Q$ of a point in the plane $z = 0$ intersects this plane in the polar $p[Q] = \mathcal{A}Q$. The points of the polar plane $\pi[Q]$ separate harmonically with Q the segment that the line of intersection cuts out of the quadric. This property is simply the generalization of what we know from the projective plane. For points Q on the plane $z = 0$, the intersection with the polar plane $\pi[Q]$, and the intersection with the quadric, this property is inherited from the three-dimensional projective space. Next, we check that the polar planes $\pi[Q] = \mathcal{A}^*Q$ of the points Q of a line g in the plane $z = 0$ have a line $p[g]$ in common that, in its turn, intersects the plane $z = 0$ in the (two-dimensional) pole $P[g] = \mathcal{B}g$ of the line g. For instance, if we choose two points on g, then their polar planes intersect in a line. This line turns out to be independent of the choice of the two points provided; of course, they remain on the line g. It is the polar $p[g]$ of the line g.

In a three-dimensional projective space, the polar figure of a line is again a line. The reason is that the three-dimensional projective space must be represented by a homogeneous linear space that is now four-dimensional. There a line is determined by *two* linear simultaneous linear equations, which can be given the form

$$h_{ik}x^k = 0$$

with an antisymmetric matrix h_{ik} of coefficients. We use the fourfold indexed permutation symbol ϵ_{iklm} (numerically identical to ϵ^{iklm} and with the reciprocity relation $\epsilon_{iklm}\epsilon^{ikrs} =$

$2(\delta^r_l\delta^s_m - \delta^r_m\delta^s_l)$). A line through two points Q and R has the array of coefficients $h_{ik}[Q, R] = \epsilon_{iklm}Q^l R^m$. The line of intersection of two planes α and β has the array of coefficients $h_{ik}[\alpha, \beta] = \alpha_i\beta_k - \alpha_k\beta_i$. Assuming the polarity $\alpha = \pi[Q]$ and $\beta = \pi[R]$, we obtain by substitution the formula

$$j_{ik} = (A^*_{ir}A^*_{ks}\epsilon^{lmrs})\, h_{lm}$$

for the polar $j = p[h]$ of the line h.

The polar line $p[g]$ intersects the plane $z = 0$ in the (two-dimensional) pole $P[g] = \mathcal{B}g$ of the line g. We can proceed with a dual construction. In this case, we first determine the planes through the line g. The poles of these planes lie, in the nondegenerate case, on the polar line constructed above.

If $\mathcal{B}$ is nondegenerate, each point is pole to a line, its polar: $p[Q] = \mathcal{A}Q$, and $\mathcal{AB} = \mathcal{E}$. In the degenerate case ($\det \mathcal{B} = 0$), $\mathcal{A}$ is degenerate too ($\det \mathcal{A} = 0$ and $\mathcal{AB} = 0$). Here, the relation between the maps $\mathcal{A}$ and $\mathcal{B}$ is incomplete.

We must establish what happens in the following cases:

1. The conic section that defines the polarity is real and nondegenerate. In the three-dimensional embedding, we can use a sphere that intersects the drawing plane in a real circle. Instead of a sphere, we can take an ellipsoid or a hyperboloid without changing the picture qualitatively. The polarity can be constructed by a real construction with a real conic section without recourse to a three-dimensional embedding. However, the real conic separates interior and exterior points as well as intersecting and nonintersecting lines. Therefore, we find three subcases.

   (a) We choose lines that intersect the absolute conic and the interior points. This gives us Klein's model of non-Euclidean geometry (Lobachevski geometry).

   (b) We choose lines that intersect the absolute conic and the exterior points. This gives us the model of the de Sitter geometry. Timelike lines intersect the absolute conic, spacelike lines do not.

   (c) We choose lines that do not intersect the absolute conic and exterior points. This gives us the model of the anti-de Sitter geometry. Timelike lines do not intersect the absolute conic, spacelike lines do.

   In each of the three subcases, the metric relations are different. The quadric can be degenerated provided the intersection with the plane is not; for instance, it could be a double cone whose intersection with the drawing plane is not in its vertex. In Chapter 7, the Lobachevski geometry was introduced in this way.

2. The conic section that defines the polarity is imaginary but nondegenerate. This case is represented in the three-dimensional embedding by a quadric that does not intersect the drawing plane. We obtain the model of elliptic (spherical) geometry.

3. The conic section that defines the polarity has only one real point. This case is represented in the three-dimensional embedding by a quadric that only touches the drawing plane. We obtain the model of the anti-Euclidean geometry. As in the first case, the quadric can be degenerate in a certain way: It could be a double cone cut by the drawing

plane only in its vertex. This point becomes the absolute pole, and all polars pass through it. In the polar pencil, an involution without real fixed rays is defined.

4. The conic section that defines the polarity degenerates into a pair of real points. This gives us the model of the pseudo-Euclidean (Minkowski) geometry. All points have a common polar, the image of the line at infinity. An involution with two fixed points is defined on this absolute polar. The fixed points are the real points of the degenerate conic.

5. The conic section that defines the polarity degenerates into a pair of imaginary points that nevertheless has a real connection (the polar common to all points). This gives us the model of the Euclidean geometry. On the polar, an involution without real fixed points is defined.

6. The conic section that defines the polarity degenerates into a double point, where it has a real tangent, the absolute polar. This gives us the model of the Galilean geometry. All points have a common polar (the image of the line at infinity), and all lines have a common pole (the real double point of the absolute conic).

7. The conic section that defines the polarity degenerates into a real pair of lines. All lines have a common pole, the intersection point that carries all polars. In the pencil of polars, an involution with two real fixed rays (the pair of lines representing the absolute conic) is defined: This is the anti-Minkowski geometry.

We now intend to formalize the nine cases. To begin with, we place the center of the quadric at the point $[x, y, z, t] = [0, 0, 1, 1]$. This allows the representation of all cases except for the one in which the intersection with the drawing plane $z = 0$ is a pair of real lines. This case can only be considered with a center of the quadric that lies in the plane, for instance $[x, y, z, t] = [0, 0, 0, 1]$. For the moment, we write

$$\mathcal{A}^* = \begin{pmatrix} \frac{1}{a^2} & 0 & 0 & 0 \\ 0 & \frac{1}{b^2} & 0 & 0 \\ 0 & 0 & \frac{1}{c^2} & -\frac{1}{c^2} \\ 0 & 0 & -\frac{1}{c^2} & -1 + \frac{1}{c^2} \end{pmatrix}, \qquad \det \mathcal{A} = -\frac{1}{a^2 b^2 c^2},$$

and

$$\mathcal{B}^* = \begin{pmatrix} a^2 & 0 & 0 & 0 \\ 0 & b^2 & 0 & 0 \\ 0 & 0 & c^2 - 1 & -1 \\ 0 & 0 & -1 & -1 \end{pmatrix}.$$

In the drawing plane, this yields

$$\mathcal{A} = \begin{pmatrix} \frac{1}{a^2} & 0 & 0 \\ 0 & \frac{1}{b^2} & 0 \\ 0 & 0 & -1 + \frac{1}{c^2} \end{pmatrix} \propto \begin{pmatrix} b^2 c^2 & 0 & 0 \\ 0 & a^2 c^2 & 0 \\ 0 & 0 & a^2 b^2 (1 - c^2) \end{pmatrix},$$

and

$$\mathcal{B} = \begin{pmatrix} a^2(-1+\frac{1}{c^2}) & 0 & 0 \\ 0 & b^2(-1+\frac{1}{c^2}) & 0 \\ 0 & 0 & 1 \end{pmatrix} \propto \begin{pmatrix} a^2(1-c^2) & 0 & 0 \\ 0 & b^2(1-c^2) & 0 \\ 0 & 0 & c^2 \end{pmatrix}.$$

If $\mathcal{A}^*$ represents a sphere, $a = b = c$, and intersects the drawing plane, $c > 1$, we obtain indefinite matrices $\mathcal{A}$ and $\mathcal{B}$ that are inverses of each other (up to the free factor because of homogeneity): $\mathcal{AB} \propto \mathcal{E}$. If the sphere does not intersect the drawing plane, $c < 1$, both $\mathcal{A}$ and $\mathcal{B}$ are definite. If the sphere touches the plane, $c = 1$, the matrix $\mathcal{B}$ is of rank 1 and $\mathcal{A}$ is of rank 2.

Now we turn our interest to the limit $b \to 0$ (with $a = c = 1$). The sphere is flattened to a disk. It is important that the plane of the disk intersects the drawing plane in a line in the finite. Otherwise we obtain in the limit the case of elliptic geometry, which we already modeled. If $c < 1$, the disk does not intersect the plane. We obtain a matrix $\mathcal{A}$ of rank 1 and a matrix $\mathcal{B}$ of rank 2 and *semidefinite*. If $c < 1$, the disk intersects the plane along a segment between two real points. The matrix $\mathcal{A}$ is again of rank 1, and $\mathcal{B}$ is of rank 2. But now $\mathcal{B}$ is *indefinite*. If the disk touches the plane, $c = 1$, both $\mathcal{A}$ and $\mathcal{B}$ are of rank 1. In all three cases, the line $y = 0$ is defined by $\mathcal{A}$. It is the image of the line at infinity, which we find absolute in the Euclidean, Minkowski, and Galilean geometries. After this consideration, only one case remains. We now assume the quadric to be a double cone with its axis in the drawing plane. To this end, we substitute $-b^2$ for b^2 in the formulas (E.1) and (E.2). Then we go to the limit $d^2 \to \infty$. This yields

$$\mathcal{A} = \begin{pmatrix} \frac{1}{a^2} & 0 & 0 \\ 0 & -\frac{1}{b^2} & 0 \\ 0 & 0 & -\frac{1}{d^2} \end{pmatrix} \propto \begin{pmatrix} b^2 & 0 & 0 \\ 0 & -a^2 & 0 \\ 0 & 0 & -\frac{a^2b^2}{d^2} \end{pmatrix} \to \begin{pmatrix} b^2 & 0 & 0 \\ 0 & -a^2 & 0 \\ 0 & 0 & 0 \end{pmatrix},$$

and

$$\mathcal{B} = \begin{pmatrix} a^2 & 0 & 0 \\ 0 & -b^2 & 0 \\ 0 & 0 & -d^2 \end{pmatrix} \propto \begin{pmatrix} -\frac{a^2}{d^2} & 0 & 0 \\ 0 & \frac{b^2}{d^2} & 0 \\ 0 & 0 & 1 \end{pmatrix} \to \begin{pmatrix} 0 & 0 & 0 \\ 0 & 0 & 0 \\ 0 & 0 & 1 \end{pmatrix}.$$

If we consider the duality of points and lines in the projective geometry of the plane, it corresponds to interchange of the roles of $\mathcal{A}$ and $\mathcal{B}$. In this sense, the names given to the anti-Euclidean, anti-Minkowski, and anti-Lobachevski geometries are justified.

We conclude the chapter by considering the sine theorem and prove it for triangles of timelike lines in the de Sitter geometry (Figure 7.23) that is drawn on a pseudosphere in a three-dimensional Minkowski geometry. All motions are Lorentz rotations here, the transformation matrices have determinant 1, and the triple product $[\vec{OA}, \vec{OB}, \vec{OC}]$ of three position vectors is invariant. Without loss of generality, we place the point A on the formal equator and B on the correlated meridian (Figure E.1). We draw the tangential unit vector $\vec{AT}_{AB}$ at A in the direction B and obtain the decomposition

$$\vec{OB} = \vec{OA}\,\Gamma[c] + \vec{AT}_{AB}\,\Pi[c], \quad \text{i.e.} \quad \vec{OB^*} = \vec{OA} + \vec{AT}_{AB}\,\Pi[c],$$

where c denotes the central angle, i.e., the length of the geodesic AB. Here, the meridional plane is *pseudo-Euclidean*, and we have $\Gamma[c] = \cosh c$ and $\Pi[c] = \sinh c$. The decomposition contains only invariantly described terms, i.e., it is itself invariant (independent of the particular orientation of the segment AB). Hence, we can put $\vec{OC^*} = \vec{OA} + \vec{AT}_{AC}\,\Pi[b]$ in general. We now evaluate the triple product $[\vec{OA}, \vec{OB}, \vec{OC}]$. It is equal to twice the volume of the pyramid $OABC$. This pyramid is equal in volume to the pyramid OAB^*C^* because the substitution of B^* for B and C^* for C does not alter the basis plane $OAB = OAB^*$ nor the height C^*G (see Figure E.2). We recall that $AB^* = \Pi[c]$ and $AC^* = \Pi[b]$, and we put $C^*G = AC^*\,\Sigma[\alpha]$. In our case, the tangential plane is pseudo-Euclidean too, and we have $\Sigma[\alpha] = \sinh\alpha$. We obtain

$$[\vec{OA}, \vec{OB}, \vec{OC}] = 2\,\Pi[b]\,\Pi[c]\,\Sigma[\alpha].$$

When all three sides of the triangle are timelike, any of the three points can be chosen to be A, and we cyclically find $\Pi[b]\,\Pi[c]\,\Sigma[\alpha] = \Pi[c]\,\Pi[a]\,\Sigma[\beta] = \Pi[a]\,\Pi[b]\,\Sigma[\gamma]$ or

$$\Pi[a] : \Pi[b] : \Pi[c] = \Sigma[\alpha] : \Sigma[\beta] : \Sigma[\gamma]. \tag{E.3}$$

Let us imagine the nine geometries in the form of such central projections. Then they differ in the characters of the meridional and tangential planes. In the case of positive curvature, the meridional planes are Euclidean ($\Pi[a] = \sin a$, Figure 7.18), and in the case of negative curvature, pseudo-Euclidean ($\Pi[a] = \sinh a$, Figure 7.20). Without curvature, they are Galilean ($\Pi[a] = a$). The tangential planes are Euclidean ($\Sigma[\alpha] = \sin\alpha$) for geometries with usual triangle inequality (Figures 7.9 and 7.10); pseudo-Euclidean ($\Sigma[\alpha] = \sinh\alpha$) for the antihyperbolic (Figure 7.28), Minkowski, and de Sitter geometries (Figure 7.26); Galilean ($\Sigma[\alpha] = \alpha$) for the anti-Euclidean, anti-Minkowski, and Galilean geometries. The function $\Pi[r]$ describes the dependence of the arc on the angle: Let r be the length of the radius. Then the arc is obtained by $b = \alpha\Pi[r]$. The function $\Sigma[\alpha]$ describes the dependence of the Sine s (i.e., the length of the perpendicular from one end of the arc on the other leg) on the arc, $s = \Pi[r]\Sigma[\alpha]$. Corresponding to the character of the meridional plane, $\Pi[a]$ can be equal to $\sin a$, a, and $\sinh a$. Corresponding to the character of the tangential plane, $\Sigma[\alpha]$ can be equal to $\sin\alpha$, α, and $\sinh\alpha$. This yields Table E.1.

E.2 The Metric

The metric of the projective-metric plane admits the representation of the distance between two points in the closed form

$$d[A, B] = \tfrac{1}{2}\ln \mathcal{D}[A, B; K_1[A \times B], K_2[A \times B]], \tag{E.4}$$

where K_1 and K_2 are the points of intersection of the connecting line AB with the absolute conic. For the angle, we find correspondingly

$$\alpha[g, h] = \tfrac{1}{2}\ln \mathcal{D}[g, h; k_1[g \times h], k_2[g \times h]].$$

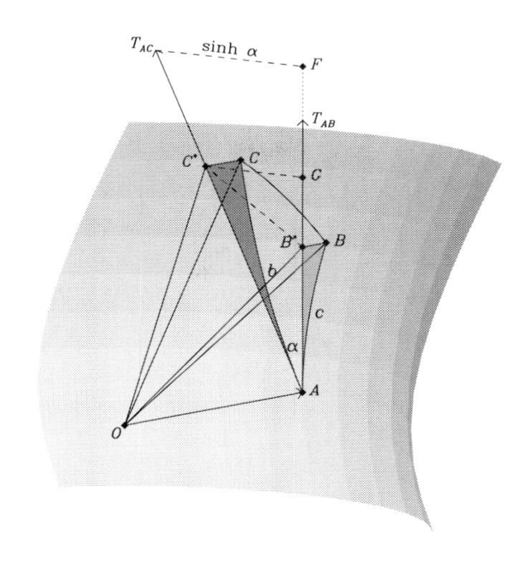

Figure E.1: *The characteristic functions. I.*

We see a part of the hyperboloid of radius 1 with a tangential plane at the point A. The side $c = AB$ of the triangle ΔABC lies on the meridian. AT_{AB} and AT_{AC} are unit vectors in the tangential plane. Here, we have $AC^* = \sinh b$, $AB^* = \sinh c$, and $C^*G/AC^* = \sinh \alpha$.

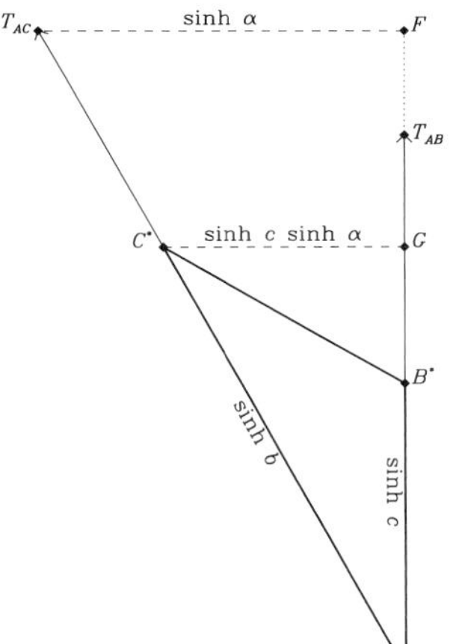

Figure E.2: *The characteristic functions. II.*

We evaluate the triangle ΔAB^*C^* in the tangential plane. The volume of the pyramid OAB^*C^* is given by $2V^* = [\vec{OA}, \vec{OB^*}, \vec{OC^*}] = \sinh b \sinh c \sinh \alpha$.

The metric degenerates if $K_1 \to K_2$. In this case, we obtain to the first order

$$K_2 = K_1 + \varepsilon V \quad \to \quad \mathcal{D}[A, B; K_1, K_1 + \varepsilon V] = 1 + \varepsilon \left(\frac{BV}{BK_1} - \frac{AV}{AK_1} \right).$$

The first-order term now yields a metric. If we give K_1 the coordinate ∞, the distance on the line is the difference of the coordinates.

We now construct the metric of the geometries that we examined in the preceding chapters. They allow us to relate the projective notions to the considerations in cosmology, which are usually in the framework of metric space–times. In the two-dimensional Euclidean plane, Cartesian coordinates give

$$\mathrm{d}s^2 = \mathrm{d}x^2 + \mathrm{d}y^2,$$

Table E.1: The characteristics of the nine geometries

	$\Pi[a] = \sin a$	$\Pi[a] = a$	$\Pi[a] = \sinh a$
$\Sigma[\alpha] = \sin\alpha$	Elliptical geometry $\mathcal{B} = + + +$ $\mathcal{A} = + + +$	Euclidean geometry $\mathcal{B} = 0 + +$ $\mathcal{A} = + 0 0$	Lobachevski geometry $\mathcal{B} = - + +$ $\mathcal{A} = - + +$
$\Sigma[\alpha] = \alpha$	Anti-Euclidean geometry $\mathcal{B} = 0 0 +$ $\mathcal{A} = + + 0$	Galilean geometry $\mathcal{B} = 0 0 +$ $\mathcal{A} = + 0 0$	Anti-Minkowski geometry $\mathcal{B} = 0 0 +$ $\mathcal{A} = - + 0$
$\Sigma[\alpha] = \sinh\alpha$	Anti-Lobachevski geometry $\mathcal{B} = - - +$ $\mathcal{A} = + + -$	Minkowski geometry $\mathcal{B} = 0 + -$ $\mathcal{A} = + 0 0$	de Sitter geometry $\mathcal{B} = + - +$ $\mathcal{A} = - + -$

and in the three-dimensional space

$$ds^2 = dx^2 + dy^2 + dz^2.$$

If we substitute polar coordinates in the plane ($x = r\cos\varphi$, $y = r\sin\varphi$), we obtain

$$ds^2 = dr^2 + r^2 d\varphi^2. \tag{E.5}$$

If we substitute spherical coordinates in space ($x = r\sin\theta\cos\varphi$, $y = r\sin\theta\sin\varphi$, $z = r\cos\theta$), we obtain

$$ds^2 = dr^2 + r^2(d\theta^2 + \sin^2\theta d\varphi^2). \tag{E.6}$$

The term inside the parentheses describes what we find on the surface of the sphere. It shows that, compared with Eq. (E.5), the expression

$$ds^2 = d\theta^2 + \sin^2\theta d\varphi^2 \tag{E.7}$$

is the line element on a homogeneous, positively curved surface.

In the three-dimensional Minkowski world, we obtained

$$ds^2 = c^2 dt^2 - dx^2 - dy^2.$$

If we substitute polar coordinates in addition to cosmological time (this is equivalent to spherical coordinates),

$$t = \tau\cosh[\chi], \qquad r = \tau\sinh[\chi],$$

we find

$$ds^2 = c^2 d\tau^2 - (c\tau)^2 (d\chi^2 + \sinh^2[\chi] d\varphi^2).$$

Distances from the spatial point $\chi = 0$ for a given cosmological time, indicated by a fixed coordinate χ, are found to increase with cosmological time τ. The line element describes Milne's explosion universe (Figure 7.20). The term inside the parentheses,

$$ds^2 = d\chi^2 + \sinh^2[\chi] d\varphi^2,$$

is the metric of a homogeneous, *negatively* curved surface. Finally, if we substitute once more the spherical coordinates (θ, φ) of the spheres around the origin of space for the angular coordinate φ of the circles around the origin of the plane, we obtain the scheme

$$ds^2 = d\chi^2 + \begin{pmatrix} \sin^2 \chi \\ \chi^2 \\ \sinh^2 \chi \end{pmatrix} (d\theta^2 + \sin^2 \theta d\varphi^2) \quad \begin{pmatrix} \text{for positive curvature} \\ \text{for curvature zero} \\ \text{for negative curvature} \end{pmatrix}$$

for the line element of homogeneous spaces.

A cosmological model is a world conceived as a homogeneous space with a time-dependent distance gauge. Its line element must be constructed in the form[1]

$$ds^2 = c^2 dt^2 - a^2[t](d\chi^2 + r^2[\chi](d\theta^2 + \sin^2 \theta d\varphi^2)).$$

The function $r[\chi]$ can be $\sin \chi$ (for positively curved spaces), χ itself (for spaces without curvature), or $\sinh \chi$ (for negatively curved spaces). The function $a[t]$ (the expansion parameter) describes the time dependence of the spatial measure with respect to the measure of time. The rate of change of a,

$$H = \frac{1}{a} \frac{da[t]}{dt},$$

is called the expansion rate. The Einstein equations of general relativity are reduced to the *Friedmann equation*, Eq. (7.3),

$$H^2 + \frac{kc^2}{a^2} = \Lambda \frac{c^2}{3} + \frac{8\pi G}{3} \varrho. \tag{E.8}$$

Here, k is the sign of the curvature, Λ is the cosmological constant, and ϱ is the (average) density of mass.

The universes of Chapter 7 are the empty Friedmann universes ($\varrho = 0$). The Minkowski world is a space without curvature and expansion ($k = \Lambda = H = 0$). The Milne universe (Figure 7.19) is a space of negative curvature in linear expansion, $k = -1, \Lambda = 0$. It originates from the Minkowski world by a different choice *and interpretation* of the coordinates but has the same abstract metric. The de Sitter universe is the surface of a pseudosphere. Depending

[1] This line element carries the names of Robertson and Walker. It was already used extensively by Friedmann.

on the choice and the interpretation of the coordinates, it appears as a flat and exponentially expanding space (Figure 7.21) with the line element[2]

$$ds^2 = c^2 dt^2 - a_0^2 \exp[2H_0 t](d\chi^2 + \chi^2(d\theta^2 + \sin^2\theta d\varphi^2)),$$

as a positively curved contracting and reexpanding space (Figure 7.23) with the line element

$$ds^2 = c^2 dt^2 - a_0^2 \cosh^2[H_0 t](d\chi^2 + \sin^2[\chi](d\theta^2 + \sin^2\theta d\varphi^2)),$$

or as a negatively curved exploding space (Figure 7.25) with the line element

$$ds^2 = c^2 dt^2 - a_0^2 \sinh^2[H_0 t](d\chi^2 + \sinh^2[\chi](d\theta^2 + \sin^2\theta d\varphi^2)).$$

Choosing the time coordinate differently, we obtain the anti-de Sitter universe (Figure 7.27) with the line element

$$ds^2 = c^2 dt^2 - a_0^2 \cos^2[H_0 t](d\chi^2 + \sinh^2[\chi](d\theta^2 + \sin^2\theta d\varphi^2)),$$

i.e., a space of negative curvature; it expands first but then collapses.

We end the appendix by deriving the line element from the cross-ratio formula. The line element of the projective-metric plane is obtained as differential form of formula (E.4). If A and B are only infinitesimally separated, the cross-ratio is near unity, and the logarithm near zero. First we determine the intersection points K_m in the form $K_m = P + \lambda_m Q$. The coefficients λ_m solve the equation

$$0 = \langle K_m, \mathcal{A} K_m \rangle = \langle P, \mathcal{A} P \rangle + 2\lambda_m \langle P, \mathcal{A} Q \rangle + \lambda_m^2 \langle Q, \mathcal{A} Q \rangle.$$

We obtain

$$\lambda_1 + \lambda_2 = -2\frac{\langle P, \mathcal{A} Q \rangle}{\langle Q, \mathcal{A} Q \rangle}, \qquad \lambda_1 \lambda_2 = \frac{\langle P, \mathcal{A} P \rangle}{\langle Q, \mathcal{A} Q \rangle},$$

$$\lambda_1 - \lambda_2 = 2\frac{\sqrt{\langle P, \mathcal{A} Q \rangle^2 - \langle P, \mathcal{A} P \rangle \langle Q, \mathcal{A} Q \rangle}}{\langle Q, \mathcal{A} Q \rangle}. \tag{E.9}$$

We now solve

$$\mathcal{D}[P, Q; K_1, K_2] = \frac{[S, P, K_1]\,[S, Q, K_2]}{[S, P, K_2]\,[S, Q, K_1]} = \frac{\lambda_1}{\lambda_2}.$$

If Q is only infinitesimally distant from P, i.e., $Q = P + dP$, the coefficient λ_1 differs from λ_2 only infinitesimally, $\lambda_2 = \lambda_1 + \mathcal{O}[dP]$, and we obtain

$$d[P, Q] = \frac{1}{2} \ln |\mathcal{D}[P, Q; K_1, K_2]| = \left| \frac{\lambda_1 - \lambda_2}{\lambda_1 + \lambda_2} \right| + \mathcal{O}_2[dP].$$

[2]The quantity H_0 is an integration constant called Hubble constant (although constant only in space, not in time). The constant a_0 denotes the radius of the pseudospheres on which we have drawn the geometry, i.e., this is a characteristic curvature radius.

Using now Vietà's theorems (E.9), we obtain

$$d[P,Q] = \sqrt{\left|\frac{\langle P, \mathcal{A}Q\rangle^2}{\langle P, \mathcal{A}P\rangle\langle Q, \mathcal{A}Q\rangle} - 1\right|} + \mathcal{O}_2[\mathrm{d}P] = \sqrt{|1 - \cos^2 d[P,Q]|} + \mathcal{O}_2[\mathrm{d}P].$$

This last expression can be found directly from consideration of the sphere and its generalization: The distance of two points on the sphere is given in homogeneous coordinates by

$$\cos d[P,Q] = \frac{\langle P, \mathcal{A}Q\rangle}{\sqrt{\langle P, \mathcal{A}P\rangle\langle Q, \mathcal{A}Q\rangle}}, \tag{E.10}$$

where we must put

$$\mathcal{A} = \begin{pmatrix} 1 & 0 & 0 \\ 0 & 1 & 0 \\ 0 & 0 & 1 \end{pmatrix}$$

as long as we are on the sphere or in elliptic geometry, respectively. On the pseudosphere, $\mathcal{A}$ must be changed correspondingly. In addition, it might be necessary to substitute for the cosine its hyperbolic partner. We note that formula (E.10) is homogeneous in the written form: P, Q, and $\mathcal{A}$ can be replaced by multiples without changing the result.

$$d^2[P,Q] \approx 1 - \cos^2 d[P,Q] = 1 - \frac{\langle P, \mathcal{A}Q\rangle^2}{\langle P, \mathcal{A}P\rangle\langle Q, \mathcal{A}Q\rangle} = \frac{\langle (P \times Q), (\mathcal{A}P \times \mathcal{A}Q)\rangle}{\langle P, \mathcal{A}P\rangle\langle Q, \mathcal{A}Q\rangle}.$$

To find the differential form, we write $Q = P + \mathrm{d}P$ and observe that $(\mathcal{A}P \times \mathcal{A}Q) \propto B[P \times Q]$. We obtain

$$\begin{aligned} \mathrm{d}s^2 &= \frac{\langle (P \times \mathrm{d}P), (\mathcal{A}P \times \mathcal{A}\mathrm{d}P)\rangle}{\langle P, \mathcal{A}P\rangle^2} \propto \frac{\langle (P \times \mathrm{d}P), \mathcal{B}(P \times \mathrm{d}P)\rangle}{\langle P, \mathcal{A}P\rangle\langle P, \mathcal{A}P\rangle} \\ &= \frac{B^{mn}\epsilon_{mij}P^i\mathrm{d}P^j\epsilon_{nkl}P^k\mathrm{d}P^l}{(A_{ik}P^iP^k)^2}. \end{aligned} \tag{E.11}$$

Now we coordinatize $P = (x, y, 1)$, $\mathrm{d}P = (\mathrm{d}x, \mathrm{d}y, 0)$ and choose $\mathcal{B}$ and $\mathcal{A}$ to be in a normal form with diagonal elements $0, 1$, or -1. Equation (E.11) is then transformed into the known metric of homogeneous planes. For the elliptic geometry, we obtain

$$\mathcal{B} = \begin{pmatrix} 1 & 0 & 0 \\ 0 & 1 & 0 \\ 0 & 0 & 1 \end{pmatrix} \quad \to \quad \mathrm{d}s^2 = \frac{\mathrm{d}x^2 + \mathrm{d}y^2 + (x\mathrm{d}y - y\mathrm{d}x)^2}{(1 + x^2 + y^2)^2},$$

for the Lobachevski geometry

$$\mathcal{B} = \begin{pmatrix} 1 & 0 & 0 \\ 0 & 1 & 0 \\ 0 & 0 & -1 \end{pmatrix} \quad \to \quad \mathrm{d}s^2 = \frac{\mathrm{d}x^2 + \mathrm{d}y^2 - (x\mathrm{d}y - y\mathrm{d}x)^2}{(1 - x^2 - y^2)^2},$$

for the de Sitter geometry

$$\mathcal{B} = \begin{pmatrix} -1 & 0 & 0 \\ 0 & 1 & 0 \\ 0 & 0 & -1 \end{pmatrix} \quad \rightarrow \quad \mathrm{d}s^2 = \frac{-\mathrm{d}x^2 + \mathrm{d}y^2 - (x\mathrm{d}y - y\mathrm{d}x)^2}{(1 + x^2 - y^2)^2},$$

and for the anti-de Sitter geometry

$$\mathcal{B} = \begin{pmatrix} 1 & 0 & 0 \\ 0 & -1 & 0 \\ 0 & 0 & 1 \end{pmatrix} \quad \rightarrow \quad \mathrm{d}s^2 = \frac{\mathrm{d}x^2 - \mathrm{d}y^2 + (x\mathrm{d}y - y\mathrm{d}x)^2}{(1 + x^2 - y^2)^2}.$$

The other geometries yield simpler, but corresponding formulas.

Exercises

1. Draw the diagrams 3.10 and 3.11 for the case of a totally inelastic collision.

2. Draw the diagrams 3.10 and 3.11 for the case of a collision of second kind.

3. Correct the calculation of the echo sounder (Figure 2.13) for the case of ideally reflected particles.

Chapter 5

4. Einstein's construction of simultaneous events uses two signals of light directed to symmetrical positions (Figure 4.7). Check that the uniform actual transport of clocks to both events has the same success.

5. Show that clocks that start with different velocities to some event *cannot* define the simultaneity (Figure 7.20).

6. Draw the Euclidean analog of the time dilation (Figure 5.13).

7. Draw the analog of the twin paradox (Figure 5.15) in the Euclidean geometry.

8. Two equal rockets move and fly by. Each calculated the motion of the other and fires a gun at its rear end when the head passes the rear end of the other. Will the bullets hit?

9. Now the guns are fired on a command from the head, when it passes the rear end of the other rocket. Will the bullets hit?

10. Draw the momentum diagram of the spontaneous emission of a tachyon without change of the rest mass of the emitting particle.

11. Try to design the momentum diagram of the emission of a particle slower than light. Why must you resign?

12. Two particles move with equal velocity toward each other. Their energy is ten times the rest of the energy. How large is the energy of the one in the rest frame of the other?

Chapter 6

13. Prove the orthocenter theorem for the Minkowski reflection (Figure 5.1).

14. Check the statement of Figure 6.5.

15. Construct the proof why the circle through the feet of the altitudes bisects the sides (Figure 6.11). Check that this property depends on the validity of the theorem for the circumference angle.

16. Draw a Feuerbach circle in the Galilean geometry.

The Geometry of Time. Dierck-E. Liebscher

ISBN: 3-527-40567-4

Chapter 9

17. Construct the length transport (Figure 9.10) in the Euclidean version and show that the shift has the other sign.

18. Choose a triangle ΔABC and an absolute pole P that is the orthocenter of the triangle. Draw the perpendicular on AP in P and mark the intersection with $a = BC$. Construct the other two such intersections and show that they are collinear. Prove that the construction represents an anti-Euclidean rule and that the relations are dual to the orthocenter theorem in the Euclidean geometry.

Appendices

19. Show that the transformations that leave the wave equation (B.1) invariant must be linear.

20. Check Eq. (C.22).

21. Show that the main-plane construction for a thick lens yields a projective map. Show that these maps when restricted to lenses on a common optical axis form a subgroup, and that any such system of lenses on a common optical axis is equivalent to one thick lens.

22. Show that the ideal refractor (two lenses with one common focus) is equivalent to an infinitely thick lens of infinite focal length.

23. Let a triangle $\Delta A_1A_2A_3$ and its orthocenter A_4 be given and show that this is not sufficient to determine the metric (resp., the absolute conic). Show that $\mathcal{B}$ is of the form $\mathcal{B} = \sum_{N=1}^{4} \lambda_N A_N \circ A_N$ so that three effective parameters remain to be chosen freely.

24. Figure D.11 seems to show that a Euclidean circle could be defined *independent* of these three free parameters of the absolute conic. Why is this a wrong impression?

25. Show that the Euclidean circumscribed circle of a triangle passes its *pseudo-Euclidean* orthocenter (Figure D.12).

26. Let us imagine the bundle of conics through a quadrangle $A_1A_2A_3A_4$. Let us define centers, altitudes, and circles through $\mathcal{B}$. The centers of the conics lie on the circle through the diagonal points if A_4 is the intersection of the altitudes of the triangle $\Delta A_1A_2A_3$.

27. Try to find a proof of the intersection theorem for the three conics of Figure D.9.

Glossary

aberration: shift of the apparent positions of distant objects found by two observers in motion with respect to each other, in particular shift of the apparent position of a star by the change in direction of the orbital motion of the earth transversal to the line of sight. When the light velocity is combined with another one, its magnitude does not change, but its direction can. A change in motion of the observer leads to a change in the apparent direction, i.e., to aberration. Aberration is an effect between two observers that depends on their relative velocity and the apparent direction of the light. It does *not* depend on the velocity of the source. On the earth's orbit, it attains a maximal value of $20,47''$. From the aberration value and the orbital velocity of the earth of approximately 30 km/s, we find that the speed of light is approximately 300 000 km/s.
Figures 4.3, 5.25, 5.22, 5.23, and D.5 39, 70, 69, 185

absolute conic section: If in the projective plane a conic is privileged and if we allow only those projective maps that leave this privileged conic invariant (even if the points on it can move sidewards), we speak of an absolute conic. Through reference to such a conic, the projective plane is transformed into a metric plane in which lengths of segments and sizes of angles are defined.
Figure 8.15 114

absolute polar: common polar for all points of the plane. An absolute polar exists if the axiom of parallels is valid, that is in the Euclidean, Galilean, and Minkowski geometries.
Figures 9.7, 9.8, and 9.9 128 ff.

absolute pole: common pole for all straight lines of the plane. The existence of an absolute pole is equivalent to the existence of a unique point on any straight line whose distance to a given point outside the line is zero. This dual to the axiom of parallels holds in the Galilean, anti-Euclidean, and anti-Minkowski geometries. In physics, it corresponds to → absolute simultaneity.
Figures 9.5, 9.6, and 9.9 126 ff.

absolute simultaneity: circumstances under which the simultaneity of two distant events can be decided independently of position, orientation, and motion of the observer. For geometry, this implies the degeneracy of orthogonality in the world. In Newtonian mechanics, absolute simultaneity is tacitly assumed. The theory of relativity shows why this assumption cannot hold, and why absolute simultaneity is a good approximation for small relative velocities.
Figures 3.15 and 3.16 33 ff.

absolute space: virtual entity that permits the determination of position, orientation, and velocity independently of material reference objects. Absolute space is a formal but fake reference object. The relativity principle asserts that positions, orientations, and velocities always refer to material objects, and that space is not absolute in this respect. However, rotation seems to remain absolute. This problem leads to → Mach's principle. 19, 47

The Geometry of Time. Dierck-E. Liebscher

ISBN: 3-527-40567-4

absolute time: time that can be defined in the case of absolute simultaneity. If some physical process would make it possible to establish a transitive relation of simultaneity, simultaneous events would form three-dimensional spaces. One could then try to set up a theory in which these spaces (together with all other laws of the theory) are preserved in changes of reference frames. The time would be absolute in this case. One of the lessons of relativity theory is that simultaneity cannot be absolute.

To the extent that one refers to material objects, specific time coordinates may be useful (→ cosmological time). 20

absolute velocity: → speed of light.

absolute zero: zero point of the thermodynamic (absolute) temperature scale. The absolute temperature is defined by the statistics of microscopic motions and represents one of the characteristics of the variance in the distribution of microscopic quantities (seen in a given state of the system in question). 20

action: product of energy and time or of length and momentum. The *action integral* attributes values to all segments of curves in the → phase space. The realized path between two points in phase space yields an extremum for the action integral. From this metrization of the phase space, all other metrization should be derived. 144, 9

acoustic signal: pre-relativistic analogon of a light signal, subject to Galilean addition of velocities. 14

adaptive optics: technology which uses the measurement of the form of the wavefront to correct the optical system for the influence of atmospheric turbulence. 67

addition theorem of velocities: formula for the *composition* of velocities. In the classical mechanics of Galileo and Newton, the composition is additive. In Einstein's theory of relativity, it is, for equal directions, the addition theorem of the hyperbolic tangent.
Figures 4.10 and 8.7, Eq. (4.1) 45, 110

aether: hypothetical substratum of the world intended to explain the propagation of light by analogy with the way pressure and shear waves explain the sound propagation. The aether is usually imagined to be a weightless fluid that permeates all space and is the seat of vibrational excitations. After the advent of the theory of relativity, no effect that must be attributed to an aether has ever been found. 40, 141

affine geometry: geometry of points and lines with a defined and invariant separation ratio. 105

annihilation of particle–antiparticle pairs: transformation of a particle–antiparticle pair into photons. This transformation is possible because the sum of the participating charges of any kind except gravitation is zero. The mass of the pair is conserved as the mass of the (purely kinetic) energy of the produced photons. The reverse process is the creation of particle–antiparticle pairs. If a energetic photon collides with a massive particle, the photon can form a particle–antiparticle pair. It is essential that the target particle removes some of the photon's momentum. The heavier it is, the less energy is lost for the pair production.

Two photons will collide with only extremely small probability, because the corresponding cross section is so small. 140

anti-Euclidean geometry:
Figure 9.5, Tables 9.1 and E.1 126, 131, 202

anti-Minkowski geometry:
Figure 9.6, Tables 9.1 and E.1 126, 131, 202

anti-Lobachevski geometry:
Figure 7.27, Tables 9.1 and E.1 103, 131, 202

apex: direction of a uniform motion. The direction is represented by a point on the (plane or spherical) field of view, the → vanishing point of the orbits.
Figure 2.16 16

apparent magnitude: measure of the intensity of the radiation of a source at the position of the observer.
Figure 2.18 17

asymptotic cone: in the locally pseudo-Euclidean geometries the cone of lightlike lines through a given origin. The locus of fixed distances from the origin approaches this cone asymptotically. That is, lightlike lines intersect infinity at the same points as the asymptotic cone. In the physical interpretation, the asymptotic cone is the → light cone.
Figure 7.19 97

atom: particle that cannot be permanently changed in its characteristics by chemical means (i.e., with energies less than 1 keV). It consists of a positively charged nucleus (which binds its constituents with energies of some MeV) and a hull in which there are bound electrons, with some energies far below 1 keV.
Chapter 2, 10 5, 139

atomic time: time measured with quantum mechanical proper frequencies. The International System (SI) uses a frequency of Cesium 133 (9 192 631 770 Hz).
Chapter 2 5

axioms, Newton's: → Newton's laws.

axiom of parallels: For each straight line g and each point A not incident with the line, there exists exactly one second line (the parallel) incident with A and not intersecting the line g.

This is the final axiom of the Euclidean geometry. After long dispute, it had to be acknowledged as independent. This independence was established by the construction of a non-Euclidean geometry (the Lobachevski geometry) that conforms to all Euclidean axioms except the axiom of parallels.
Table 9.1 121, 131

axioms of projective geometry: axioms that regulate the algebraic relations between points and lines, which are thereby implicitly described by the axioms. In this sense, projective geometry is a particular system of algebraic relations.

To begin with, points are the objects of projective structures for the moment, → lines are subsets of the set of points. We suppose more than one point and more than one line to exist. A line is supposed to contain at least two points. Its definition completed by the condition that it can be extended and shortened. The line through two distinct points is supposed to be unique, as is the common point of two distinct lines.

A → projective map assigns lines to lines and points to points while preserving the incidence. Consequently, the pattern of points in figures like the harmonic pattern is preserved can be recognized after projective transformations (Figures 8.8 and 8.9). On this basis, the cross-ratio can

be defined in a coordinate-free way and is also preserved. With the construction of Figure C.2, homogeneous coordinates are defined in which the projective maps take the form of linear maps.
Chapter 8, Appendix C 105, 165

axioms of reflections: In the plane, a group of motions $\mathcal{B}$ is generated by a system $\mathcal{S}$ of reflections g (involution maps with $gg = 1$) called straight lines. If the product of two straight lines g, h is again an involution (i.e., $ghgh = 1$), it is called a point. In this case, the lines are said to be perpendicular, and we write $g \mid h$.

1. Any two points A, B are supposed to determine a connecting line $g = (A, B)$ (i.e., $AgAg = 1$ and $BgBg = 1$, denoted by $A, B \mid g$).
2. If two points are connected by two lines, $A, B \mid g, h$, either the points or the lines coincide.
3. If three lines are concurrent, i.e., $a, b, c \mid A$, a line d exists with $abc = d$.
4. If three lines a, b, c have a common perpendicular, $a, b, c \mid g$, a line d exists with $abc = d$.
5. There exist three lines g, h, j for which g is perpendicular to h, but j is neither perpendicular to g nor to h.

Although we can always imagine points and lines of the plane, the axioms refer to an abstract group that does not necessarily act on a separate object space but may act on itself.
Appendix A, Section D.2 145, 180

azimuthal projection: Figure 7.9 ff. 90 ff.

bisection: basic construction in geometry, → harmonic separation.
Figure 8.9 111

Bohr, N.: 1885–1962, physicist, Nobel prize 1922. One of the founders of the quantum theory, invented the first quantum model of the atom. We consider the *Bohr radius*, i.e., the radius of the smallest orbit around a proton in which an electron can revolve in accordance with Bohr's quantum conditions. The Bohr radius is still an appropriate measure for all atomic distances in the final quantum mechanics. It is determined by Planck's constant h, the elementary electric charge e, and the rest mass of the electron m_e to be
$r_{\text{Bohr}} = h^2 m_e^{-1} e^{-2}$. 15

Bolyai, J. 1802–1860, mathematician, constructed contemporarily with N. Lobachevski the first non-Euclidean geometry. 139

Brunelleschi, F.: 1377–1446, architect, was the first of the artists of renaissance to demonstrate perspective images.
Figure 8.1 106

Cartesius (Descartes, R.): 1596–1650, philosopher and mathematician, founder of analytic geometry. *Cartesian coordinates* are referred to rectangular axes. In curved spaces, they exist only in local approximations. In Cartesian coordinates, the metric of space is the unit matrix (up to the sign of its diagonal elements).

Cayley, A.: 1821–1895, mathematician, founder of algebraic geometry. The Cayley–Klein geometries are subject of this book.

causal order: (partial) ordering of the events of a world. If two events can be distinguished as cause and effect, this order requires that the effect happens later than the cause (→ tachyons). The existence of a causal order is called *causality*. However, the notion causality is sometimes used synonymously for deterministic causality, i.e., for the expectation that a given preparation of a system makes it possible to calculate uniquely at least the near future.

center of mass: virtual point at which the dipole moment of mass vanishes. The velocity of the center of mass multiplied by the total mass is equal to the total momentum of the given system.
Figures 3.10 and 3.11 31, 31

circle: locus of points of fixed distance from a center in a plane.
Chapter 6, Section D.3, 75, 183
Figures 3.7, 2.8, 6.3, and 6.4 27, 11, 77

clock: measures the flow of time by counting the periods of appropriate processes. Their stability (uniformity) depends of the ratio of the internal forces (which determine the periodic process) to the external forces (which can accelerate or deform the clock).
Chapter 2 5

clock paradox: → twin paradox.

collinearity: Three points are collinear when they lie on a common straight line (→ axioms of projective geometry).
Figure 2.10, Chapter 8 12, 105

collision: interaction that can be evaluated without detailed consideration of the time interval of interaction. In collisions, the equality of all conserved quantities before and after the collision is used. All other quantities are calculated statistically in the form of scattering cross sections, which allow conclusions to be drawn about the details of acting forces (→ perfectly elastic collision, → totally inelastic collision).
Figures 2.6, 3.7, 3.8, and 5.6 9, 27, 55

Compton effect: change in frequency of light scattered on free electrons. This change is described completely as collision of photons (of momentum $p = E/c$ and energy $E = h\nu$) with electrons.
Figure 5.10 57

conformal map: locally shape-preserving map, i.e., map that preserves the angular relations at every point (→ aberration).
Figures 5.24 and D.5 70, 185

congruence: 1. general equivalence of form after transformation by the operations of a given group of motions or symmetries. In particular, congruence is equivalence with respect to translations and rotations in space.
2. $(n-1)$-parametric family of curves in an n-dimensional space.

conic (section): planar curve of second degree. A conic section is intersected by any straight line in at most two points. It preserves this property in projective maps. A conic section is determined by five given points or other elements.

Conic sections can be represented as solutions of quadratic equations and are the simplest geometrical figures after points and lines.
Figures 8.6 and C.6 109, 174

conic (section), absolute: → absolute conic section.

constancy of the speed of light: absolute independence of the speed of light of the direction of propagation. 43

contingency of geometry: feasibility and necessity to decide the applicability of geometry by experiment and observation. 46

coordinates: numbers representing the position of a point with respect to other points or lines.
Appendix B 155

coordinates, homogeneous: → homogeneous coordinates.

cosmological constant: term in the Einstein equations that can be interpreted as ground state of the curvature of the world, or density of the quantum vacuum. It was used by Einstein to find a static model for the universe, just as Poincaré had used it to find a static model for the electron.
Equations (7.3) and (E.8) 95, 203

cosmological red-shift: Light emitted from a distant source appears red-shifted because of the expansion of space itself. This effect is loosely interpreted as the → Doppler effect of the recession of surrounding galaxies. 18

cosmological time: In general relativity, coordinates acquire physical properties only in relation to objects embedded in space–time. If we do not refer to such objects, for instance if the space–time is empty, all foliations of the world in sequences of spaces define time coordinates of equal status. In cosmology, we assume the existence of a foliation of approximately homogeneous spaces. This is a particular foliation, and the corresponding time is the cosmological time. If the universe is empty, more than one such foliation can exist: The Minkowski world and the de Sitter world are examples. If the universe contains matter, at most one such foliation exists. The lines of constant position are timelike lines that are chosen to represent the average motion of the matter. The cosmological time is the proper time of this motion.

cosmology: theory of the global consistency (cosmos) of physics and its testability on the observable part of the universe, which is assumed to realize the cosmos. Its main basis is the cosmological principle, which is the assumption that the observed part of the universe is typical and that the universe behind the horizon is identical in its physical laws, its matter density, constitution, and distribution. The homogeneity of the universe is well supported by the isotropy of the → microwave background radiation, though the *scale* of the homogeneity is still in question. In strictly homogeneous models of the universe, the world can be envisaged as a homogeneous space uniformly expanding (or contracting) in time. This time is called the → cosmological time. Any motion can be decomposed into the motion produced by the expansion and a peculiar motion. The space coordinates can be chosen so that their change in time indicates a peculiar motion. These are the expansion-reduced or comoving coordinates. The expansion of the universe is ruled by the → Friedmann equation (Eqs. (7.3) and (E.8)).
Section 7.2 92 ff., 203

Coulomb, C. de: 1736–1806, physicist. He explored the electrostatic interaction and found its law, the *Coulomb force* whose strength between two particles is (like gravitation) inversely proportional to the square of the distance and acts in the direction of the connecting line. 19

cross-ratio: characteristic quantity of projective geometry. Its invariance determines the group of projective transformations.
Eq. (8.1), Figure 8.5 108, 109

cross product: antisymmetric bilinear product of two vectors in three-dimensional space yielding the plane spanned by the two. 168

curvature: deviation from the Euclidean geometry.
Figures 7.1, 7.2, 7.4, and 7.11 85, 87, 91

degeneracy: coincidence or vanishing of quantities generically different or different from zero. A conic section degenerates when one of the principal axes vanishes (degeneracy to a line), or becomes infinite (degeneracy to a parallel pair of lines), or when both vanish (degeneracy to a point or a pair of intersecting lines). 117 ff., 197 ff.

Desargues, G.: 1591–1661, architect and engineer, found the first basic theorem of modern projective geometry.
Figures 8.3 and 8.4 108

de Sitter, W.: 1872–1934, astronomer, constructed, for instance, the first (empty) universe that satisfies the Einstein equations.
Figure 7.21 ff. 99 ff.

direct product: formal linear map that assigns to all vectors a vector of fixed direction. 169

Doppler, C.J.: 1803–1853, physicist, famous for his finding of the *Doppler effect*, an apparent change in wavelength of sound or light caused by motion of the source, observer, or both. Waves emitted by a moving object as received by an observer will be blue-shifted (compressed) if the object is approaching, red-shifted (elongated) if receding. It occurs both in sound and light. How much the frequency changes depends on how fast the object is moving toward or away from the receiver, → cosmological red-shift.
Figures 2.14 and 5.11 14, 61

dual construction: in projective geometry, the construction in which points are interchanged with lines, point rows with ray pencils, tangents with contact points, and so on. Dual constructions are typical for the projective geometry of the plane because in homogeneous coordinates points as well as lines are defined by a coordinate triple, and linear maps of points on points are equivalent to the inverse maps of lines on lines. Consequently, points can be interchanged with lines in all theorems if at the same time intersection is interchanged with connecting line and collinearity with pencil property. 116

duality: possibility of interchanging pairs of complementary notions, as for instance point and line in a projective plane (→ dual construction).

Dürer, A. 1471–1528, painter.
Figure 8.2 107

ecliptic: the plane containing the earth's orbit (strictly the orbit of the center of mass of the earth–moon system). The projection of the ecliptic on the apparent sky is the apparent mean orbit of the sun. Eclipses can occur when the lunar orbit around the earth crosses the ecliptic. The ecliptic is inclined to the equator at an angle of $23^\circ 27'$.
Figure 2.15 16

Einstein, A.: 1879–1955, physicist, one of the creators of the quantum theory (Nobel prize 1921) and the (special) theory of relativity, founder of the general theory of relativity.

Einstein's equations: → general relativity.

Einsteinian geometry: geometry of a locally pseudo-Euclidean world modified by the curvature of the world. 163

electrodynamics: theory of the behaviour of the electromagnetic field, discovered as unified description of the electric and the magnetic field by →Maxwell. Visible light is electromagnetic waves in the wave-length range between 350 and 700 nm.
Section 5.2 54

elementary particles: here the particles of the subnuclear level, such as protons, neutrons, electrons, and photons. On this level, the elementary particles can be classified as baryons, mesons, and leptons. For baryons (protons, neutrons, and various hyperons) and leptons (electrons, muons, neutrinos) number conservation laws exist and characterize the baryon and lepton number as charges. The conservation of the baryon number implies that the lightest baryon (proton or antiproton) is stable; the conservation of the electric charge ensures the stability of the electron (and the positron, respectively). All other particles decay into lighter ones, and the typical unit for the lifetime of weakly decaying particles is 10^{-10} s. The strongly decaying particles live typically only 10^{-23} s (the time light takes to pass across the particle) and are called resonances because they reveal their existence in large cross sections of other particles at energies corresponding to the resonances' rest mass. Mesons are particles without baryon or lepton charge and are all unstable unless massless like the photon (which represents a species of its own). The most famous meson is the π meson predicted to mediate the nuclear forces between proton and neutron. Mesons and baryons can be imagined to be composed of subelementary particles called quarks. The composition rule was vindicated by the detection of the predicted Ω^- (Figure 2.7), which unexpectedly was found to decay only weakly.

In relativity, the apparent slowdown of the decay of the muon in flight is a qualitative demonstration of time dilation. Figure 2.7 10

elliptic geometry:
Figure 9.1, Tables 9.1 and E.1 122, 131, 202

energy: fundamental quantity of physics, universal measure for both motion and the capacity to motion that is strictly conserved in isolated systems. The dependence of the energy on the general coordinates (positions and momenta) of the state of a system completely determines the physical motions. 29, 58 ff.

ephemeris time: time describing the planetary motion without removable perturbations. The ephemeris time is determined implicitly by appropriate observations. If all orbits are known, the ephemeris time is determined by the validity of energy conservation. 15

equation of motion: instrument used to identify integral real motions as solutions of a local (differential) law. 9

equivalence of mass and energy: Inertia of energy. The theory of relativity shows why all energy contributes to the inertia of a body. The proportionality of energy and inertial mass contains as factor the square of an absolute velocity c. This absolute velocity is the speed of light in vacuo (at least there is no counterindication).
Eq. (5.4) 58

equivalence of inertial and gravitational mass: In the equations of motion in a gravitational field, the gravitational charge (i.e., the gravitational mass) and the inertial mass (i.e., the coefficient of the velocities in the momentum conservation law) cancel. Hence, both are given in the same units. Their strong proportionality implies that the gravitational field can be represented by a → metric in a world of variable curvature. 92

Euclid: 330–275 BC, mathematician, author of the oldest text with a complete axiomatization of geometry.

Euclidean geometry: Simplest geometry of space with positive definite metric and zero curvature. Chapter 9 121

eV: electron volt. Unit of energy in nuclear physics, 1 eV= $1.60201\ 10^{-19}$ J.
The corresponding mass is $\approx 1.8 \times 10^{-36}$ kg, the corresponding temperature ≈11 600 K.

event: point in a space–time (world) defined by the position coordinates and the time. Section 2.1 5

extremum principle: strategy to identify real motions and structures as extremum of values attributed both to real and virtual motions and structures, → action. 9

Fermat, P. de: 1601–1665, mathematician, one of the creators of analytic geometry. Fermat found the first integral principle in the history of physics (*Fermat's principle*). The path of an observed light ray between two points A and B is the shortest connection of the two points if the geometrical length $\mathrm{d}s = \sqrt{\mathrm{d}x^2 + \mathrm{d}y^2 + \mathrm{d}z^2}$ of its infinitesimal segments is multiplied by the local refractive index (or divided by the local phase velocity of light): The light ray from A to B realizes a minimum of the integral

$$S = \int_A^B n\,\mathrm{d}s.$$

In → mechanics, the analog is the principle of Maupertuis and Jacobi: In a time-independent potential, the path of a particle is the shortest connection if the geometric length is multiplied by the expression $\sqrt{E_{\mathrm{tot}} - E_{\mathrm{pot}}}$ formed from the total energy E_{tot} and potential energy E_{pot}[position]: The length of a path is given by

$$S = \int \sqrt{E_{\mathrm{tot}} - E_{\mathrm{pot}}}\sqrt{E_{\mathrm{kin}}}\,\mathrm{d}t.$$

The kinetic energy E_{kin} is a homogeneously quadratic function of the velocity coordinates, so that the time t remains an undetermined free parameter.

This principle is valid, in general, as Hamilton's principle, which characterizes and determines the general motion as realization of an extremum of an → action integral. 8, 142

Feuerbach circle: Characteristic circle in geometries with valid axiom of parallels that connects midpoints of the sides and altitudes of a triangle with the feet of the altitudes. Projectively, it is an eleven-point conic.
Figures 6.11 and 8.12 81, 112

fine-structure constant, Sommerfeld's: → Sommerfeld.

flat angle: angle with sides forming one straight line. In the Euclidean geometry, it has the size π or $180°$.

force: cause of the change of the → momentum of an object. Deduction of a force from the hypothesis of a corresponding action on other objects yields an equation of motion that must be solved and whose solutions must be tested. → Newton's laws. 19 ff.

four-vector: indicates a four-component vector in space–time, in contrast to the three-component vectors of ordinary space. Any direction in a space–time diagram corresponds to a four-vector. The identification of the fourth component (time component) of a formerly three-component ordinary vector is one of the tasks of relativistic kinematics. The fourth component of the velocity is the clock rate (ratio of time increment in the frame of reference to proper time increment of the moving object). It is trivially equal to one in Galilean geometry. The fourth component of the momentum is the inertial mass (its ratio with the rest mass is equal to the clock rate), which is also the total energy (→ equivalence of mass and energy).
Appendix B 155

freely falling frames: construction of inertial frames in homogeneous gravitational fields.
Figure B.2 157

Friedmann, A.A.: 1888–1925, mathematician and physicist. The *Friedmann equation* is the fundamental equation of cosmology.
Eqs. (7.3) and (E.8) 95, 203

Galilean geometry: geometry of the space–time of classical mechanics.
Figures 3.16 and 3.17, Tables 9.1 and E.1 34, 131, 202

Galilean group: group of → transformations that in Newtonian mechanics convert the coordinates of inertial reference systems into each other and leave the laws of point mechanics formally invariant.
Chapter B 155

Galilei, G.: 1564–1642, astronomer and physicist, founder of the physics of modern times. In particular, he stated the uniformity of motion which is screened from external influence (later Newton's first law). 35

Gauß, K.F.: 1777–1855, mathematician and astronomer. 91

general theory of relativity: Simplest gravitation theory consistent with the equivalence of inertial and gravitational mass, developed by A. Einstein. The coefficients of the wave equation are no longer constant. Gravitation theory becomes a theory for the metric of the world. General relativity requires a world with curvature. This curvature describes the gravitational field. Locally (i.e., in the domain of its definition) special relativity remains valid.

The metric field is governed by *Einstein's equations*. They state that the curvature, averaged pointwise over all orientations of two-dimensional surface elements at the given point, is proportional to the energy–momentum tensor. 92 ff.

geodesic: connection between two points of extremal length. In locally Euclidean geometries, geodesics are shortest connections. In locally pseudo-Euclidean geometries, timelike geodesics are longest connections (→ twin paradox). The geodesic is the generalization of the straight line of plane geometries to general metric geometries.
Figures 7.4, 7.5, and 7.6 87 ff.

geometry: theory of the relations between different structures, forms, or positions.

geometry, affine: → affine geometry.

geometry, Einsteinian: → Einsteinian geometry.

geometry, Euclidean: → Euclidean geometry.

geometry, metric: → metric.

geometry, non-Euclidean: → non-Euclidean geometry.

geometry, projective: → projective geometry.

geometry, pseudo-Euclidean: → pseudo-Euclidean geometry.

geometry, Riemannian: → Riemannian geometry.

geometry, spherical: → spherical geometry.

Giotto di Bondone: 1267–1337, painter, cited for his attempts to get the effect of space in his works although he failed to find the correct laws of the perspective. 107

gravitational field: name for the experience that at each point in space there is a → gravitational force on a (test) body.
Section 7.2 92

gravitational force: proportional to mass, not screenable, but extremely weak force of long range. The source strength and charge in the gravitational field are proportional to the → inertial mass. There exist only positive masses. Therefore, the gravitational field cannot be screened off. The gravitational field is very weak: Compared with the electrostatic force between two protons, the gravitational attraction is only 10^{-36}. While all other forces can be screened or are of short range, gravitation adds up to play the dominant role for the celestial bodies and in the universe in general. In the theory of general relativity, the concept of gravitational force must be abandoned and is replaced with the concept of curvature. 19

gravitational lens: astrophysical phenomenon (image distortion and magnification) observed at cosmic objects behind foreground gravitational sources. Gravitational lensing is a manifest indication of gravitational light deflection.
Figure 7.13 94

gravitational mass: charge in the gravitational field. A body responds the more to a given gravitational field, the larger is its gravitational mass. In everyday life, the gravitational mass is found by weighing. In contrast to electrostatics, in which the *specific* electric charge (the ratio of the charge to the inertial mass) can vary from object to object, the specific gravitational charge is universal. This fact is called → equivalence of gravitational and inertial mass. It is the basis of general relativity.

The notion of gravitational mass can itself be further refined into two concepts: The passive charge, which measures the reaction to a given gravitational field, and the active charge, which produces the gravitational field. Proportionality of the two is the simplest assumption that will satisfy Newton's third law and is usually supposed to be valid. 59

gravitation potential: → potential of the gravitational field. In the case of a swarm of point particles of mass M_A (G gravitational constant, $\vec{r}_A$ positions), it is given by

$$\Phi = \frac{G}{c^2} \sum_A \frac{M_A}{\mid \vec{r} - \vec{r}_A \mid}.$$

141

great circle: intersection of a sphere with a plane that passes through the center of the sphere and is thus perpendicular to the surface at each point. On a curved surface, a line is a geodesic if at each point of it the plane in which the line is curved is perpendicular to the surface at that point. Great circles are such geodesics.
Figure 9.4 124

group: set with operations defined so that to any two elements a and b a product $c = ab$ is assigned that is again an element of the group. There exists a unit element e that reproduces all other elements in multiplication, $ea = ae = a$. In addition, the product is invertible, i.e., a reciprocal a^{-1} exists for all elements ($aa^{-1} = a^{-1}a = e$). In multiple products, brackets can be freely moved. However, the sequence of factors must not be changed. In general, ab and ba are different. Group elements are advantageously represented by → matrices and the group operation as matrix product.
21, 105

group of motions: Appendix A 145

gyroscope: instrument that uses the conservation of angular momentum to measure changes of orientation. 86

Hamilton, W.R.: 1805–1865, mathematician and physicist, author of the fundamental integral principle of mechanics (→ action integral). The integrand is called Hamiltonian and descibes the total energy as function of the coordinates and momenta of the system under consideration.

harmonic pattern: configuration of lines and points that solely by their incidence have a cross-ratio $\mathcal{D} = -1$. Figure 8.8 110

harmonic range: configuration of four points in *harmonic separation*, i.e., with the cross-ratio $\mathcal{D} = -1$.
Figures 8.8 and 8.9 110, 111

Helmholtz, H.: 1821–1894, physician and physicist, answered → Riemann with a thesis *On the facts that form the basis of geometry* [74].

Hilbert, D.: 1862–1943, mathematician, cited here for his contribution to the axiomatic foundation of geometry.
Chapter 10 139

hodograph: description of a motion in the space of velocities. 183

homogeneous coordinates: coordinates based on the interpretation of points as the intersection of the projective space in question with central rays in a space with a supplementary dimension and an origin outside the projective space. Correspondingly, lines are interpreted as generated by central planes, and so on. The ray coordinates include an irrelevant factor. In homogeneous coordinates, the projective group is the special linear group. Plücker was the first to introduce them as barycentric coordinates of a triangle: Each point of the plane can be the center of mass of a given triangle if the vertices A_i have appropriate weights m_i. These are homogeneous coordinates.
Figure C.1 167

homogeneous expansion: expansion without center. Figures 7.17 and 7.19 96, 97

horizon: boundary line of observability (particle horizon) or reachability (event horizon).
In projective geometry: Image of the line at infinity.
Figure 8.3 108

Huygens, Ch.: 1629–1695, physicist and astronomer, leading proponent of a theory that explained light as pulses in a medium (this theory was extended to a wave theory by Euler). Huygens stated that in a collision between different bodies "motion" (his term for momentum) is not lost or gained (known, in other form, as Newton's third law).
Figure 3.9 28

hyperbolic geometry: → non-Euclidean geometry.

hyperons: → elementary particles.

hypersurface: configuration of dimension $n - 1$ in an n-dimensional space that can be represented in general by an equation $f[x_1, ..., x_n] = 0$. If the coordinates are those of a linear space, a hyperspace of the function f is linear.

ideal line: projective completion of the plane by the points at infinity, which are projectively a line.
Figure 8.3 108

incidence: relation between geometric elements of different character that in the case of a point A and a line g signifies that the point A belongs to the point row carried by the line and the line belongs to the ray pencil carried by the point (→ axioms of projective geometry).
Figure A.2 148

inertial frame: inertial → reference frame.

inertial mass: coefficient with which velocities are multiplied in order to obtain a conserved quantity in the form of a weighted sum of velocities. The weighted velocity is called the momentum. Any force that changes velocity does so more strongly the smaller is the inertial mass. Hence, the inertial mass can be interpreted as a resistance to accelerations. The conservation law of momentum contains the conservation law of mass if written in a four-dimensional space–time. 92

instantaneous rest frame: → rest frame.

intersection of altitudes: The altitudes of a triangle intersect at one point. This is equivalent to the intersection theorem of perpendicular bisectors and central to the notion of perpendicularity.
Figures 6.5, 9.11, A.7, and D.2 78, 132, 152, 179

involution: map $\mathcal{I} : x \to \mathcal{I}[x]$ that yields the initial state if applied twice ($\mathcal{I}[\mathcal{I}[x]] = x$) but is not the identity. Involutions can be understood as reflections, although the reflections of everyday language are only a special case of involutory maps.

Projective involutions on a line are determined by the definition of two pairs $A, \mathcal{I}[A]$ and $B, \mathcal{I}[B]$ of points, which can also be fixed points, of course.

isotropic: 1. independent of direction. In particular, light propagation is isotropic independently of the motion of the observer. This contradicts the additive composition of velocities and is the starting point for relativity theory.

2. → lightlike.

isotropy of light propagation: property of light propagation that is recognized and used in relativity theory and according to which the speed of light is independent of direction and keeps this property when it is combined with other velocities. 41

Jacob's staff: instrument used to determine angles between lines of sight to different stars. Often assumed to be invented by Ibn Sina (Avicenna, 980–1037), probably already known to Ptolemy as Hipparch's Dioptra.
Figure 2.10 12

Jacobi, C.G.J.: 1804–1851, mathematician, cited here for his contribution to analytical mechanics.

Kant, I.: 1724–1804, philosopher, cited here for his belief that the Euclidean geometry is a priori, not dependent on experience. 10

Kepler, J.: 1571–1630, astronomer and mathematician. One of the founders of modern astronomy, he formulated the famous three laws of planetary motion. They comprise a correct quantitative formulation of the heliocentric theory that the planets revolve around the sun. *Kepler's first law:* A planet orbits the sun in an ellipse with the sun at one focus. *Kepler's second law:* A ray directed from the sun to a planet sweeps out equal areas in equal times. *Kepler's third law:* the square of the period of a planet's orbit is proportional to the cube of that planet's semimajor axis; the constant of proportionality is the same for all planets. 15

Klein, F.: 1849–1925, mathematician, cited here for his Erlangen program for geometry, and for his model of non-Euclidean geometry.
Chapter 10 139

Klein's model: model of non-Euclidean geometry consisting of the segments of a circle and the points in it. It can be interpreted as the projection of the time shell from the center to the plane.
Figure 7.12 91

length contraction: projection effect in relativity.
Figures 5.18 and 5.19 66, 66

length transfer: basic geometric construction.
Figures 8.11 and 8.10 112

length unit: given classically by the extension of a solid body. Because of the particularly precise methods that can be used to reproduce the speed of light, the length unit is now derived from the unit of time.
→ Bohr.

light clock: Virtual construction of a clock that uses solely light propagation and parallel transport and is thus independent of conceptually more complicated processes.
Figures 5.14 and 5.16 62

light cone: cone of → lightlike world-lines through a given event. The light cones separate the (absolute) future from the (absolute) past (the inner region of the double cone) and the relative present (the outer region). The events in the inner part lie → timelike to the vertex, the events outside, → spacelike.
Figures 4.1 and 7.19 38, 97

light ray: Spatial projection (orbit) of the world-line of a photon, also used for this world-line itself.

light signal: classical equivalent to the photon in its average propagation, but unstable due to dilution.

light velocity: → speed of light.

lightlike: relative position of two events for which one lies on the light cone of the other. If the interval between two events is lightlike, one of them can be reached by a light signal that passed through the other or was sent by it to the other. A → vector is lightlike if its direction coincides with such an interval. Lightlike vectors have a null norm. The paradigm for a lightlike vector is the velocity or the momentum of a particle of zero → rest mass. 50

light-ray quadrilateral: quadrilateral with lightlike sides. Lightray quadrilaterals are used to construct reflection and orthogonality in the Minkowski geometry.
Figure 4.9 45

line: → axioms of projective geometry.
Appendix A 145

line element: representation of the length of a short interval connecting two events as generalization of Pythagoras's theorem, i.e., as quadratic form in the coordinate differences of the two events. If a point P has the coordinates x^k, $k = 1, ..., n$, and $Q = P + \mathrm{d}P$ the coordinates $x^k + \mathrm{d}x^k$, the line element is written as

$$\mathrm{d}s^2 = \sum_{ik} g_{ik}[x]\mathrm{d}x^i \mathrm{d}x^k.$$

The arc length of a general curve $x^k[\lambda]$, $0 < \lambda < 1$, is then given by the integral

$$s = \int_0^1 \sqrt{\sum_{ik} g_{ik}[x] \frac{\mathrm{d}x^i}{\mathrm{d}\lambda} \frac{\mathrm{d}x^k}{\mathrm{d}\lambda}} \mathrm{d}\lambda.$$

Appendix E 195

line at infinity: → ideal line.

linear map: structure-preserving map of a linear vector space. The image of a linear combination is the linear combination of the images. Linear maps are most simply represented by → matrices.

Lobachevski, N.I.: 1792–1856, mathematician, constructed contemporarily with J. Bolyai the first non-Euclidean geometry. 139

Lorentz, H.A.: 1853–1928, physicist, Nobel prize 1902. Founder of the classical theory of the electron. His name was given to the *Lorentz group*. It is the group of transformations that relate the different inertial reference frames in relativity theory. Invariance with respect to the Lorentz group is a principal demand on all theories of elementary phenomena.
Section B.2 156

Lorentz contraction: → length contraction.

loxodrome: curve of fixed inclination to a given point.
Figure 7.3 86

Mach, E.: 1838–1916, physicist and philosopher. Einstein gave the name *Mach's principle* to the loosely defined conviction that the laws of inertia are due to the existence of and interaction with the masses in the surrounding universe. Mechanics alone should not, for instance, be able to distinguish rotational motion of an isolated body from its rest. Mach's principle admits different physical interpretations. Hence, neither its appropriate definition nor its applicability has yet decided [101]. 142

magnitude, apparent: → apparent magnitude

map: assignment of the objects of a mapped set to the objects of an image set. Maps can be defined in different degrees of generality and can be restricted by different contexts. In geometry, maps are applied to geometric objects; in the simplest case, points are mapped to points and the map of all other figures can be derived. In this volume, we often consider a polarity, which is a map of points on flat hypersurfaces (straight lines in the plane, flat planes in the space) and vice versa.
Appendix A 145

map, conformal: → conformal map.

map, linear: → linear map.

map, projective: → projective map.

mass: → equivalence of → inertial and → gravitational mass; → variation of mass with velocity.

mass, gravitational: → gravitational mass.

mass, inertial: → inertial mass.

mass defect: difference between the sum of the rest masses of individual components and the rest mass of a bound system formed by them. The mass defect is proportional to the binding energy. It can be measured for binding energies in the atomic nucleus. 59

mass shell: spacelike surface in the four-dimensional (and pseudo-Euclidean) momentum space, locus of the endpoints of momentum vectors with a fixed rest mass.

In the classical dynamics of particles, the momentum four-vector always ends on the mass shell of the particle. In quantum theory, intermediate and mediating particles may have momenta off the mass shell for short time intervals.
Section 7.1 83

matrix: rectangular array of components characterized by its row number m and column number n, which define the type (m, n). The position in the array is characterized by two indices $1 \leq i \leq m$ and $1 \leq k \leq n$. The multiple of a matrix is found by writing of a matrix with multiples of each individual component: $(\lambda A)_{ik} = \lambda A_{ik}$. Two matrices A and B of the same type can be added to yield a matrix of the same type by adding corresponding components: $C_{ik} = A_{ik} + B_{ik}$. Two matrices A and B can be multiplied if the column number n_1 of the first factor is equal to the row number m_2 of the second factor. The product is a matrix of the type m_1, n_2. Its components C_{ik} are the linear combinations $C_{ik} = \sum_l A_{il} B_{lk}$.

Maupertuis, P.-L.M.: 1698–1759, cited here for being the first to formulate a principle of least action for mechanics.

Maxwell, J.C.: 1831–1879, physicist, cited here for his formulation of electrodynamics (the Maxwell equations), which was later shown to be invariant with respect to the → Lorentz group.

mechanics: theory of the motion of material objects subject to given forces, which are not necessarily explained in mechanics. Mechanics is based on → Newton's laws.
Chapter 3 21

medium: continuum that can carry local excitations that propagate by local interactions and thereby create wave phenomena. 37

mesons: → elementary particles.

metric: definition of distances between points. In a differentiable manifold, it is sufficient to define the distances of infinitesimally adjacent points. Then the simplest definition is given by a → line element. A metric allows the measurement of a vector's length given as a norm by some homogeneous expression quadratic in the vector's components. The norm can be negative in locally pseudo-Euclidean worlds. The coefficients of the → line element form a matrix called the metric tensor.
Appendices B.3 and E.2 161, 200

metrical plane: plane that includes the definition of a distance between its points.
Chapter 9, Appendix E 121, 195

MeV: mega electron volt. → eV.

Michelson, A.A.: 1852–1931, physicist, found the failure of the additive composition of frame velocities with the speed of light through use of his interferometer (beginning in 1881 in Potsdam).
Figure 4.5 40

microphysics: physics of small systems. The smallness of a system is decided by the product of the → momenta of its parts and the lengths of their paths in the system. Such a product is an → action that is quantized by the smallest value given by → Planck's constant h. A system is small if the actions in it are so small that the existence of the quantum of action is felt.

microwave background radiation: Homogeneously distributed electromagnetic black-body radiation in the universe of temperature 2.73 K and relative inhomogeneity 10^{-5}. This radiation has been effectively decoupled from ponderable matter since the neutralization of the primordial high-temperature plasma, and was formerly the main component of the heat bath of the universe. The temperature of the background radiation is inversely proportional to the expansion parameter of the universe. 141

Milne, E.A.: 1896–1950, astronomer.
Figures 7.19 and 7.20 97

Minkowski, H.: 1864–1909, mathematician, constructed the relativistic geometry of the *Minkowski world*, the flat world of space and time whose group of motions reconciles the relativity of velocity with the existence of an absolute velocity. Its geometry carries the same name.
Figure 5.1, Chapter 5 49

Minkowski circle: locus of points equidistant by pseudo-Euclidean measure to some centre.
Figure 6.4 77

mirror: here a formal object that remains pointwise fixed in the corresponding → reflection map.
Figure 3.1 23

Mössbauer effect: effect of the reduction of recoil in γ-emitting nuclei by cooling below the acoustic excitation temperature of the embedding crystal structure. Then the big mass of the crystal receives the recoil momentum, and the recoil velocity drops to extremely small values.
Figure 4.5 40

molecule: smallest part of a chemical compound and a bound system of atoms. The binding energy of the atoms in a molecule ($\approx$10 eV) is far smaller than the binding energy of the constituent parts of an atomic nucleus (1 MeV).

momentum: measure of the motion that is defined by a conservation law that can be tested in collisions without deeper knowledge of the acting forces. The momentum is the velocity weighted (i.e., multiplied) with the inertial mass.
Figures 3.10 and 3.11 31

momentum diagram: representation of the space–time components of the momentum vectors. The affine relation to the ordinary space–time diagrams reveals the geometric path to the definition of mass and the variation of mass with velocity.
Figures 5.4 ff. 53

motion: in *physics* mainly a change of position and orientation with time, in geometry mainly the result of this process. The motion of physical objects is subject to equations of motion, based on → Newton's laws. The geometric motions are usually combined to form a → group. In space–time, a geometric motion can represent the superposition of a physical process in space with a common overall velocity.

moving-cluster parallax: Figure 2.16 16

muon: → elementary particles.

neutron: electrically neutral → elementary particle with spin $s = \frac{1}{2}\hbar$, mass $m_n c^2 = 938$ MeV and magnetic moment $\mu = -1,9131\ e\hbar(2m_p)^{-1}$. With respect to strong interactions, it is identical to the proton apart from its isospin orientation. The differences are due to weak and electromagnetic interaction and yield the different electric charge as well as a small mass difference. The mass difference causes the instability of the free neutron, which decays with a lifetime of 887 s into a proton and two lighter particles (leptons).

Newton, I.: 1642–1727, mathematician, physicist, astronomer, and philosopher. Founder of classical mechanics and, at the same time as Leibniz, the calculus. He formulated *Newton's laws: Lex prima*: A force-free body moves uniformly on a straight line. *Lex secunda*: The uniform motion is disturbed by forces, which act in the direction of the change of the → momentum. *Lex tertia*: The forces between two bodies are equal and opposite.

The third law implies that the → center of mass of a body can realize an ideal point mass and that the point mass is also a useful approximation for extended bodies. In addition, the third law permits the determination of the (inertial) mass.
Chapter 3 21

Newtonian mechanics: mechanics based on the three laws of → Newton. The first (Galileo's) and third (Huygens') law can be used also in relativistic mechanics, the second (Newton's law of gravitation) determines the geometry to be Galilean.

non-Euclidean geometry: geometry without axiom of parallels, in particular hyperbolic geometry.
Figure 7.12 ff., Tables 9.1 and E.1 91 ff., 131, 202

orthocentre: intersection of the altitudes of a triangle. The statement of the existence of this intersection is known as othocentre theorem.
Figure 9.11 132

orthogonal: → perpendicular

pair creation: → annihilation of particle–antiparticle pairs.

Pappos of Alexandria: about 320 BC, found, for instance, the invariance of the cross-ratio (Figure 8.5) and the so-called Pappos theorem (Figure C.3). 109, 172

paradox: phenomenon that is apparently contradictory, surprising, or unexpected because of inappropriate analysis.
Chapter 5 49

parallax: denomination for the distances in the universe. It refers to the fact that the simplest determinations are performed by angle measurements in a triangle of known base length.
Figures 2.15, 2.16, and 2.18 16 ff.

parallel transport: transport of a direction without local change. A definition of parallel transport is necessary if vectors at different points are to be compared. It is necessarily nontrivial in curved spaces. If the space is metric, it can be most simply defined by the requirement of constant fixed angles with a geodesic curve (geodesic parallel transport, → geodesics). There are natural choices that are different from geodesic transport, for instance, transport by a magnetic needle.
Figures 7.1 and 7.4 ff. 85 ff.

parallels: straight lines that do not intersect in the finite (→ axiom of parallels). 121

Pascal, B.: 1623–1662, mathematician, physicist, and philosopher. A fundamental theorem about → conics carries his name.
Figure C.5 174

pencil: Lines lie in a pencil if they are concurrent, i.e., pass through a common point, the origin (carrier, vertex) of the pencil, or if they have a common perpendicular. A pencil of rays is a congruence of lines that all meet at the same point (the carrier, vertex, or origin of the pencil).
→ homogeneous coordinates, Figure A.4 150

perfectly elastic collision: → collision in which the total kinetic energy is conserved, i.e., in which the participating rest masses are not changed.
Eq. (3.2) 29

periphery: here used synonymously with circumference, see Appendix D.4 187

perpendicular: particular relative position of two intersecting lines. Two lines are perpendicular if successive reflections on them yield a rotation through a flat angle, i.e., if the composition of the reflections is an involutive map (with the intersection point fixed).

The notion of perpendicularity of lines is to some degree equivalent to the notion of reflections. They cannot be derived from other properties but must to be defined appropriately. A central property of the choice of the definition of perpendicular lines is the theorem of the → intersection of altitudes.
Figures 3.4 and 8.19 24, 116

perspective map: projective map in which all connecting lines between mapped points and their images intersect at one point, the center of perspective.
Figures 8.1, 8.2, and 8.3 106 ff.

phase space: space of states of a system described by general position coordinates and conjugate momenta. 9

photon: quantum of energy in an oscillator component of the electromagnetic field. Its energy is proportional to the frequency, $E = h\nu$. If this energy is comparable to or higher than the energy of particles that react with the electromagnetic field, the photon can itself be interpreted as an → elementary particle, with the above energy and momentum $p = h\nu/c$. In classical electrodynamics, the photon can be interpreted (badly) as a group of superposed electromagnetic waves that have zero amplitude except for some small region in space. Such a wave group has an arbitrary energy but its momentum is always $p = E/c$.

Planck, M.: 1858–1947, physicist, Nobel prize 1918. One of the founders of quantum theory, discovered in the spectrum of heat radiation the very first law that allowed the determination of the quantum h of action. This quantum of action is $h = 6.626 \times 10^{-34}$ J s.

plane, metrical: → metrical plane.

Plücker: 1801–1868, mathematician, cited here for his method of defining → homogeneous coordinates.

Poincaré, H.: 1854–1912, mathematician, cited here for his formulation of the relativity principle. The Lorentz group, when complemented with ordinary translations, is called Poincaré group.
Chapter 10 139

points: → axioms of projective geometry.
Figure A.1 148

polar coordinates: coordinates that represent the position by the distance from a center and the orientation of the connecting line.
Figures 7.9 and 7.10 90, 90

polar to a point: line defined for a point P with respect to a conic $\mathcal{K}$. On any line through the point P that intersects the conic twice (E and F), the intersection Q with the polar is the fourth harmonic, i.e., $\mathcal{D}[P, Q; E, F] = -1$. If the point P is an exterior point, we can draw the tangents to the conic, and we obtain the polar as the line connecting the contact points.
Figure 8.17 115

polar triangle: triangle in which each vertex is the pole of the opposite side. In metric geometry, proper polar triangles exist only in elliptic geometry.
Figure D.3 182

polar, absolute: → absolute polar

polarity: in the plane, a map between points and straight lines. The polarity associates a line (the polar) with each point and a point (the pole) with each line.

In space, points and planes are mapped to each other, and lines are mapped to lines. In a general n-dimensional linear space, a polarity is a linear incidence-preserving map that associates with each linear submanifold r another linear submanifold $P[r]$ having dimension $\dim P[r] = n - \dim r$.
Figure 9.3 ff. 124 ff.

pole of a line: point defined for a line g with respect to a conic $\mathcal{K}$. For any point on the line l from which two tangents (u and v) can be drawn to the conic, the connection h to the pole is the fourth harmonic ray, i.e., $\mathcal{D}(g, h; u, v) = -1$. If the line g intersects the conic twice, we can draw the tangents at the two points of intersection, and we obtain the pole as the point at which the two tangents meet.
Figure 8.16 114

pole, absolute: → absolute pole

potential: quantity introduced to yield in simple cases (gravitational field, electrostatic field) the field strength as gradient of a descent. The function describing the formal height is the potential. Usually, it is normalized to have the value zero at infinity. 19

power: → energy released in the unit of time.
Figure 2.18 17

product, direct: → direct product

projective coordinates: → homogeneous coordinates.

projective geometry: → axioms of projective geometry.
Chapter 8, Appendix C 105, 165

projective map: simultaneous map of points to points and lines to lines that preserves incidence. Projective maps leave the cross-ratio of four points on a straight line and that of four rays of a plane pencil unchanged.
Section C.2 169

projective theory: strategy for unifying the gravitational and the electromagnetic field using higher-dimensional space–times which are projected somehow onto the usual four-dimensional space-time. 135

proper motion: apparent motion, corrected for parallax and aberration, across the sky perpendicular to the line of sight, measured in angle per time.
Figure 2.16 16

proper time: the time that passes in an instantaneous rest system. It is measured by a clock comoving with the object in question. Formally, it is the arc length of timelike curves in space–time. 51

proton: lightest of the heavy → elementary particles (baryons) and stable if baryon number is a conserved quantity. The proton has a positive elementary charge e, a spin of $\frac{1}{2}\hbar$ (like the neutron) and a magnetic moment of $\mu = 2,793\, e\hbar(2m_p)^{-1}$. Its mass is 937 MeV. Together with neutrons, protons form atomic nuclei in the mutual balance of strong attraction and electrostatic repulsion.

pseudo-Euclidean geometry: geometry in which the axiom of parallels holds and the square of distance is indefinite (→ Minkowski).
Chapter 5 49

Pythagoras: 582–496 BC, mathematician and philosopher, famous for a fundamental theorem of Euclidean geometry, presumably of earlier origin. With appropriate reinterpretation, this theorem can also be applied to the Minkowski geometry.
Figures 3.5, 5.2, and 8.3 25, 52, 108

quadric: hypersurface of a projective space determined by a homogeneous quadratic equation. In the projective plane, a quadric is a conic section. 121

quantum mechanics: reformulation of classical mechanics to conform to the quantization of action, found by → M. Planck. Momentum and position can no longer be measured simultaneously with arbitrary precision (Heisenberg's uncertainty relation). Consequently, classical paths no longer exist. Instead, one obtains interfering probability waves whose values are subject in the simplest case to a Schrödinger equation. The uncertainty relation implies that motion cannot cease to exist even in a ground state. This is the reason for the existence of a zero-point energy. 39

quasar: quasistellar source (QSS), quasistellar object (QSO). A starlike object of very large red-shift that is usually a strong source of radio waves; presumed to be extragalactic and highly luminous.
Figure 7.13 94 ff.

radial velocity: The speed at which an object is moving away or toward an observer. By observing the → Doppler shift of spectral lines, one can determine this velocity; however, these spectral lines cannot be used to measure the → proper motion.
Figure 2.16 16

radiation: continuous and free propagation of → energy and mass with large velocities (particles with rest mass) or with the absolute velocity (light), respectively. Used without attribute, the notion is specialized by the context. The intensity of radiation is directly proportional to the power of the source and inversely proportional to the square of its distance (strictly to the surface of a sphere drawn around the source with the corresponding radius).
Figure 2.18 17

range, harmonic: → harmonic range

reference frame: combination of clocks and rulers to obtain locally a characterization of all events and all → vectors by coordinates. In general, this is necessary for the quantitative analysis of physical motions. A *linear* reference frame maps the addition of vectors to the addition of the corresponding coordinates. In mechanics, force-free motion is represented by linear relations of the coordinates of a linear frame. An *inertial* reference system admits in addition the formulation of → Newton's laws with isotropic inertial masses. Inertial reference frames exist in a metric space–time. 26

reflection: in general a map that restores the initial state if performed twice, i.e., an involution. Specifically, reflections are the involutory elements of a group G called a group of motions: $\varrho \in G$ is a reflection if and only if $\varrho \cdot \varrho = 1$ but $\varrho \neq 1$. The question of when the product of two reflections is again a reflection is the central issue of the generated geometry.

The abstract definition of reflections is independent of the considered geometric objects. It only deals with the algebraic relations that they satisfy.
Figures 3.1, 3.2, 5.1, 8.18, and 8.15
23, 51, 115

refractive index: simplest representation of a non-uniform light propagation, useful for isotropic materials where it indicates the ratio of the light velocity in vacuo with the light velocity in the material. 8

relativity: relation of a statement to external objects or circumstances that change the statement if they themselves change. The problem that gave rise to the relativity theory is the consistency of the relativity of velocity with the existence of an absolute velocity (the → speed of light).
Figure 3.9 28

relativity of simultaneity: dependence of statements about simultaneity of spatially separated events on the state of motion of the observer. The relativity of simultaneity is a characteristic feature of the relativity theory and the source of most of the misunderstandings about it.
Figures 4.1, 4.7, and 4.8 38, 42

relativity principle: requirement that the construction of a theory implements a priori that some notions are only definable and measurable with respect to external circumstances and that, accordingly, they must not enter a theory of closed systems.

In the (special) relativity theory, this applies mainly to velocity. The position, orientation, and velocity of a closed system are relative and cannot be found by strictly internal observations.

The relativity principle is valid in Newton mechanics as well as in relativistic mechanics. The difference lies in the composition law for velocities and in the group of motions that realizes the relativity principle.
Figure 3.9 28

Relativity theory: theory that transfers the invariance of the wave equation (for light) to all other physical phenomena. In the case of mechanics, one obtains the → special relativity theory, which is not able to include gravitation. Because of the equivalence principle for inertial and gravitational mass, the gravitational field is represented by the coefficients of the wave equation. One obtains a theory for the metric of space–time, → general relativity theory.

representation of a group: structure-preserving (homomorphic) map of a group into the special group of quadratic matrices of given dimension (into the group of regular linear operators of a vector space, respectively).

resonances: → elementary particles.

rest frame: inertial → reference frame in which the considered object is at rest. For an object moving generally, one can define a separate instantaneous rest frame at each event of its world-line. If the object is accelerated, it is at rest in this instantaneous rest frame only for the defining event. 45

rest mass: mass of an object in its instantaneous rest system. Whereas the inertial mass is conserved for a closed system, the sum of the rest masses of its constituents can vary through the interchange of internal energies with kinetic energies. 54

Riemann, B. 1826–1866, mathematician, cited here for his consideration of curved spaces of arbitrary dimension, beginning with his work *On the hypotheses that form the basis of geometry* (→ Helmholtz). He gave his name to *Riemannian geometry*, i.e., the geometry of a locally Euclidean space modified by curvature.

163, 155

rotation: motion with one finite fixed point.
Figure 9.16 ff. 135 ff.

Rydberg, J.: 1854–1919, physicist, helped to develop spectral analysis to a state that permitted the detection of the underlying physical laws. The *Rydberg constant*, i.e., the typical size of the frequency differences in the main structure of the line spectrum of an atom, is named after him. This constant is a measure for the tightness of bound states in atomic systems. The binding energy of the hydrogen electron in its ground state is (for an idealized infinitely heavy proton) equal to

$$h\,\mathrm{Ry}_\infty = 2\pi^2 m_e e^4 h^{-2} = 2.18 \times 10^{-18}\ \mathrm{J}.$$

scalar product: 167

simultaneity, absolute: → absolute simultaneity.

Sine theorem: In a triangle of the Euclidean plane, the sines of the angles are proportional to the opposite sides. In fact, the form of this theorem characterizes the geometries of the plane in general: If instead of the length a of a side we use the circumference $\Pi[a]$ of a circle with the radius a, the sine theorem in the form [36]

$$\Pi[a] : \Pi[b] : \Pi[c] = \sin\alpha : \sin\beta : \sin\gamma$$

summarizes the elliptical, Euclidean, and Lobachevski geometries. If instead of the sine we write the ratio Σ of the length of the projecting perpendicular to the projected side, we obtain for all nine geometries of the plane

$$\Pi[a] : \Pi[b] : \Pi[c] = \Sigma[\alpha] : \Sigma[\beta] : \Sigma[\gamma].$$

There are geometries with $\Pi[a]$ equal to $\sin a$, a, or $\sinh a$ just as $\Sigma[\alpha]$ can be equal to $\sin\alpha$, α, or $\sinh\alpha$. There are constructions that realize all nine combinations.
Figures E.1 and E.2, Table E.1 201, 202

Sommerfeld, A.: 1868–1951, physicist, made important contributions to the theory of atomic spectra and atomic structure. He gave his name to *Sommerfeld's fine-structure constant.* This is a dimensionless constant that characterizes the fine structure of atomic spectra. The fine-structure constant α can be interpreted as the ratio of the atomic unit of velocity to the speed of light. The atomic unit is the product of the → Rydberg constant and → Bohr radius,

$$v_{\mathrm{atom}} = 2r_{\mathrm{Bohr}}\mathrm{Ry}_{\infty} = \alpha c.$$

The value is $\alpha = e^2/(hc) \approx 1/137$. 43

sound waves: pressure (and shear) waves that are audible in the frequency region between 30 and 30 000 Hz.
Figure 4.2 38

space: → world.

space, absolute: → absolute space.

space of constant curvature: Chapter 7 83

spacelike: relative position of two events for which one lies outside the light cone of the other. If the interval between two events is spacelike, neither can be reached by a signal from the other. They are causally not connected. A → vector is spacelike if it has the same direction as such an interval. Spacelike vectors have negative norm. The paradigm of a spacelike vector is the acceleration of a particle. 50

special theory of relativity: the physical theory of space and time developed by Albert Einstein, based on the postulates that the form of all the laws of physics (mechanics and electrodynamics in particular) is the same in all inertial frames of reference, which can move at uniform velocity with respect to each other, and that the speed of light is independent of direction, regardless of how fast or slow the source or the observer are moving. Consequences of the theory are the relativistic mass increase of rapidly moving objects and the equivalence of mass and energy; it was generalized to → general relativity.

speed of light: in the relativity theory synonymously used for the absolute velocity, which is not changed in composition with other velocities. The synonym is appropriate because the speed of light (in vacuum) is assumed to be this velocity, i.e., photons are assumed to be massless. If the rest mass of photons should turn out to be positive, the relativity theory would still be valid but the speed of light would lose the crown of being the absolute velocity. For the consistency and applicability of the relativity theory, it is not necessary for any object to exist at all that moves with the absolute velocity. The geometrical relations alone decide whether such a velocity exists. The consistency and applicability of the relativity theory can be tested even without particles that move with the absolute velocity.

Ordinary particles with nonvanishing rest mass always move slower than light in vacuum. A speed faster than that of light is observed only if the speed of light is smaller in the given circumstances than that in vacuum, i.e., smaller than the absolute velocity. In this case, an analog of the Mach cone in acoustics is observed as the Cherenkov effect. Velocities faster than the *absolute* velocity may exist for hypothetical elementary particles but have *never* been observed. In addition, they lead to problems of consistency with the well-observed → causality. The hypothetical particles that move faster than the absolute velocity are called → tachyons.

In the SI, the speed of light relates the length unit to the time unit and is defined to be equal to 299 792 458 m/s.

sphere: (in our context) locus of the points of fixed distance from a center in space.
Figure 7.1 85

spherical excess: excess of the sum of the angles of a geodesic triangle on a curved surface over the flat angle.
Figure 7.1 85

spherical geometry: geometry of the surface of a → sphere. If centrally projected onto a plane, it yields elliptical geometry.

stereoscopic aberration: method to complement the map of the apparent sky by → aberration for a map of the full space.
Figure 5.25 70

summation convention: convention that is used in formulas with indexed components of tensors. If in some term a character is used for both an upper and a lower index, summation over this pair of indices goes without saying.
Section B.3 161

tachyon: hypothetical particle that moves faster than light. The norm of the momentum of a tachyon is negative, its momentum spacelike. The hypothesis of the existence of tachyons is in conflict with universal → causal order.
Figures 5.26 and 5.27 73

theorem of altitudes: The altitudes of a triangle meet at one point.
Figures 6.5, 9.11, A.7, and D.2 78, 132, 152, 179

theorem of circumference angles: The angles at the points on the circumference of a circle sustended by a fixed chord of it are all equal.
Figures 6.8, 6.9, 6.10, and 9.13 79, 133

theorem of perpendicular bisectors: The perpendicular bisectors of a triangle meet at one point, the center of the circumcircle. This is the geometric equivalent of the transitivity of equality.
Figures 6.1, 9.12, A.5, and D.4 76, 132, 151, 182

Thomas precession: deviation from parallel transport of a spacelike vector (in particular, the angular momentum of a gyroscope), produced by the constraint of orthogonality to the (four-component) velocity vector. The Thomas precession is an effect of special relativity, i.e., it arises in a world without curvature. It can be understood as due to the curvature of the → velocity space [77]. The Thomas precession produces a part of the fine structure of spectral lines.
Figures 7.12 and D.6 91, 185

time: order relation between configurations in space.
Chapter 2 5

time dilation: in relativity theory a projection effect between timelike lines measurable by clock rates.
Figures 5.13 and 5.14 62

time shell: spacelike surface in four-dimensional (and pseudo-Euclidean) space–time, locus of the endpoints of position vectors with a fixed norm (proper time).
Figure 7.11 91

timetable: graphic presentation of the motion in space by a curve in the space–time, → world-line.

time, absolute: $\rightarrow$ absolute time.

timelike: relative position of two events in which one lies inside the light cone of the other. If the interval between two events is timelike, one can be reached from the other by a massive body moving through space. A $\rightarrow$ vector is timelike if it has the same direction as such an interval. Timelike vectors have positive norm. Examples of a timelike vector are the velocity and momentum of a particle with positive rest mass.
50

totally inelastic collision: $\rightarrow$ collision in which the kinetic energy referred to the $\rightarrow$ center of mass is transformed completely into internal energy and in which the collision product moves on with the constant velocity of the center of mass.
Eq. (3.1) 29

transformation: change of form of quantities usually subject to a substitution of coordinates or other variables.

In group theory, automorphism of a group $G = \{g\}$ produced by multiplication with a given element $a \in G$, i.e., $T_a[g] = a^{-1}ga$.
Appendix A 145

transitivity region: region accessible by a point via transformations by the elements of the transformation group. If $\mathcal{T} = \{T\}$ denotes the transformation group, the transitivity region of the point P is the set of all points of the form $\{T[P], T \in \mathcal{T}\}$. If all points belong to one transitivity region, the group is said to be transitive.
Appendix A 145

triangle inequality: axiomatic requirement of a definite metric that the distance $d[A, B]$ between two points A and B is never larger than the sum of the distances to a third point, $d[A, B] \leq d[A, C] + d[C, B]$. In this form, the triangle inequality holds in the elliptic, Euclidean, and Lobachevski geometries. In locally pseudo-Euclidean geometries, the triangle inequality has a different form ($\rightarrow$ twin paradox).

triple product: volume of the parallelepiped as a function of the three independent edges and their orientation, both given by a vector in space. 168

twin paradox: apparent paradoxical conclusion drawn from the symmetry of $\rightarrow$ time dilation.
Figures 5.15 and 5.16 64

vanishing line: image of the line at infinity. Its points are *vanishing points*, which are defined by the direction of a pencil of parallel lines.
Figure 8.3 108

variation of mass with velocity: fundamental result of the relativity theory, corollary to the $\rightarrow$ equivalence of mass and energy.
Figure 5.4 53

vector: objects that are defined by their algebra (*vector algebra*). This is a structure of operations, including the definition of a (commutative) addition between the vectors and a distributive and associative multiplication with numbers.

In this volume, vectors are used in the simple intuitive sense. A vector is determined by its length or norm and its direction. It is described by as many components as we have dimensions of the space or the world. Its norm is determined by the same formula that is used for the square

of distance of infinitesimally separated points. Momenta and field strengths are vectors. While momentum is associated with a moving object, a field strength is, in principle, associated with all the points of space and varies with location. We then speak of a vector field. The action of motions on a vector decomposes into rotations around its point of definition, which are just as simple as the rotations of space, and into translations of the point of definition. These translations are called parallel transport and require detailed investigation in spaces with curvature.
Section C.1 165

velocity space: space of relative velocities that parametrize Galilean or Lorentz transformations. In the Galilean geometry, the velocity space is Euclidean. In the relativity theory, it is still homogeneous but negatively curved. In the two-dimensional plane, the relative velocities fill a circle with radius c (absolute velocity) that reproduces Klein's model of the non-Euclidean geometry.
Figures D.5 and D.6 185

virtual displacement: When the state of a physical system is described by redundant coordinates, these coordinates are subject to conditions that may vary with time. The (infinitesimally small) changes of the state that are allowed by the conditions at a given instant are called virtual displacements. The virtual displacements weight the generalized forces to find equilibria, for instance. 56

wave: excitation that propagates through space by microscopic coupling. The idealized equation for this propagation is the wave equation. It immediately defines a metric of space–time. The relativity principle implies that all the metric tensors definable by the propagation of free waves coincide with the metric determined by mechanics.

wave group: pulse of excitation that is to be regarded as a superposition of monochromatic waves of different wavelengths. If the propagation velocity depends on the wavelength, the velocity of a wave group (group velocity) differs from the phase velocity. In general, the energy is transported with the group velocity. That is the reason why a wave group (or wave packet) can be regarded as equivalent to a particle.

Weyl, H.: 1885–1955, mathematician, cited here for his contribution to relativity theory by the analysis of unified field theories. 10

world: notion combining space and time. Both notions are fundamental and correspond to the elementary experience that objects are geometrically arranged. Space is the ensemble of the virtual or real arrangements. Physical motion is the change of these arrangements, which are thereby ordered too. This order is time. It is a task of physics to give a measure to all the arrangements, and it is a task of mathematics to find the appropriate rules of calculation.

In the relativity theory, the world is the formal product of space and time. This product becomes so tightly bound together by the local Minkowski geometry of the events and world-lines that quantum constructions, which require the existence of a privileged time, get characteristic problems.
Chapter 2 5

world-line: curve in a world that describes the history of the positions of an object in a space–time.
Figures 2.1, 2.2, 2.3, 2.4, and 2.8 7, 11

zero, absolute: → absolute zero.

References

[1] F. Klein, Über die geometrischen Grundlagen der Lorentz-Gruppe, Phys. ZS **12**, 17–27 (1911).

[2] F. Klein, Bemerkungen über die Beziehungen des deSitterschen Koordinatensystems B zu der allgemeinen Welt konstanter Krümmung, Proc. Amsterdam **21**, 614–615 (1919).

[3] H.K. DeVries, *Die vierte Dimension* (Teubner, Leipzig, 1926).

[4] D.-E. Liebscher, *Relativitätstheorie mit Zirkel und Lineal* (Akademie-Verlag, Berlin, 1991), 2. Aufl.

[5] J. Mroczkowski, *Zastosowanie metod geometrii wykreślnej w szczególnej teorii względności* (Wyd. Politechniki Wrocławskiej, Wrocław, 1986).

[6] R. Resnick, *Introduction to Special Relativity* (J. Wiley, London, 1979).

[7] R.B.Salgado, http://physics.syr.edu/courses/modules/LIGHTCONE/

[8] L.D. Landau and Ju.B. Rumer, *Was ist die Relativitätstheorie* (Akad. Verlagsges. Geest & Portig, Leipzig, 1962); *What is relativity* (Mir editorial, Moskva 1978).

[9] T.A. Moore, *A Travellers Guide to Space–Time* (McGraw-Hill, New York, 1996).

[10] E.F. Taylor and J.A. Wheeler, *Space–Time Physics* (Freeman, San Francisco, 1966).

[11] H. Ruder and M. Ruder, *Die spezielle Relativitätstheorie* (Vieweg-Verlag, Braunschweig, 1993).

[12] M. Born, *Die Relativitätstheorie Einsteins* (Springer, Berlin, 1920); *Einstein's Theory of Relativity* (Dover, New York, 1962).

[13] A.P. French, *Special Relativity* (MIT, Cambridge, 1966).

[14] B.F. Schutz, *A First Course in General Relativity* (Cambridge University Press, Cambridge, 1985).

[15] A. Einstein, *Grundzüge der Relativitätstheorie* (Akademie-Verlag, Berlin, 1969), 5. Aufl.

[16] J. Heidmann, *Introduction à la Cosmologie* (Presses universitaires de France, Paris, 1973).

[17] D.-E. Liebscher, *Cosmology* (Springer, Heidelberg, 2005).

[18] H.J. Treder, *Relativität und Kosmos* (Braunschweig, Berlin, Oxford, 1968).

[19] R. Bereis, *Darstellende Geometrie* (Akademie-Verlag, Berlin, 1964).

[20] W. Blaschke, *Projektive Geometrie* (Wolfenbüttel, 1947).

[21] H.S.M. Coxeter, *Projective Geometry* (Springer, New York et al., 1963).

[22] H.S.M. Coxeter, *Introduction to Geometry* (Wiley, New York, 1966).

[23] K. Doehlemann, *Projektive Geometrie in synthetischer Behandlung* (Sammlung Göschen, W. de Gruyter, Berlin and Leipzig, 1922, 1924), Vols. 72, 876.

[24] F. Schilling, *Projektive und nichteuklidische Geometrie* (Teubner, Leipzig and Berlin, 1931).

[25] A. Beutelspacher and U. Rosenbaum, *Projective Geometry. From Foundations to Applications* (Cambridge University Press, Cambridge, 1998).

[26] G. Buchmann, *Nichteuklidische Elementargeometrie* (Teubner, Stuttgart, 1975).

[27] B. Klotzek and E. Quaisser, Nichteuklidische Geometrie (Deutscher Verlag der Wissenschaften, Berlin 1978).

[28] G. Bär, *Eine Einführung in die analytische und konstruktive Geometrie* (Teubner, Stuttgart, 1996).

[29] W. Benz, *Real Geometries* (BI Wissenschaftsverlag Mannheim, 1994).

[30] H.S.M. Coxeter, *Non-Euclidean Geometry* (University of Toronto Press, Toronto, 1961).

[31] B. Klotzek, *Geometrie* (Deutscher Verlag der Wissenschaften, Berlin, 1971).

[32] E. Quaisser and H.J. Sprengel, *Räumliche Geometrie* (Deutscher Verlag der Wissenschaften, Leipzig, 1981).

[33] E. Quaisser, *Bewegungen in der Ebene und im Raum* (Deutscher Verlag der Wissenschaften, Berlin, 1983).

[34] B. Pareigis, *Analytische und projektive Geometrie für die Computer-Graphik* (Teubner, Stuttgart, 1990).

[35] C. Seelig (Hrsg.), *Helle Zeit—dunkle Zeit. In memoriam Albert Einstein* (Zürich, 1956).

[36] I.M. Jaglom (I.M. Yaglom), *Princip otnositelnosti Galileja i neevklidova geometrija* (Izd. Nauka, Moskva, 1969); *A Simple Noneuclidean Geometry and its Physical Basis* (Springer, New York et al., 1979).

[37] S. Antoci, Devil's Advocate Online Service, private communications, 1997.

[38] R.B. Palmer, D. Radojicic, R.R. Rau, C. Richardson, N.P. Samios, I.O. Skillicorn, and J. Leitner, Precision measurement of the Ω^- and Ξ^o masses, Phys. Lett. B **26**, 323–326 (1968).

[39] J. Hevelius, *Machina Coelestis* (Bd. I, Danzig, 1673).

[40] H. Bondi, *Relativity and Common Sense* (Heinemann, London, 1962).

[41] J. Ehlers, F.A.E. Pirani, and A. Schild, The geometry of free fall and light propagation, *Papers in Honour of J.L. Synge*, edited by L. O'Raifeartaigh (Oxford University Press, Oxford, 1972), pp. 63–84.

[42] N.D. Mermin, Relativistic addition of velocities directly from the constancy of the velocity of light, Am. J. Phys. **51**, 1130–1131 (1983).

[43] H. Poincaré, *Science and Method* (Dover, New York, 1952); *Science and Hypothesis* (Dover, New York, 1952).

[44] R. Eötvös, V. Pekar, and E. Fekete, Beitrag zum Gesetz der Proportionalität von Trägheit und Gravität, Ann. Phys. (Lpz.) **68**, 11–66 (1922).

[45] V.B. Braginski and V.I. Panov, Verification of the equivalence of inertial and gravitational mass, J. Exp. Theor. Phys. **34**, 463–466 (1972).

[46] R.F. Marzke and J.A. Wheeler, Gravitation as geometry: I. The geometrodynamical standard meter, Grav. Rel. **29**, 40–64 (1964).

[47] C.W. Misner, K.S. Thorne, and J.A. Wheeler, *Gravitation* (Freeman, San Francisco, 1971).

[48] J.B. Barbour, *The End of Time. The Next Revolution in Physics* (Weidenfeld and Nicolson, London, and Oxford University Press, New York, 1999).

[49] F. Bachmann, *Der Aufbau der Geometrie aus dem Spiegelungsbegriff* (Springer, Heidelberg, 1959).

[50] G. Choquet, *L'enseignement de la géométrie* (Hermann, Paris, 1964); *Geometry in a Modern Setting* (Hermann, Paris, 1969).

[51] L. Lange, *Die geschichtliche Entwicklung des Bewegungsbegriffs und ihr voraussichtliches Endergebnis. Ein Beitrag zur historischen Kritik der mechanischen Prinzipien* (W. Engelmann, Leipzig, 1886).

[52] P.G. Tait, Note on reference frames, Proc. R. Soc. Edinburgh 1883–1884, 743–745 (1884).

[53] H.J. Treder, *Relativität der Trägheit* (Akademie-Verlag, Berlin, 1972).

[54] Ch. Huygens, De motu corporum ex percussione, *Opuscula Posthuma* (Manuscript 1656, Ostwalds Klassiker Bd. 138, Leipzig, 1903).

[55] G. Galilei, *Dialogo di Galileo Galilei, Linceo matematico straordinario dello studio di Pisa e filosofo, e matematico primario del Serenissimo Gran Duca di Toscana. Dove nei congressi di quattro giornate si discorre sopra i due massimi sistemi del mondo tolemaico, e copernicano* (Landini, Fiorenza, 1632) (translated in [56]).

[56] M.V. Berry, *Principles of Cosmology and Gravitation* (Adam Hilger, Bristol, 1989).

[57] G. Holton, Resource letter special relativity theory, Am. J. Phys. **30**, 462–469 (1962).

[58] D.C. Champeney, G.R. Isaak, and A.M. Khan, An aether drift experiment based on the Mössbauer effect, Phys. Lett. **7**, 241–243 (1963).

[59] M.P. Haugan and C.M. Will, Modern tests of special relativity, Phys. Today **40**(5), 69–76 (1987).

[60] M. T. Murphy, Webb, J.K., Flambaum, V.V., Dzuba, V.A., Churchill, C.W., Prochaska, J.X., Barrow, J.D., Wolfe, A.M., Possible evidence for a variable fine structure constant from QSO absorber lines. Motivations, analysis, and results, Mon. Not. R. Astron. Soc. 327, 1223 (2001)

[61] D.F. Mota, Variations of the fine structure constant in space and time, *Preprint* astro-ph/0401631, 2004.

[62] A. Einstein, Zur Elektrodynamik bewegter Körper, Ann. Phys. (Lpz.) **17**, 891–921 (1905).

[63] A. Einstein, *Relativity: The special and general theory*, (H. Holt, New York 1920).

[64] W. Dietze (ed.), *Limericks* (Edition Leipzig, Leipzig 1977).

[65] H. Bondi, Negative Mass Within General Relativity, Rev. Mod. Phys. **29**, 423 (1957).

[66] A. Einstein, Ist die Trägheit eines Körpers von seinem Energieinhalt abhängig? Ann. Phys. (Lpz.) **18**, 639–641 (1905).

[67] F. Rohrlich, An elementary derivation of $E = mc^2$, Am. J. Phys. **58**, 348–349 (1990).

[68] S.P. Boughn, The case of identically accelerated twins, Am. J. Phys. **57**, 791–799 (1989).

[69] T. Dray, The twin paradox revisited, Am. J. Phys. **58**, 822–825 (1990).

[70] R. Shaw, Length contraction paradox, Am. J. Phys. **30**, 72 (1962).

[71] H.-Ch. Freiesleben, *Beiträge zum Problem der astronomischen Aberration* (Thomas & Hubert, Weida, 1926).

[72] H.A. Atwater, Non-simultaneity in the aberration of starlight, Am. J. Phys. **42**, 1022–1024 (1974).

[73] M. Santander, The Chinese south-seeking chariot: a simple mechanical device for visualizing curvature and parallel transport, Am. J. Phys. **60**, 782–787 (1992).

[74] H. Reichardt, *Gauß und die Anfänge der nichteuklidische Geometrie. Mit Originalarbeiten von J. Bolyai, N.I. Lobatschewski, F. Klein* (Teubner, Leipzig, 1985).

[75] H. Urbantke, A note on elementary geometry and special relativity, Eur. J. Phys. **5**, 119 (1984).

[76] M.C. Escher, *Graphik und Zeichnungen* (Moos, München, 1984).

[77] L.H. Thomas, The kinematics of an electron with an axis, Phil. Mag. **3**, 1–22 (1927).

[78] E. Gausmann, R. Lehoucq, J.-P. Luminet, J.-P. Uzan, and J. Weeks, Topological lensing in spherical spaces, Class. Quantum Grav. **18**, 1–32 (2001) (gr-qc/0106033).

[79] J.-P. Luminet, J. Weeks, A. Riazuelo, R. Lehoucq, and J.-P. Uzan, Dodecahedral space topology as an explanation for weak wide-angle temperature correlations in the cosmic microwave background, Nature **425**, 593–595 (2003) (astro-ph/0310253).

[80] P.G. Roll, R. Krotkov, and R.H. Dicke, The equivalence of inertial and passive gravitational mass, Ann. Phys. (NY) **26**, 442–517 (1964).

[81] L. Russo, *La rivoluzione dimenticata* (Feltrinelli, Milano, 1996), pp. 79–86; *The forgotten revolution* (Springer, Heidelberg 2004)

[82] A. Dürer, *Opera* (Johan Jansen, Arnem, 1525).

[83] P. Jordan, *Schwerkraft und Weltall* (Vieweg-Verlag, Braunschweig, 1952).

[84] T.V. Kaluza, Zum Unitätsproblem der Physik, Sitzungsber. K. Preuss. Akad. Wiss. Phys. Math. Kl., 966–972 (1921).

[85] O. Klein, Quantentheorie und fünfdimensionale Relativitätstheorie, Z. Phys. **37**, 895–906 (1926).

[86] E. Schmutzer, *Projektive Einheitliche Feldtheorie* (H. Deutsch, Frankfurt 2004).

[87] G. Gabrielse, D. Phillips, W. Quint, and H. Kalinowsky, Special relativity and the single antiproton: Fortyfold improved comparison of $\bar{p}$ and p charge-to-mass ratios, Phys. Rev. Lett. **74**, 3544–3547 (1995).

[88] G. Salmon, *A Treatise on Conic Sections* (Longman, Brown, Green and Longmans, London, 1855), 3rd edn.

[89] U. Bleyer and D.-E. Liebscher, Mach's principle and local causal structure, in J.B. Barbour and H. Pfister (eds.): *Mach's Principle: From Newton's Bucket to Quantum Cosmology* (Birkhäuser, Boston 1995), pp. 293–307.

[90] P.L.S. Lagrange, *Théorie des functions analytiques* (Impr. de la République, Paris, 1797).

[91] A. Einstein, Geometrie und Erfahrung. Sitzungsber. Preuss. Akad. Wiss. 1–8 (1921).

[92] E.M. Schröder, Gemeinsame Eigenschaften euklidischer, galileischer und minkowskischer Ebenen, Mitt. Math. Ges. Hamburg **10**, 185–217 (1974).

[93] E.M. Schröder, Über die Grundlagen der affin-metrischen Geometrie, Geometria dedicata **11**, 415–442 (1981).

[94] E.M. Schröder, Fundamentalsätze der metrischen Geometrie, J. Geom. **27**, 36–59 (1986).

[95] H. Struve, Singulär projektiv-metrische und Hjelmslevsche Geometrie, Dissertation, Uni. Kiel, 1979.

[96] H. Struve and R. Struve, Endliche Cayley-Kleinsche Geometrien, *Arch. Math.* **48**, 178–184 (1987).

[97] H. Struve and R. Struve, Zum Begriff der projektiv-metrischen Ebene, ZS f. math. Logik u. Grundl. d. Math. **34**, 79–88 (1988).

[98] J. Steiner, *Vorlesungen über synthetische Geometrie, I* (3.ed. Leipzig, 1887); *Vorlesungen über synthetische Geometrie, II* (3.ed. Leipzig, 1898).

[99] R.H. Dicke, Experimental tests of Mach's principle, Phys. Rev. Lett. **7**, 359–360 (1961).

[100] S.K. Lamoreaux, J.P. Jacobs, B.R. Heckel, F.J. Raab, and E.N. Fortson, New limits on spatial anisotropy from optically pumped Hg^{201} and Hg^{199}, Phys. Rev. Lett. **57**, 3125–3135 (1986).

[101] J.B. Barbour and H. Pfister (eds.) *Mach's Principle: From Newton's Bucket to Quantum Gravity* (Birkhäuser, Boston, 1995).

[102] R.C. Tolman, Non-Newtonian mechanics, the mass of a moving body, Phil. Mag. **23**, 375–380 (1912).